Pipeline Electrical and Instrumentation

Level Two

Trainee Guide
Third Edition

PEARSON

Boston Columbus Indianapolis New York San Francisco Amsterdam
Cape Town Dubai London Madrid Milan Munich Paris Montreal Toronto Delhi
Mexico City Sao Paulo Sydney Hong Kong Seoul Singapore Taipei Tokyo

NCCER

President: Don Whyte
Vice President: Steve Greene
Chief Operations Officer: Katrina Kersch
Director of Product Development: Daniele Dixon
Pipeline Mechanical Project Managers: Patty Bird, Elizabeth Schlaupitz, Justin Street, Chris Wilson
Senior Development Manager: Mark Thomas
Senior Production Manager: Tim Davis
Quality Assurance Coordinator: Karyn Payne
Desktop Publishing Coordinator: James McKay
Permissions Specialist: Adrienne Payne
Production Specialist: Adrienne Payne
Editors: Graham Hack, Karyn Payne, Debie Hicks, Jackie Strawbridge, Adrienne Payne

Writing and development services provided by Topaz Publications, Liverpool, NY

Lead Writers/Project Managers: Don Congdon, Veronica Westfall
Desktop Publisher: Joanne Hart
Art Director: Alison Richmond
Permissions Editors: Andrea LaBarge, Briana Morgan
Writers: Don Congdon, Tony Clay, Thomas Burke

Pearson

Director of Alliance/Partnership Management: Andrew Taylor
Editorial Assistant: Collin Lamothe
Program Manager: Alexandrina B. Wolf
Digital Studio Project Managers: Heather Darby, Tanika Henderson, Jose Carchi
Director of Marketing: Leigh Ann Simms
Senior Marketing Manager: Brian Hoehl
Composition: NCCER
Printer/Binder: LSC Communications
Cover Printer: LSC Communications
Text Fonts: Palatino and Univers

Credits and acknowledgments for content borrowed from other sources and reproduced, with permission, in this textbook appear at the end of each module.

ScoutAutomatedPrintCode

PEARSON

ISBN-13: 978-0-13-471650-3
ISBN-10: 0-13-471650-7

Preface

To the Trainee

NCCER is pleased to release the 3rd edition of Pipeline Training and Qualification with updates that will engage you and give you the best training in the industry. The curriculum has been redesigned to better align with the learning objectives, and also includes new end-of-section review questions to compliment the Module Review. Changes to the NCCER's covered task (CT) modules include updates to the latest technology and standards, as well as alignment to the American Petroleum Institute's *API Recommended Practice for Pipeline Operator Qualification (OQ) 1161*, 3rd edition (2014). Each CT module contains task-specific Abnormal Operating Conditions (AOCs) to which the trainee must be prepared to respond. General AOCs for both Field (Liquid and Gas) and Control Center settings are also available as single modules.

CT modules are available as traditional printed levels and as individual module eBooks. Modules are completed by successfully passing the proctored Module Examination, and completing the Performance Profile under the supervision of an NCCER-certified evaluator. These completions are tracked within the NCCER Registry for training delivered through an Accredited Training Sponsor. Covered task module completions in the Registry are also documented in ISNetworld.

The modules are designed to be used for Operational Qualification, and they are each a vital part of NCCER's pipeline workforce development curricula. They provide in-depth training on the fundamental knowledge and skills required for pipeline work, as well as technical training that increases the depth and breadth of a pipeline worker's capability within several specialty areas.

We invite you to visit the NCCER website at **www.nccer.org** for information on the latest product releases and training, as well as online versions of the *Cornerstone* magazine and Pearson's NCCER product catalog.

Your feedback is welcome. You may email your comments to **curriculum@nccer.org** or send general comments and inquiries to **info@nccer.org**.

NCCER Standardized Curricula

NCCER is a not-for-profit 501(c)(3) education foundation established in 1996 by the world's largest and most progressive construction companies and national construction associations. It was founded to address the severe workforce shortage facing the industry and to develop a standardized training process and curricula. Today, NCCER is supported by hundreds of leading construction and maintenance companies, manufacturers, and national associations. The NCCER Standardized Curricula was developed by NCCER in partnership with Pearson, the world's largest educational publisher.

Some features of the NCCER Standardized Curricula are as follows:

- An industry-proven record of success
- Curricula developed by the industry for the industry
- National standardization providing portability of learned job skills and educational credits
- Compliance with the Office of Apprenticeship requirements for related classroom training (*CFR 29:29*)
- Well-illustrated, up-to-date, and practical information

NCCER also maintains the NCCER Registry, which provides transcripts, certificates, and wallet cards to individuals who have successfully completed a level of training within a craft in NCCER's Curricula. *Training programs must be delivered by an NCCER Accredited Training Sponsor in order to receive these credentials.*

Additional Pipeline Resources

The following additional Pipeline Resources are available from NCCER's website at **www.nccer.org/pipeline**:

1. A complete copy of NCCER's Covered Task (CT) list
2. Access to:
   a. An online bookstore to order print and digital versions of NCCER's pipeline curricula.
   b. Instructor resources (Lesson Plans and PowerPoints)
   c. Exams for Operator Qualification
   d. A site to custom build curriculum with pipeline and/or other NCCER-developed modules

Pipeline Career Pathway Training

Entry-level Pipeline Craft Trainee → **Pipeline Fundamentals** → **Level 1**

The newly established pipeline career pathway represented on this page provides the recognition pipeline professionals deserve through standardized training and industry-recognized credentials.

Levels of training have been established by combining training modules into specific focus areas within the industry. This allows trainees to progress through a standardized program and earn a professional credential.

In addition, Levels Two and Three are comprised of covered task training modules that can lead to operator qualifications. *Pipeline Corrosion Control is the only exception which actually provides covered task training starting in Level One.*

The blue boxes denote the professional credential that can be earned with successful completion of each program.

A trained workforce is a safe and qualified workforce!

Pipeline Fundamentals

- **Core Curriculum**
- **Introduction to the Pipeline Industry**

  and

  a.) Field Abnormal Operating Conditions
  Module: AOCFG-17

  or

  b.) Control Center Abnormal Operating Conditions
  Module: AOCCC_17

Level 1

- **Field and Control Center Operations**
- **Pipeline Maintenance and Mechanical**
- **Electrical and Instrumentation**
- **Pipeline Corrosion Control** (CT)

Level 2: Craft-Specific Training with Operator Qualification*	Level 3: Craft-Specific Training with Operator Qualification*	Trained and Qualified Pipeline Professional Credential
Pipeline Operations (Control Center/ Gas/Liquid) CT		Pipeline Operations Technician
Pipeline Mechanical CT	Pipeline Mechanical CT	Pipeline Mechanical Technician
Pipeline Maintenance CT	Pipeline Maintenance CT	Pipeline Maintenance Technician
Pipeline Electrical and Instrumentation CT		Pipeline Electrical and Instrumentation Technician
Pipeline Corrosion Control CT		Pipeline Corrosion Technician

CT *Each of these titles have corresponding operator qualifications. The Covered Task list is available online at www.nccer.org/pipeline-program*

Special Features

In an effort to provide a comprehensive and user-friendly training resource, this curriculum showcases several informative features. Whether you are a visual or hands-on learner, these features are intended to enhance your knowledge of the construction industry as you progress in your training. Some of the features you may find in the curriculum are explained below.

Introduction

This introductory page, found at the beginning of each module, lists the module Objectives, Performance Tasks and Trade Terms. The Objectives list the knowledge you will acquire after successfully completing the module. The Performance Tasks give you an opportunity to apply your knowledge to real-world tasks. The Trade Terms are industry specific vocabulary that you will learn as you study this module.

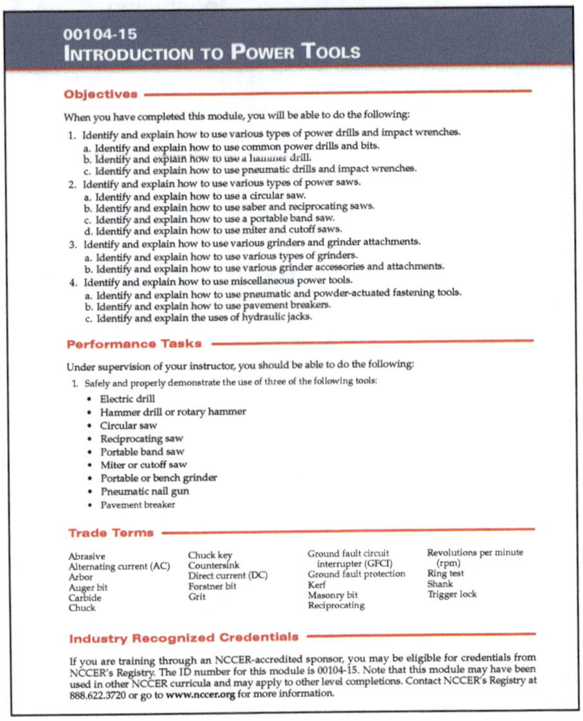

00104-15
INTRODUCTION TO POWER TOOLS

Objectives

When you have completed this module, you will be able to do the following:

1. Identify and explain how to use various types of power drills and impact wrenches.
   a. Identify and explain how to use common power drills and bits.
   b. Identify and explain how to use a hammer drill.
   c. Identify and explain how to use pneumatic drills and impact wrenches.
2. Identify and explain how to use various types of power saws.
   a. Identify and explain how to use a circular saw.
   b. Identify and explain how to use saber and reciprocating saws.
   c. Identify and explain how to use a portable band saw.
   d. Identify and explain how to use miter and cutoff saws.
3. Identify and explain how to use various grinders and grinder attachments.
   a. Identify and explain how to use various types of grinders.
   b. Identify and explain how to use various grinder accessories and attachments.
4. Identify and explain how to use miscellaneous power tools.
   a. Identify and explain how to use pneumatic and powder-actuated fastening tools.
   b. Identify and explain how to use pavement breakers.
   c. Identify and explain the uses of hydraulic jacks.

Performance Tasks

Under supervision of your instructor, you should be able to do the following:

1. Safely and properly demonstrate the use of three of the following tools:
   - Electric drill
   - Hammer drill or rotary hammer
   - Circular saw
   - Reciprocating saw
   - Portable band saw
   - Miter or cutoff saw
   - Portable or bench grinder
   - Pneumatic nail gun
   - Pavement breaker

Trade Terms

Abrasive	Chuck key	Ground fault circuit interrupter (GFCI)	Revolutions per minute (rpm)
Alternating current (AC)	Countersink	Ground fault protection	Ring test
Arbor	Direct current (DC)	Kerf	Shank
Auger bit	Forstner bit	Masonry bit	Trigger lock
Carbide	Grit	Reciprocating	
Chuck			

Industry Recognized Credentials

If you are training through an NCCER-accredited sponsor, you may be eligible for credentials from NCCER's Registry. The ID number for this module is 00104-15. Note that this module may have been used in other NCCER curricula and may apply to other level completions. Contact NCCER's Registry at 888.622.3720 or go to **www.nccer.org** for more information.

Trade Features

Trade features present technical tips and professional practices based on real-life scenarios similar to those you might encounter on the job site.

Bowline Trivia

Some people use this saying to help them remember how to tie a bowline: "The rabbit comes out of his hole, around a tree, and back into the hole."

Figures and Tables

Photographs, drawings, diagrams, and tables are used throughout each module to illustrate important concepts and provide clarity for complex instructions. Text references to figures and tables are emphasized with *italic* type.

Notes, Cautions, and Warnings

Safety features are set off from the main text in highlighted boxes and categorized according to the potential danger involved. Notes simply provide additional information. Cautions flag a hazardous issue that could cause damage to materials or equipment. Warnings stress a potentially dangerous situation that could result in injury or death to workers.

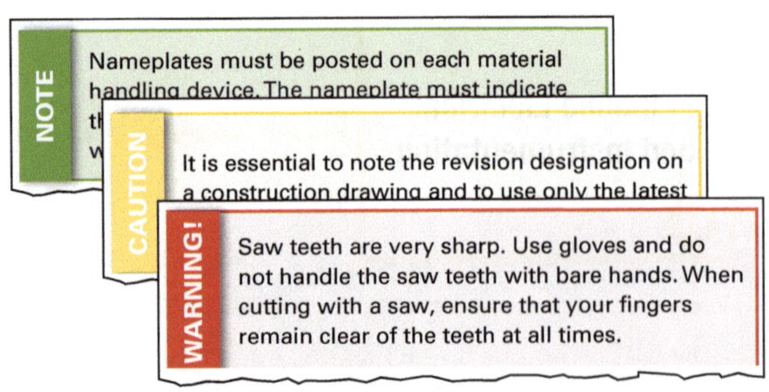

Case History

Case History features emphasize the importance of safety by citing examples of the costly (and often devastating) consequences of ignoring best practices or OSHA regulations.

Case History

Requesting an Outage

An electrical contractor requested an outage when asked to install two bolt-in, 240V breakers in panels in a data processing room. It was denied due to the 24/7 worldwide information processing hosted by the facility. The contractor agreed to proceed only if the client would sign a letter agreeing not to hold them responsible if an event occurred that damaged computers or resulted in loss of data. No member of upper management would accept liability for this possibility, and the outage was scheduled.

The Bottom Line: If you can communicate the liability associated with an electrical event, you can influence management's decision to work energized.

Going Green

Going Green features present steps being taken within the construction industry to protect the environment and save energy, emphasizing choices that can be made on the job to preserve the health of the planet.

GOING GREEN

Reducing Your Carbon Footprint

Many companies are taking part in the paperless movement. They reduce their environmental impact by reducing the amount of paper they use. Using email helps to reduce the amount of paper used,

Did You Know

Did You Know features introduce historical tidbits or interesting and sometimes surprising facts about the trade.

Did You Know?

Safety First

Safety training is required for all activities. Never operate tools, machinery, or equipment without prior training. Always refer to the manufacturer's instructions.

Step-by-Step Instructions

Step-by-step instructions are used throughout to guide you through technical procedures and tasks from start to finish. These steps show you how to perform a task safely and efficiently.

Perform the following steps to erect this system area scaffold:

Step 1 Gather and inspect all scaffold equipment for the scaffold arrangement.

Step 2 Place appropriate mudsills in their approximate locations.

Step 3 Attach the screw jacks to the mudsills.

Trade Terms

Each module presents a list of Trade Terms that are discussed within the text and defined in the Glossary at the end of the module. These terms are presented in the text with **bold, blue** type upon their first occurrence. To make searches for key information easier, a comprehensive Glossary of Trade Terms from all modules is located at the back of this book.

During a rigging operation, the **load** being lifted or moved must be connected to the apparatus, such as a crane, that will provide the power for movement. The connector—the link between the load and the apparatus—is often a sling made of synthetic, chain, or **wire rope** materials. This section focuses on three types of slings:

Section Review

Each section of the module wraps up with a list of Additional Resources for further study and Section Review questions designed to test your knowledge of the Objectives for that section.

Additional Resources

Materials Handling Handbook, The American Society of Mechanical Engineers (ASME) and The International Material Management Society (IMMS), Raymond A. Kulwiec, Editor-in-Chief. 1985. New York, NY: Wiley-Interscience.

Manufacturing Facilities Design & Material Handling, Matthew P. Stevens, Fred E. Meyers. 2013. West Lafayette, IN: Purdue University Press.

1.0.0 Section Review

1. For material handling tasks, it is just as important to be mentally fit as it is to be _____.
   a. physically fit
   b. physically aggressive
   c. closely supervised
   d. over 200 pounds

2. Which of the following is a type of knot that is often used to join the ends of two ropes in non-critical, low-strain applications?
   a. Bowline
   b. Clove hitch
   c. Half hitch
   d. Square knot

Review Questions

The end-of-module Review Questions can be used to measure and reinforce your knowledge of the module's content.

Review Questions

1. Identification tags for slings must include the _____.
   a. type of protective pads to use
   b. type of damage sustained during use
   c. color of the tattle-tail
   d. manufacturer's name or trademark

2. The type of wire rope core that is susceptible to heat damage at relatively low temperatures is the _____.
   a. fiber core
   b. strand core
   c. independent wire rope core
   d. metallic link supporting core

3. Synthetic slings must be inspected _____.
   a. once every month
   b. visually at the start of each work week
   c. before every use
   d. once wear or damage becomes apparent

4. An alloy steel chain sling must be removed from service if there is evidence that _____.
   a. the sling has been used in different hitch configurations
   b. replacement links have been used to repair the chain
   c. the sling has been used for more than one year
   d. strands in the supporting core have weakened

5. A piece of rigging hardware used to couple the end of a wire rope to eye fittings, hooks, or other connections is a(n) _____.
   a. eyebolt
   b. hitch
   c. shackle
   d. U-bolt

6. A lifting clamp is most likely to be used to move loads such as _____.
   a. steel plates
   b. piping bundles
   c. concrete blocks
   d. plastic tubing

7. Chain hoists are able to lift heavy loads by utilizing a _____.
   a. rope and pulley system
   b. rigger's strength
   c. stationary counterweight
   d. gear system

8. Before attempting to lift a load with a chain hoist, make sure that the _____.
   a. hoist is secured to a come-along
   b. load is properly balanced
   c. tag lines are properly anchored
   d. tackle is connected to its power source

9. A hitch configuration that allows slings to be connected to the same load without using a spreader beam is a _____.
   a. double-wrap hitch
   b. choker hitch
   c. bridle hitch
   d. basket hitch

10. To make the emergency stop signal that is used by riggers, extend both arms _____.
    a. horizontally with palms down and quickly move both arms back and forth
    b. directly in front and then move both arms up and down repeatedly
    c. vertically above the head and wave both arms back and forth
    d. horizontally with clenched fists and move both arms up and down

00106-15 Introduction to Basic Rigging Module Six 31

NCCER Standardized Curricula

NCCER's training programs comprise more than 80 construction, maintenance, pipeline, and utility areas and include skills assessments, safety training, and management education.

Boilermaking
Cabinetmaking
Carpentry
Concrete Finishing
Construction Craft Laborer
Construction Technology
Core Curriculum: Introductory Craft Skills
Drywall
Electrical
Electronic Systems Technician
Heating, Ventilating, and Air Conditioning
Heavy Equipment Operations
Highway/Heavy Construction
Hydroblasting
Industrial Coating and Lining Application Specialist
Industrial Maintenance Electrical and Instrumentation Technician
Industrial Maintenance Mechanic
Instrumentation
Insulating
Ironworking
Masonry
Millwright
Mobile Crane Operations
Painting
Painting, Industrial
Pipefitting
Pipelayer
Plumbing
Reinforcing Ironwork
Rigging
Scaffolding
Sheet Metal
Signal Person
Site Layout
Sprinkler Fitting
Tower Crane Operator
Welding

Maritime

Maritime Industry Fundamentals
Maritime Pipefitting
Maritime Structural Fitter

Green/Sustainable Construction

Building Auditor
Fundamentals of Weatherization
Introduction to Weatherization
Sustainable Construction Supervisor
Weatherization Crew Chief
Weatherization Technician
Your Role in the Green Environment

Energy

Alternative Energy
Introduction to the Power Industry
Introduction to Solar Photovoltaics
Introduction to Wind Energy
Power Industry Fundamentals
Power Generation Maintenance Electrician
Power Generation I&C Maintenance Technician
Power Generation Maintenance Mechanic
Power Line Worker
Power Line Worker: Distribution
Power Line Worker: Substation
Power Line Worker: Transmission
Solar Photovoltaic Systems Installer
Wind Turbine Maintenance Technician

Pipeline

Pipeline Fundamentals
Pipeline Corrosion Control
Pipeline Electrical and Instrumentation
Pipeline Maintenance
Pipeline Mechanical
Pipeline Operations (Control Center, Gas, Liquid)

Safety

Field Safety
Safety Orientation
Safety Technology

Supplemental Titles

Applied Construction Math
Tools for Success

Management

Construction Workforce Development Professional
Fundamentals of Crew Leadership
Mentoring for Craft Professionals
Project Management
Project Supervision

Spanish Titles

Acabado de concreto: nivel uno (*Concrete Finishing Level One*)
Aislamiento: nivel uno (*Insulating Level One*)
Albañilería: nivel uno (*Masonry Level One*)
Andamios (*Scaffolding*)
Carpintería: Formas para carpintería, nivel tres (*Carpentry: Carpentry Forms, Level Three*)
Currículo básico: habilidades introductorias del oficio (*Core Curriculum: Introductory Craft Skills*)
Electricidad: nivel uno (*Electrical Level One*)
Herrería: nivel uno (*Ironworking Level One*)
Herrería de refuerzo: nivel uno (*Reinforcing Ironwork Level One*)
Instalación de rociadores: nivel uno (*Sprinkler Fitting Level One*)
Instalación de tuberías: nivel uno (*Pipefitting Level One*)
Instrumentación: nivel uno, nivel dos, nivel tres, nivel cuatro (*Instrumentation Levels One through Four*)
Orientación de seguridad (*Safety Orientation*)
Paneles de yeso: nivel uno (*Drywall Level One*)
Seguridad de campo (*Field Safety*)

Acknowledgments

This curriculum was revised as a result of the farsightedness and leadership of the following sponsors:

American Petroleum Institute
Central NDT
Construction Craft Academy
CP Masters
EMS USA, Inc.
Enbridge Pipelines, Inc.
Enhance Co.
Enterprise
Flint Energy Services, Inc.
Jomax Construction
Kinder Morgan, Inc.
L.E. Bell Construction Co. Inc.
Marathon Pipe Line
Midwestern Contractors
OneOK
Plains All-American Pipeline
Praxair Services
Protech EIS
Safety Advantage LLC
Southland Safety
Sunoco Logistics, L.P.
Sunoco Pipeline
Thorco Holdings, LLC
URS/Flint Energy

This curriculum would not exist were it not for the dedication and unselfish energy of those volunteers who served on the Authoring Team. A sincere thanks is extended to the following:

Gary Affolter
Carl Allebach
Travis Ashby
Steve Barker
Art Birx
James Carroll
Leslie Carroll
Gordon Carson
Bill Emery
Paul Fontenot II
Paula Gates-Rippe
Trobi Hall
Sid Harrell
David Hoffman
Kimberly-Joy Harris
Les Harrison
Lisa Hartnup
Benjamin Johnson
Dan Lake
Tom Magness
Wallace McCracken
Tim Mills
Steve Montoy
Michael Nordstrom
Tom Pickthall
Donnie Pittsinger
William Rado
David Reeder
Tom Reeder
Randall Rogers
Dan Rohret
Dan Ruane
DeWayne Schapansky
Mark Seitz
Blaine Shellhammer
Lynn Simpson
Cameron Strother
Charles Whiten
Rick Wilenzick
Michael Williams
Erin Worthington
Donavon Wright
Will Wright

NCCER Partners

American Council for Construction Education
American Fire Sprinkler Association
Associated Builders and Contractors, Inc.
Associated General Contractors of America
Association for Career and Technical Education
Association for Skilled and Technical Sciences
Construction Industry Institute
Construction Users Roundtable
Design Build Institute of America
GSSC – Gulf States Shipbuilders Consortium
ISN
Manufacturing Institute
Mason Contractors Association of America
Merit Contractors Association of Canada
NACE International
National Association of Women in Construction
National Insulation Association
National Technical Honor Society
National Utility Contractors Association
NAWIC Education Foundation
North American Crane Bureau
North American Technician Excellence
Pearson
Prov
SkillsUSA®
Steel Erectors Association of America
U.S. Army Corps of Engineers
University of Florida, M. E. Rinker Sr., School of Construction Management
Women Construction Owners & Executives, USA

NCCER Business Partners

ISNetworld Data Provider
Judgment Index
NACB North American Crane Bureau, Inc.
Pearson
Prov

Contents

AOCCC-17

Abnormal Operating Conditions - Control Center

Describes how control center personnel recognize and properly react to abnormal operating conditions (AOCs) that may occur during pipeline operations. Trainees will learn about the federal regulations and agencies governing the operation of gas and liquid pipelines.. (Module ID AOCCC-17; 5 Hours)

AOCFG-17

Abnormal Operating Conditions - Field and Gas

Describes how field personnel recognize and properly react to abnormal operating conditions (AOCs) that may occur during pipeline operations. Trainees will learn about the federal regulations and agencies governing the operation of gas and liquid pipelines. (Module ID AOCFG-17; 5 Hours)

CT25_1-17

Inspect, Test, and Calibrate Pressure Switches

Presents skills and knowledge for inspecting, testing, and calibrating pressure switches in pipelines. Pressure switches are common equipment related to process control, and their correct function is critical to pipeline operation. (Module ID CT25_1-17; 7.5 Hours)

CT25_2-17

Inspect, Test, and Calibrate Pressure Transmitters

Presents skills and knowledge for inspecting, testing and calibrating pressure transmitters in pipelines. Pressure transmitters are common equipment related to process control, and their correct function is critical to pipeline operation. (Module ID CT25_2-17; 7.5 Hours)

CT26_0-17

Verify or Set Protection Parameters for Programmable Controllers and/or Other Instrumentation Control Loops

Presents basic methods for tuning and troubleshooting process controllers. To perform these task, technicians need to have a good understanding of the basic components and functions of the various control loops that are widely used in the pipeline industry. This module reviews the different types of control loops, their functions, and the methods that are used to tune them. It explains how to verify and set protection parameters. It also covers how to perform continuity checks and troubleshoot loops, including troubleshooting an oscillating process. (Module ID CT26_0-17; 15 Hours)

CT30_0-17

Test Overfill Protective Devices

Introduces trainees to overfill protection devices (OPDs), their design, and usage. After instruction covering the skills required, trainees will test OPDs to verify that they are functioning within specifications. (Module ID CT30_0-17; 5 Hours)

CT31_0-17

Inspect and Calibrate Overfill Protective Devices

Introduces trainees to overfill protection devices (OPDs), their design, and usage. After instruction covering the skills required, trainees will calibrate OPDs to ensure that they are functioning within specifications. (Module ID CT31_0-17; 7.5 Hours)

CT44_3-17

Inspect, Test, and Maintain Flow Computer for Hazardous Liquid Leak Detection

Provides an overview of computational pipeline management systems with particular focus on the role that flow computers play in leak detection. In this module, trainees receive an overview of how flow computers communicate with other components of the CPM system, and how to operate and maintain digital flow computers. (Module ID CT44_3-17-17; 7.5 Hours)

CT44_4-17

Inspect, Test, and Perform Corrective and Preventative Maintenance of Tank Gauging for Leak Detection

Describes the importance of tank gauging to the petroleum industry in the receipt and transfer of liquid products. The methods and equipment used to accurately track and measure tank levels are discussed. Symptoms of abnormal conditions associated with monitoring and performing maintenance on tank gauging equipment and appropriate pipeline E&I technician responses are reviewed. The general approach to performing tests and maintenance on tank gauging equipment is discussed. (Module ID CT44_4-17; 7.5 Hours)

CT44_5-17

Prove Flow Meters for Hazardous Liquid Leak Detection

Describes the various techniques and devices used for calibrating, or proving, pipeline flow meters. The trainee is then instructed on the various competencies required for recognizing and reporting abnormalities associated with flow meter proving, as well as the general approach to conducting flow meter proving in order to aid detection of hazardous fluids leakage. (Module ID CT44_5-17; 7.5 Hours)

CT44_6-17

Maintain Flow Meters for Hazardous Liquid Leak Detection

Introduces trainees to a variety of pipeline flow meter technologies and describes the principles upon which they operate. They learn the key advantages and disadvantages of the instruments and how these influence the need for servicing. The trainee is then instructed in the various competencies required for recognizing and reporting abnormalities associated with flow meter operation and maintenance, as well as the general approach to conducting flow meter maintenance. (Module ID CT44_8-17; 7.5 Hours)

CT44_7-17

Inspect, Test and Maintain Gravitometers/ Densitometers for Hazardous Liquid Leak Detection

Identifies and describes the operating principles of the various instruments used to measure pipeline product density. Monitoring product density is essential to custody transfer process and to alert operators of abnormal conditions that may be occurring during a product transfer. Trainees will learn how to service density measuring equipment and recognize abnormal operating conditions relating to petroleum product gravitometers and densitometers. (Module ID CT44_7-17; 7.5 Hours)

CT44_8-17

Inspect, Test and Maintain Temperature Transmitters for Hazardous Liquid Leak Detection

Describes the important methods of measuring and communicating temperature in a process pipeline. The principal temperature sensors and data transmission methods are discussed. Trainees will learn the general approaches to servicing temperature measuring and transmission equipment as well as recognizing abnormal operating conditions relating to petroleum product temperature monitoring systems. (Module ID CT44_8-17; 7.5 Hours)

CT55_0-17

Maintain Fixed Gas Detection Equipment

Presents an introduction to representative categories of analytical equipment associated with a typical pipeline. These instruments monitor and analyze properties of the pipeline contents, as well as detect the presence or absence of specific elements in or around the pipeline equipment. Analytical equipment can also be used by pipeline personnel to test atmospheres and as personal protective devices to indicate when hazardous conditions exist. (Module ID CT55_0-17; 25 Hours)

AOCCC-17

Abnormal Operating Conditions – Control Center

Overview

Pipeline operators and contractors have programs, procedures, training, safety equipment, and warning devices for practically every facet of pipeline operations. Situations that are outside of normal operating conditions may arise at any time, and control center personnel must know what to do to help keep pipeline operations safe for everyone. In this module, personnel will learn how to recognize, properly react to, and properly report AOCs that may occur during pipeline operations.

Trainees with successful module completions may be eligible for credentialing through NCCER's National Registry. To learn more, go to **www.nccer.org** or contact us at **1.888.622.3720.** Our website has information on the latest product releases and training, as well as online versions of our *Cornerstone* magazine and Pearson's product catalog.

Your feedback is welcome. You may email your comments to **curriculum@nccer.org,** send general comments and inquiries to **info@nccer.org**, or fill in the User Update form at the back of this module.

This information is general in nature and intended for training purposes only. Actual performance of activities described in this manual requires compliance with all applicable operating, service, maintenance, and safety procedures under the direction of qualified personnel. References in this manual to patented or proprietary devices do not constitute a recommendation of their use.

AOCCC-17
Abnormal Operating Conditions – Control Center

Objectives

When you have completed this module, you will be able to do the following:

1. Explain how to recognize abnormal operating conditions in a control center for both gas and liquid pipelines.
   a. Explain how to recognize abnormal facility conditions and pipeline system damage.
   b. Explain how to recognize the activation of a safety device.
   c. Explain how to recognize a communications failure and a control system failure.
   d. Explain how to recognize power interruptions.
   e. Explain how to recognize a fire, explosion, and natural disaster occurring with a pipeline.
   f. Explain how to recognize hazardous liquid or gas encounters.
   g. Explain how to recognize an unexplained flow rate/pressure change.
2. Describe the proper reaction to abnormal operating conditions in a control center for both gas and liquid pipelines.
   a. Describe the proper reaction to abnormal facility conditions and pipeline system damage.
   b. Describe the proper reaction to the activation of a safety device.
   c. Describe the proper reaction to a communications failure and a control system failure.
   d. Describe the proper reaction to power interruptions.
   e. Describe the proper reaction to a fire, explosion, and natural disaster occurring with a pipeline.
   f. Describe the proper reaction to hazardous liquid or gas encounters.
   g. Describe the proper reaction to an unexplained flow rate/pressure change.

Performance Tasks

This is a knowledge-based module; there are no performance tasks.

Trade Terms

Abnormal operating condition (AOC)
American National Standards Institute (ANSI)
Breakout tank
Computational pipeline monitoring (CPM)
Disbondment
Encroachment
Highly volatile liquid (HVL)
Lower explosive limit (LEL)
Maximum allowable operating pressure (MAOP)
Maximum operating pressure (MOP)
Naturally occurring radioactive materials (NORMs)
One-Call notification system
Pipeline and Hazardous Materials Safety Administration (PHMSA)
Pipeline facility
Pipeline operator
Redundant system
Resistance temperature device (RTD)
Right-of-way (ROW)
Self-contained breathing apparatus (SCBA)
Supervisory control and data acquisition (SCADA)
Surge pressure
Uninterruptible power system (UPS)

Industry-Recognized Credentials

If you're training through an NCCER-accredited sponsor, you may be eligible for credentials from NCCER's Registry. The ID number for this module is AOCCC-17. Note that this module may have been used in other NCCER curricula and may apply to other level completions. Contact NCCER's Registry at 888.622.3720 or go to **www.nccer.org** for more information.

Contents

Topics to be presented in this module include:

1.0.0 Recognizing Abnormal Operating Conditions....................................1
1.1.0 Recognizing Abnormal Pipeline Facility Conditions....................................2
1.1.1 Pipeline System Damage....................................5
1.1.2 Damaged Pipeline Supports....................................5
1.1.3 Malfunctions Due to Human Error....................................5
1.2.0 Recognizing the Activation of a Safety Device....................................5
1.2.1 Pressure-Limiting Devices....................................6
1.2.2 Seal Failure Alarms....................................6
1.2.3 Overfill Protection Devices....................................6
1.2.4 Pressure Switches....................................6
1.2.5 Activation of a Pressure Relief System....................................6
1.2.6 High Pump-Case Temperature/Pressure Alarm....................................6
1.3.0 Recognizing a Communications Failure....................................7
1.3.1 Failure of Personnel Communication Tools....................................7
1.3.2 System Communications Failures....................................7
1.3.3 Control System Failures....................................8
1.4.0 Recognizing Power Interruptions....................................8
1.5.0 Recognizing Fire, Explosion, or Natural Disaster....................................8
1.5.1 Fire....................................8
1.5.2 Explosion....................................9
1.5.3 Natural Disaster....................................9
1.6.0 Recognizing the Unexpected Release of Hazardous Gas or Liquid....................................10
1.7.0 Recognizing Unexplained Flow Rate/Pressure Changes....................................10
2.0.0 Reacting to Abnormal Operating Conditions....................................11
2.1.0 Reacting to Abnormal Pipeline Facility Conditions....................................11
2.1.1 Pipeline System Damage....................................11
2.1.2 Damaged 0Pipeline Supports....................................12
2.2.0 Reacting to the Activation of a Safety Device....................................12
2.2.1 Activation of a Pressure Relief System....................................12
2.2.2 High Pump-Case Temperature/Pressure Alarm....................................12
2.3.0 Reacting to a Communications Failure....................................12
2.3.1 Failure of Personnel Communication Tools....................................13
2.3.2 System Communications Failures....................................13
2.4.0 Reacting to Power Interruptions....................................13
2.5.0 Reacting to a Fire, Explosion, or Natural Disaster....................................14
2.5.1 Emergency Shutdowns....................................14
2.5.2 Evacuations....................................14
2.6.0 Reacting to the Unexpected Release of Hazardous Gas or Liquid....................................14
2.7.0 Reacting to Unexplained Flow Rate/Pressure Changes....................................15
2.7.1 Unexplained Flow Rate Changes....................................16
2.7.2 Unexplained Pressure Changes....................................16
Appendix A Weather and Natural Disaster Information Sources....................................24
Appendix B Organizations of Interest to the Pipeline Industry....................................25

Figures and Tables

Figure 1 Control center console 2
Figure 2 Pipeline on pipe supports 5
Figure 3 Relief valve 6
Figure 4 Resistance temperature devices (RTDs) on a pump case 7
Figure 5 Flame detector 9
Figure 6 Self-contained breathing apparatus 15

Table 1A Abnormal Operating Conditions: Examples and General Reactions 3
Table 1B Abnormal Operating Conditions: Examples and General Reactions 4

SECTION ONE

1.0.0 RECOGNIZING ABNORMAL OPERATING CONDITIONS

Objective

Explain how to recognize abnormal operating conditions in a control center for both gas and liquid pipelines.

a. Explain how to recognize abnormal facility conditions and pipeline system damage.
b. Explain how to recognize the activation of a safety device.
c. Explain how to recognize a communications failure and a control system failure.
d. Explain how to recognize power interruptions.
e. Explain how to recognize a fire, explosion, and natural disaster occurring with a pipeline.
f. Explain how to recognize hazardous liquid or gas encounters.
g. Explain how to recognize an unexplained flow rate / pressure change.

Trade Terms

Abnormal operating condition (AOC): A condition within a pipeline system, identified by the operator, that may indicate (1) a malfunction of a component or (2) a deviation from normal operations that may indicate a condition exceeding design limits or result in a hazard to persons, property, or the environment.

American National Standards Institute (ANSI): An organization that adopts standards for a variety of products, services, and processes in the United States to ensure consistency.

Breakout tank: A tank used to (a) relieve surges in a hazardous liquid pipeline system or (b) receive and store hazardous liquid transported by a pipeline for reinjection and continued transportation by pipeline.

Computational pipeline monitoring (CPM): Technology that provides monitoring tools used by control center personnel to detect temperature, flow, and pressure irregularities in a pipeline.

Disbondment: Separation of the protective coating from the pipeline.

Encroachment: Activity carried out by and/or equipment belonging to outside parties on the pipeline right-of-way. A particular encroachment might be authorized or not, so each situation requires specific handling.

Highly volatile liquid (HVL): a hazardous liquid which will form a vapor cloud when released to the atmosphere and which has a vapor pressure exceeding 276 kPa (40 psia) at 37.8°C (100°F).

Lower explosive limit (LEL): The lowest concentration of a gas or a vapor in the atmosphere capable of producing fire in the presence of an ignition source. Different products have different LELs.

Maximum allowable operating pressure (MAOP): The maximum pressure at which a pipeline or segment of a pipeline may be operated (gas).

Maximum operating pressure (MOP): The maximum pressure at which a pipeline or segment of a pipeline may be normally operated (liquid).

One-Call notification system: A centralized communication system established by utilities, government agencies, and operators of underground facilities. This system provides one telephone number for contractors and the general public to call to notify of their intent to use equipment on or near a pipeline. It is also in place for any individual to call if pipeline damage is discovered. The One-Call system is generally set up on a state-by-state basis.

Pipeline and Hazardous Materials Safety Administration (PHMSA): An agency of the US Department of Transportation responsible for developing and enforcing regulations for pipeline operations in the United States.

Pipeline facility: New and existing pipe; rights-of-way; and any equipment, facility, or building used in the transportation of hazardous liquids or carbon dioxide, or new and existing pipelines; rights-of-way; and any equipment, facility, or building used in the transportation of gas or in the treatment of gas during the course of transportation.

Pipeline operator: The company that operates the pipeline, as opposed to an individual person who works for the company that operates the pipeline.

Redundant system: A system that duplicates the operations of a system so that, in the event of a failure, a backup system is available.

Resistance temperature device (RTD): A device used on pumps that reads the temperature of the pump case and the product in the pipeline. It will sound an alarm when the temperature reaches a critical level and shut down the pump automatically if needed.

Right-of-way (ROW): The area along the pipeline which has been legally permitted to be used by the pipeline operator to construct, operate, and maintain the pipeline. The ROW extends the entire length of the pipeline.

Supervisory control and data acquisition (SCADA): A computer-based system or systems used by a controller in a control room that collects and displays information about a pipeline facility and may have the ability to send commands back to the pipeline facility.

Surge pressure: Pressure produced by a change in velocity of the moving stream that results from shutting down a pump station or pumping unit, closure of a valve, or any other blockage of the moving stream.

Uninterruptible power system (UPS): A system that provides continuity of power when the normal power supply is interrupted. A UPS provides AC power to the system it serves.

An **abnormal operating condition (AOC)** is defined under federal regulations as a condition identified by the **pipeline operator** that may indicate a malfunction of a component or a deviation from normal operations that may indicate a condition exceeding design limits or result in a hazard to persons, property, or the environment. AOCs are defined as such by federal pipeline safety law 49 *United States Code (U.S.C.)* 60102 and pipeline safety regulations 49 *Code of Federal Regulations (CFR)* 192 (for transportation of gas) and 49 *CFR* 195 (for transportation of liquids). The federal agency that regulates pipeline maintenance is the **Pipeline and Hazardous Materials Safety Administration (PHMSA)** in Washington, DC.

In the control room, control center personnel may become aware of AOCs from reports directly from field personnel, by reports from third parties, or by alerts from computer systems in the control room. Some pipeline equipment is fitted with safety devices that automatically alert control room systems to problems. The control center console (see *Figure 1*) consists of computer systems and monitors that allow control center personnel to observe pipeline operations and take actions as needed.

Types of AOCs, along with examples and proper reactions, are listed in *Table 1*. Note that the reactions are general.

1.1.0 Recognizing Abnormal Pipeline Facility Conditions

Any problem with the physical structure of the pipeline is defined as an abnormal **pipeline facility** condition. A pipeline facility includes the pipeline itself; **right-of-way (ROW)**; and any equipment, facility, or building used in the transportation of a liquid or gas. Examples of components of a pipeline facility include compressors for gas pipelines and pump stations for liquid pipelines (this module refers to both as units). The terms *pipeline* or *pipeline system* mean all parts of a pipeline facility through which a hazardous liquid or gas moves in transportation, including, but not limited to, line pipe, valves, and other equipment connected to line pipe; pumping or compressor units and the fabricated assemblies associated with them; metering and delivery stations; and regulator stations. The following are examples of abnormal pipeline facility conditions that can be reported to the control center:

- Exposed pipeline that should be covered
- Damaged pipeline
- Missing or damaged pipeline supports
- Corrosion
- Coating separating from the pipeline (**disbondment**)
- Missing signs or markers
- Pressure reduction for various reasons
- Reports of unauthorized activity and/or equipment on the pipeline right-of-way (unauthorized **encroachment**)
- Sunken floating tank roof

65102-13_F01.EPS

Figure 1 Control center console.

Table 1A Abnormal Operating Conditions: Examples and General Reactions (1 of 2)

Abnormal Operating Condition	Examples	Appropriate Reaction
Abnormal pipeline facility condition	• Exposed pipe • Evidence of corrosion • Incorrect ANSI pressure rating of a component or material type • Disbondment (coatings separating from the pipe) • Unauthorized activity and/or equipment on the right-of-way (unauthorized encroachment) • Incorrect data or records caused by equipment malfunction • Mechanical/hydraulic system failures • Fire and explosion • Pipeline system damage, such as: – Missing or damaged pipeline supports – Damaged or inoperable pipeline components (pipe, valves, pumps, compressors, meters, tank gauges) due to external force – Gouges, dents, pitting, wrinkles, buckles, or cracks – Cracked or corroded flange or bolts – Sunken floating tank roof – Damaged or malfunctioning pontoon – Defective tank roof seal – Damaged tank shell – Damaged pipe due to natural disaster	Follow standard operating procedures and make notifications per company policy. Isolate and/or shut down the pipeline segment or facility if necessary. This may require coordination with field personnel. Complete the required documentation.
Activation of a safety device	• Emergency or abnormal shutdown alarm • High-pressure shutdown alarm • Activation of a pressure relief valve • Activation of a smoke or fire detector • Activation of a hazardous atmosphere alarm • Activation of an overfill protection device • High sump-level shutdown alarm • High tank-level shutdown alarm • High pump-case pressure alarm • Seal failure alarm	Follow standard operating procedures and make notifications per company policy. Isolate and/or shut down the pipeline segment or facility if necessary. This may require coordination with field personnel. Complete the required documentation.
Communications failures	Personnel communication tools: • Lost use of radio communications • Lost use of landline telephone • Lost use of cell phone System communications: • Facility outages • SCADA communication outage • System programming errors/incorrect command logic • Loss of satellite communications • Loss of microwave signals • Stale data (communications going from the control center to the field, but not being received by the control center from the field)	Switch to alternate equipment or backup systems. Follow standard operating procedures and make notifications per company policy. Complete the required documentation.

- Incorrect data or records caused by equipment malfunction
- Incorrect **American National Standards Institute (ANSI)** pressure rating of a component or material type
- Malfunctions due to human error

Control center personnel may be notified of an abnormal facility condition by field personnel. A report may also come in to the control room as

Table 1B Abnormal Operating Conditions: Examples and General Reactions (2 of 2)

Abnormal Operating Condition	Examples	Appropriate Reaction
Power interruptions	• Transformer failure • Downed power lines • Broken leads or wires • Incorrect wire termination • Station or terminal power problems (due to motor overload or a tripped breaker, for example)	Follow standard operating procedures. Switch to an alternate control system or power source if necessary. Notify the power company of the outage. Make additional notifications per company policy. Complete the required documentation.
Fire, explosion, or natural disaster	• Electrical fire • Ignition of a hazardous vapor • Unintentional detonation • Weather conditions, such as an ice storm • Fire, explosion, or pipeline movement caused by natural disaster, such as: – Lightning strike – Flooding – Landslide – Earthquake – Tornado	Follow company emergency action plans. Call local emergency personnel if appropriate. Isolate and/or shut down the pipeline segment or facility if necessary. Limit access to the area as much as possible. Make notifications per company policy. Complete the required documentation.
Unexpected release of a hazardous gas or liquid	Reports of: • Vapor clouds or frost balls • Smell of hydrocarbon release • Dead vegetation/contaminated soil • Dead wildlife • Rainbow sheen on water Hazardous products include: • Hydrogen sulfide • Carbon dioxide • Anhydrous ammonia • Propane • Natural gas • Refined fuels/products Note that the controlled transfer, sampling, venting, draining, and handling of hazardous products are not considered AOCs.	Follow company emergency action plans. Call local emergency personnel if appropriate. Isolate and/or shut down the pipeline segment or facility if necessary. Make notifications per company policy. Complete the required documentation.
Unexplained flow rate / pressure change	Any change from the expected flow or pressure, such as: • Increased pressure • Decreased pressure • No pressure • High flow limit • Low flow limit More often than not, irregularities are detected through a computational pipeline monitoring (CPM) system	Follow standard operating procedures and make notifications per company policy. Isolate and/or shut down the pipeline segment or facility if necessary. Complete the required documentation.

a result of a call made to the **One-Call notification system** or from a person calling the pipeline operator directly. When a report comes in, personnel must obtain complete information so that they can quickly find the problem and address it. Many pipeline operators use a form to capture the following information:

- Caller's name and telephone number
- Date and time of the call
- Was anyone injured? Was 911 notified by the caller?
- Type of product involved, if relevant and if known
- Location of the problem, including latitude and longitude, if possible
- Nearby landmarks such as highways, streams, mile markers, or any other structure
- The nature of the damage and any conditions observed at the site

One of the most important things to remember as a control center operator is to always follow company policy and procedures with any abnormal operating condition.

1.1.1 Pipeline System Damage

The control center may receive the indication of pipeline damage on the control center console as an unexplained change in operating conditions. Pipeline damage or exposed pipe may also be discovered by field personnel or a third party who then reports the damage to the control center. Third-party examples include the aerial patroller of the pipeline ROW who notifies the pipeline operator, or an individual landowner who calls the One-Call notification system.

One way that control center personnel notice damage is by monitoring the pressure and flow changes on the control center console. Control center personnel use monitoring tools, referred to as **computational pipeline monitoring (CPM)** systems, to detect pressure, flow, and temperature irregularities in a pipeline. These irregularities may indicate a release. Most CPM systems provide an alarm and display other data related to pressures, flows, and temperatures that personnel use to detect problems.

Physical signs of damage to the pipeline and its components may include gouges, cracks, pitting, dents, buckles, and inoperable valves, pumps, or tank gauges. Flanges or bolts may be cracked or show significant corrosion. Tank shells or pipeline supports may show damage. A sunken roof or a damaged pontoon also indicates damage. Less obvious signs of damage could include a rainbow sheen on the water near submerged pipelines, which may indicate a release. The release of a **highly volatile liquid (HVL)** may be discovered by observing a frost ball or vapor cloud along the ROW of a liquid pipeline. While the control center personnel are not able to see physical damage to a pipeline, unless viewed through a camera system, they often receive notice of these types of damage indications.

1.1.2 Damaged Pipeline Supports

Control center personnel may receive reports about damage to pipeline supports (see *Figure 2*) from maintenance or inspection personnel who are traversing the pipeline. Damage can include visible cracking, sagging, or buckling of a support, or bolts or other components of the support that have broken or become loose. This type of damage can be a serious problem and can require a shutdown. Sometimes pipeline operations can continue while repairs are made.

65102-13_F02.EPS

Figure 2 Pipeline on pipe supports.

1.1.3 Malfunctions Due to Human Error

Malfunctions are not always due to mechanical or other system failures; they can occur simply due to human error. Control center personnel should be able to recognize the following possible malfunctions as a result of human error:

- Deviation from the design limits
- Valves that are not sequenced correctly
- Cavern upset
- Units or valves that are not aligned properly
- Units that are set incorrectly, which may cause overpressure in the line
- Overflow of a tank

1.2.0 Recognizing the Activation of a Safety Device

Safety devices are part of the pipeline system designed to sense and alert personnel of any potential danger and/or hazardous condition. Hazardous conditions can include changes in pressure that might exceed pipeline design limits as well as the release of harmful gases. Safety devices include, but are not limited to, pressure relief systems, pressure limiting devices, high pump-case-pressure alarms, seal failure alarms, overfill protection devices, tank-level alarms, main-line pressure shutdown alarms, and hazardous atmosphere alarms.

Generally when a safety device activates, a sensor on the device sends a signal to the control center. An alarm may sound in the control room, and the signal appears as an alarm on the control center console. All safety alarms must be verified and the problem corrected before clearing the alarm.

1.2.1 Pressure Limiting Devices

Pressure-limiting devices include control valves. These valves are monitored and controlled by the control center. If the pressure in the pipeline is different than it should be, the control valves are adjusted as needed.

1.2.2 Seal Failure Alarms

Seal failure alarms are activated when a leak is detected in the seal on a pump. There are different types of seals that can activate seal failure alarms when not working properly. One common type of seal is a tandem seal. Tandem seals are a combination of two seals with a barrier liquid between the seals in a chamber that is not pressurized. When the barrier fluid rises to a certain level from process liquid entering the chamber the alarm is activated.

Another type of seal commonly used is a double seal. Double seals consist of two seals that are mounted back-to-back. Usually a seal failure results in product leaking into the atmosphere. This could cause a hazardous atmosphere alarm to activate.

1.2.3 Overfill Protection Devices

Overfill protection devices activate if product reaches a certain level in the tank. As a result of activation, the system may shut off product flow, restrict product flow, or signal an alarm. Two types of overfill protection device are a high and a high-high level alarm that detects the level of product when it reaches a predetermined level. The alarm may also activate a pump shutdown or valve to divert the product flow to prevent a release. Another device related to the overfill protection device is the low-low level alarm. This alarm activates when the product level reaches a predetermined low level. The pump does not shut down with a low-low alarm activation, but monitoring may be required to ensure the levels do not become dangerously low and damage the pump.

1.2.4 Pressure Switches

Pressure switches activate when electrical contacts within the switch close when the pressure reaches a set level. The switch may be designed to make contact on pressure rise or on pressure fall. Pressure switches can have either automatic or manual resets. The automatic resets will reset when the pressure returns to normal operating range. Manual resets need to be reset by personnel.

1.2.5 Activation of a Pressure Relief System

The activation of relief devices along the pipeline system is a time-sensitive occurrence. Relief valves can discharge into tanks or vessels that are designed and constructed in compliance with American Petroleum Institute (API) *Standard 650* or *API 2510* for pressure vessels. Relief valves can also open into the atmosphere. Relief valves (see *Figure 3*) can be installed on a pipeline or **breakout tank**. The purpose of a pressure relief device is to protect the pipeline or tank from overpressure. Breakout tanks can also be used to temporarily absorb **surge pressure** from the main-line and manifold pressure-relief valves to lower pressure on the line. Pressure-relief devices must be provided for each low-pressure and high-pressure breakout tank. There are different specifications for low-pressure and high-pressure tanks.

Some valves can be manually operated by field personnel, and some are remotely operated by control center personnel. When a relief valve is activated, an alarm may sound.

1.2.6 High Pump-Case Temperature/Pressure Alarm

Main-line pumps are fitted with a **resistance temperature device (RTD)** that give temperature readings for the pump case and product. These devices are preset to signal an alarm in the control center

65102-13_F03.EPS

Figure 3 Relief valve.

when the temperature reaches a critical level, and will shut the pump down automatically if some action cannot be taken to lower the temperature. When a pump-case temperature reaches such a critical level and automatically shuts down, it is considered an AOC. *Figure 4* shows RTDs on a pump case.

1.3.0 Recognizing a Communications Failure

Communications refers to the tools that personnel use to communicate with each other and with systems. Communications failures include the failure of communications among personnel, or the failure of communications between the control center and equipment on the pipeline.

1.3.1 Failure of Personnel Communication Tools

Pipeline personnel use landline telephones, cell phones, and radios to communicate with one another on the job. A breakdown in personnel communications equipment may be easy to recognize. Personnel may not hear messages or be able to make phone calls. Personnel must keep their equipment in good working order and test their equipment regularly to ensure that it is working properly.

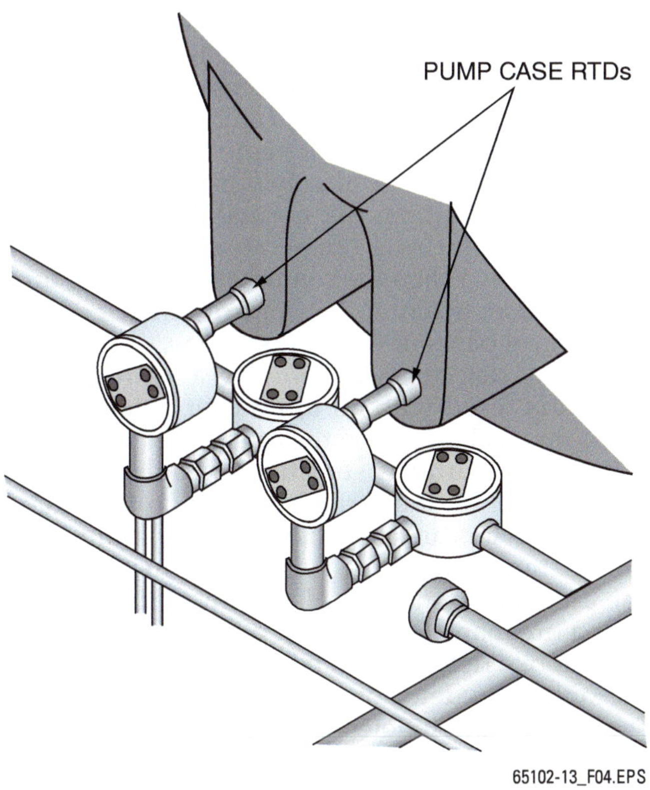

Figure 4 Resistance temperature devices (RTDs) on a pump case.

1.3.2 System Communications Failures

System communications refers to the computer equipment located in the control center and on some pipeline equipment. Personnel in the control center use the **supervisory control and data acquisition (SCADA)** system to manage pipeline operations. The system receives signals from and sends signals to pipeline equipment that may be many miles away. The SCADA system allows control center personnel to monitor pressure, volume, and temperature and keep products flowing safely and on schedule. If the system fails, control center personnel lose the ability to communicate with equipment on the pipeline. Some systems are set up to switch to a **redundant system**—a backup that comes on line if the main system fails.

Some failures of communications systems show up as warnings or messages on the control center console. Another potential sign of a communications failure is stale data showing on the computer monitor. A system failure may include any of the following events:

- Random facility outages
- Partial or complete facility outages
- Loss of the SCADA system
- Loss of satellite communications
- Loss of microwave signals
- Loss of a signal point within the system
- Stale data

If a communication failure occurs, the control center computer signals control center personnel of a problem so that an investigation can begin and the problem can be isolated, if needed.

If an individual remote terminal unit (RTU) loses communication with the control center, control center personnel must follow appropriate company procedures. In an emergency shutdown situation, if communication is lost to a main-line station that must be shut down, control center personnel must use other means to shut down the station. One method is to allow the suction pressure to decrease below the low-suction set point and call the appropriate field personnel.

The discovery that a large group of RTUs is not responding reveals a serious problem. Some SCADA systems are set up to switch to a backup communication system so that an attempt to control the situation can begin as soon as possible. Some companies are equipped with two backup computer systems in case the SCADA fails. Control center personnel should request a switch to one of the backup systems if available.

Control center personnel must consult with appropriate personnel to try to estimate how long the communications system will be in failure. The console monitors will reveal how many systems are affected. If control center personnel are advised that the system will be in failure for a specified amount of time, they will initiate appropriate procedures.

1.3.3 Control System Failures

Mechanical and hydraulic failures, also known as control system failures, are related to system communications failures. For example, the SCADA system sends a signal to a piece of equipment, but the equipment does not respond to the signal. In some cases, the problem is a result of a programming or command logic error. In other cases, equipment may not work because of a mechanical, electrical, or hydraulic problem. Examples include frozen valves, electrical component burnouts, and compressor breakdowns. A valve may have a mechanical failure that prevents it from being operated remotely by the control system. Sometimes valve actuators, electrical components, or connections burn out. This prevents the equipment from responding to a command by the system to start up or shut down.

1.4.0 Recognizing Power Interruptions

Electric power is essential to pipeline operations. Nearly every part of the pipeline depends on electricity in some way. A loss of power impacts the abilities of the control center or facility operations to safely control and monitor the pipeline facility. Power interruptions that result from bad weather or a problem at the local utility company, such as transformer failure or downed power lines, can be common.

Some pipeline operators have an **uninterruptible power system (UPS)** that turns on if electrical power is off for more than a few seconds. UPS systems supply emergency alternating current (AC) power to pipeline components to keep the pipeline operating until power is restored. Instead of electricity, these systems run on engines fueled with gasoline, diesel, or natural gas that are designed to start up if power is off for more than a few seconds. When power fails, an AC power failure alarm may be sent to the control center. To ensure that the backup system has turned on, control center personnel check to see whether they are receiving updated information from the pipeline.

If power is lost, electronic indicators may be lost. The SCADA system may switch off and pumps or compressors may shut down. In addition, lights, heat, and air conditioning may fail. An area-wide power outage may cause a loss of telephone communications. Local power outages may shut down only a portion of the pipeline system. Power outages can also cause pumps or compressors at the receiving terminal to stop operating, causing a liquid or gas shipment to overflow in tanks or be unable to be shipped further.

Some equipment is protected from power failure by battery systems. These systems run on direct current (DC). The following types of equipment can operate on DC power:

- Fire detection equipment and alarms
- Halon dispenser systems
- Fire suppression systems
- Emergency lighting systems
- SCADA
- Alarm panels
- Control modules
- Process control panels
- Specialized compressor units

1.5.0 Recognizing Fire, Explosion, or Natural Disaster

Fire, explosion, or natural disaster can occur at any time. Some events may be anticipated, but many are not. Control center personnel must be trained to recognize these events.

1.5.1 Fire

Detection devices installed along the pipeline and in pipeline facility buildings alert personnel in case of a fire. These devices send an alarm when their sensors detect heat or flame. In most cases, control center personnel will only see or hear the alarm, not the fire itself. Detection devices include thermal detectors, flame detectors, and smoke detectors.

A thermal detector is a heat-sensitive unit. It is fitted with electrical contacts that stay open as long as conditions are normal. When the unit senses rapidly rising temperatures, the contacts start to close. When the contacts connect, the unit may send an alarm to the control center and the fire control panel in the workstation.

A flame detector (see *Figure 5*) is a light-sensitive unit. It is fitted with an ultraviolet detector that turns on an alarm when the unit senses light from a flame. The unit is sensitive enough to detect a flame even in the earliest stages of a fire. Some flame detectors are connected to a firefighting system that may dump fire suppression material into the area. Personnel can also dump this material over a fire manually by using pull switches.

A smoke detector can be a light-sensitive unit, an ionization detector, or a combination of both types. Ionization detectors are more sensitive to fires in the early smoldering stage, while light detectors are more sensitive to the flaming stage of fires than ionization detectors.

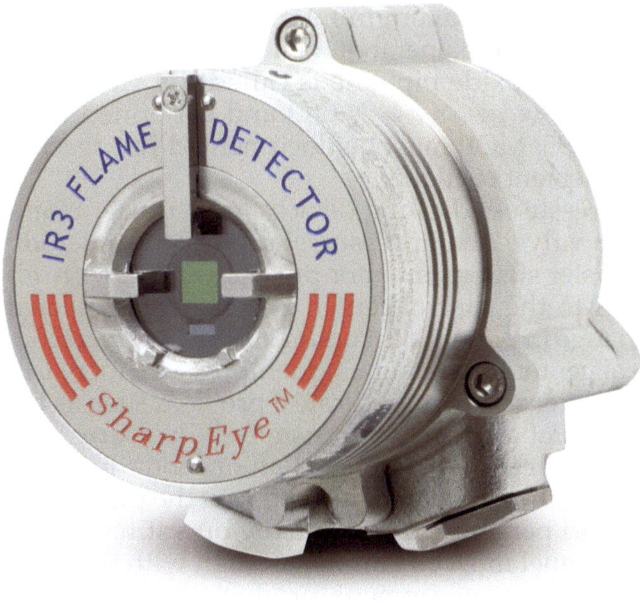

65102-13_F05.EPS

Figure 5 Flame detector.

An ionization smoke detector produces ionization in air; when there is smoke, a difference is detected. If any smoke particles pass into the chamber of the detector, the ions attach to the particles, and become less able to carry the current. An electronic circuit detects the current drop, and sounds the alarm. The light detector operates by detecting light beams. In the absence of smoke, the light passes in front of the detector in a straight line. When smoke enters the chamber across the path of the light beam, light is scattered by smoke particles, directing it at the sensor that triggers the alarm.

Monitoring devices that detect a **lower explosive limit (LEL)** produce alarms to indicate that there is a risk of explosion. The LEL is the lowest concentration of a gas or a vapor in the atmosphere capable of producing fire in the presence of an ignition source. Different products have different LELs. The monitoring device must be set to detect the particular LEL in an enclosed area. Hazardous atmosphere detectors also need to be used in an enclosed area where an LEL has gone off before anyone re-enters the area.

1.5.2 Explosion

Pipeline operators install a number of safety devices to warn personnel about conditions that could cause explosions. However, operators cannot control every circumstance, such as intentional damage, that could lead to an explosion on a pipeline. An explosion on a pipeline registers in the control center as an unexplained change in pressure. An explosion may occur in any enclosed space where hazardous vapors can accumulate to dangerous levels. Examples of such spaces include pipeline facility buildings.

If natural gas leaks into an enclosed space, the odorant added to the gas sometimes alerts people to the problem. However, gas can sometimes accumulate undetected. For example, a building may be vacant or the odors may be too far away from people inside the building. In addition, products may not have an odorant added during transport. Be aware that a spark from electrical equipment can cause the gas to explode.

Whenever a call comes into the control center in which the caller describes a smell related to a pipeline, always make note of it for reporting purposes.

WARNING!

Personnel should never rely on their sense of smell to detect a natural gas leak. Products do not always have a detectable odor.

1.5.3 Natural Disaster

Natural disasters include earthquakes, hurricanes, tornadoes, landslides, wildfires, or flooding. These events may cause soil movement, which may result in forces sufficient enough to cause pipeline failure or product release. Even heavy rains resulting in saturated soil conditions may cause soil movement in some areas.

Pipeline operators monitor for weather that can develop into a natural disaster. Control center personnel may monitor for earthquakes with seismic indicators. In addition, they may obtain real-time information on the internet. Up-to-date earthquake information is available from the National Earthquake Information Center, which is located in Denver, Colorado. Personnel may track weather alerts and warnings through the National Weather Service website. In case of a power failure, personnel can continue to receive information through the National Weather Service's National Weather Radio or other electronic devices. A list of weather and natural disaster information sources is included in *Appendix A*.

1.6.0 Recognizing the Unexpected Release of Hazardous Gas or Liquid

Control center personnel will usually receive reports about an unexpected release from field personnel. The field personnel may notice the physical signs of a release such as a pool of liquid on the ground, a frost ball, vapor cloud or mist in the air, dead or discolored vegetation, or a rainbow sheen on water. Other signs include dead fish in waterways or other dead wildlife. Know company policies and procedures for each type of release and be prepared to act quickly and decisively to protect people, the pipeline, and the environment.

1.7.0 Recognizing Unexplained Flow Rate / Pressure Changes

Flow rate and pressure are important indicators of both normal and abnormal operating conditions. Minor changes in pressure may occur normally on the pipeline. However, significant unexplained pressure changes are considered AOCs. Unexplained pressure changes include absences of, increases in, and decreases in pressure. The maximum pressure a gas pipeline or segment of a gas pipeline may be safely operated is called the **maximum allowable operating pressure (MAOP)** and is defined in 49 *CFR* 192. MAOP is based on the materials transported and the design of the system. The maximum internal pressure permitted during the operation of a hazardous liquid or carbon dioxide pipeline is called the **maximum operating pressure (MOP)** and is defined in 49 *CFR* 195. MOP must be documented if it reaches 110 percent or higher.

Pressure-indicating devices located near mainline valves, compressor stations, and custody transfer stations monitor pressure and send readings to the control center. When an unexpected change in pressure occurs, an alarm may alert control center personnel. Sudden pressure increases or decreases might indicate a problem such as the failure of a valve or a transmitter. Pressures may also increase when inspection or monitoring devices called pigs are launched into the pipeline or if field personnel are manually operating valves. Personnel must be familiar enough with the pipeline system, equipment, product characteristics, and environmental factors to quickly identify any important changes in pressure and flow.

Additional Resources

It is assumed that you will learn to perform these procedures under the direct supervision of your crew leader, foreman, or instructor. For independent study, refer to the recommended Additional Resources at the back of this module.

1.0.0 Section Review

1. Which of the following is an AOC?
   a. Aboveground pipe on pipe supports
   b. Damaged pipeline
   c. Valves operating properly
   d. Authorized encroachment

2. Safety devices include _____.
   a. hazardous atmosphere alarms
   b. tank shells
   c. SCADA
   d. compressors

3. SCADA serves as a _____.
   a. regulatory agency
   b. notification system for all utilities
   c. device worn to provide breathable air where the atmosphere is hazardous
   d. command system that assists control center personnel

4. An uninterruptible power system (UPS) supplies _____.
   a. temperature readings
   b. emergency AC power
   c. emergency DC power
   d. pressure readings

5. A flame detector is designed to activate when it detects _____.
   a. an increase in temperature
   b. light
   c. smoke
   d. pipeline pressure

6. Dead vegetation or dead wildlife may be a sign of an unexpected release of a hazardous gas or liquid.
   a. True
   b. False

7. The maximum operating pressure of a gas pipeline is called _____.
   a. MOP
   b. MAOP
   c. surge pressure
   d. compression

SECTION TWO

2.0.0 REACTING TO ABNORMAL OPERATING CONDITIONS

Objective

Describe the proper reaction to abnormal operating conditions in a control center for both gas and liquid pipelines.

a. Describe the proper reaction to abnormal facility conditions and pipeline system damage.
b. Describe the proper reaction to the activation of a safety device.
c. Describe the proper reaction to communications failure and a control system failure.
d. Describe the proper reaction to power interruptions.
e. Describe the proper reaction to a fire, explosion, and natural disaster occurring with a pipeline.
f. Describe the proper reaction to hazardous liquid or gas encounters.
g. Describe the proper reaction to an unexplained flow rate / pressure change.

Trade Terms

Naturally occurring radioactive materials (NORMs): Materials with radioactive elements that can be extracted during oil and gas extraction from the earth and can get into product streams and become transported in the pipeline.

Self-contained breathing apparatus (SCBA): A device worn to provide breathable air for an individual in an area where the atmosphere is hazardous.

Time is of the essence when AOCs occur, and control center personnel are expected to take decisive action. Personnel must always follow established company-specific policies and procedures. Personnel must make use of their training, skills, and good judgment to maintain safe operation until solutions are found.

2.1.0 Reacting to Abnormal Pipeline Facility Conditions

Some abnormal pipeline facility conditions may require immediate action. Others may require scheduling repairs at a later date. In most cases, if control center personnel are notified of an abnormal pipeline facility condition, they must notify the appropriate personnel so that repairs can be made safely and in a timely manner. If notified of a condition by a third-party call, keep in mind that a caller's description may not be entirely accurate. The problem may be other than what was described. If a report is confirmed and the condition does not require shutdown, make appropriate notifications as required by company policy. Reports of exposed pipe and damaged pipeline supports are examples of conditions that may require immediate action.

2.1.1 Pipeline System Damage

If a report of exposed pipe comes in to the control center, control center personnel should check the data received from the monitors and SCADA and send field personnel to investigate according to company policy. It is possible that communications were destroyed at the time the damage occurred and thus no alarm was signaled. The pipeline may need to be shut down until the investigation is complete, according to company policies and procedures. The decision to do so should be based on the assessment provided by trained and qualified personnel. The sooner qualified personnel can arrive on the scene, the more quickly a practical assessment of conditions can be obtained and control of the situation can begin.

Some incidents may be serious enough to require an emergency shutdown of the pipeline. One example of a serious incident is the loss of control of a vehicle on the ROW, extensively gouging or denting a large section of pipeline or damaging a valve. Another common example is third-party damage to pipeline from excavation. Whenever pipeline damage results in a release and/or explosion, control center personnel should receive a report from field personnel. In some cases, control center personnel may be able to drain a portion of the pipeline into a delivery system to lessen the amount of product that could be released from the pipeline.

Pipeline shutdowns do not have to occur only because of obvious damage to the pipeline. A blocked valve that causes increased pressure in the line can be reason for a shutdown. However, if pressures remain within the normal operating parameters, a shutdown may not be necessary, but the control center personnel must report the status change to the appropriate personnel.

Once the pipeline has resumed normal operations, the pipeline operator will conduct an internal investigation. This investigation focuses on evaluating control systems and mechanical equipment, as well as personnel performance. Personnel actions must be evaluated to determine whether company policies and procedures were followed. The work must also be evaluated to determine the effectiveness of existing company procedures. If the operator suspects that the damage was intentional, the appropriate personnel may contact regulatory and law enforcement agencies that may then conduct a criminal investigation.

2.1.2 Damaged Pipeline Supports

The proper reaction to the discovery of damaged pipeline supports varies depending on the severity of the damage. If the damage to the support is critical or indicative of a possible release, control center personnel may receive a report from field personnel. Notification is especially important if the pipe is in danger of excessive movement. Follow company policy precisely to investigate the condition and dispatch qualified personnel to the scene.

2.2.0 Reacting to the Activation of a Safety Device

Control center personnel must respond to safety device alarms quickly and calmly. They must be able to dispatch the appropriate personnel and inform them of the potential for a hazardous environment. Control center personnel may need to shut down the pipeline, if necessary. Field personnel may need to reset alarms. Company procedures must always be followed.

It is important to ensure that an alarm is not a false alarm. In case of a false alarm, the control center should notify affected personnel and have the alarm equipment checked. Generally, no further action is required when an alarm is false.

For example, if responding to a seal failure alarm, personnel should investigate the validity of the alarm, check the station sump level, and notify the appropriate personnel. This may not necessarily result in an AOC, but it may require control center personnel to document the alarm and notify the appropriate personnel to check and possibly replace the seal. Control center personnel should never attempt to restart a pump before investigating the cause of the alarm and responding to the alarm according to company policy.

During an emergency, personnel must focus on three key areas: ensuring personnel safety, reducing the damage to equipment, and protecting the environment. The general rule is to put people first. The goal is also to prevent small problems from turning into larger incidents. All reactions must be made safely.

WARNING!

For safety reasons, only authorized personnel who have had the proper training may use PPE.

2.2.1 Activation of a Pressure Relief System

Each relief event should be recognized and must be documented by control center personnel. If a relief device is activated, control center personnel must immediately minimize the situation and verify the device is operating correctly. If the alarm is signaled and the pressure remains high, personnel must take immediate action to reduce pressure. A pressure reduction is usually achieved by shutting down the upstream pumping or compressor station or the pipeline system.

2.2.2 High Pump-Case Temperature/Pressure Alarm

Some main-line pumps are set to shut down on low pressure. Pipeline operators fit pumps with differential pressure switches and temperature switches to protect the equipment from damage caused by low pressure, no pressure, or no flow; to avoid exceeding maximum operating pressure (MOP)/maximum allowable operating pressure (MAOP), which means no more than 110 percent MOP/MAOP; and to ensure that the temperatures do not increase in the pipeline facility. Pumps also have high discharge-pressure alarms that will shut down the pump when the discharge pressure exceeds a preset limit. Many times, this indicates a downstream line obstruction that is not allowing the product to flow properly from the pump.

2.3.0 Reacting to a Communications Failure

The proper reaction to a communications failure depends on whether the failure involves a communications tool that personnel use or the failure of system communications.

2.3.1 Failure of Personnel Communication Tools

When communication tools fail, personnel should first check to see if the problem is a simple one that can be quickly and easily fixed. For example, some telephones have a volume control that may accidentally be turned off. The telephone may seem to be dead, but a simple adjustment to the volume corrects the problem. Other simple solutions include replacing a battery or recharging a unit.

Some communications problems are more serious. Weather or problems at the local utility company can knock out communications completely. Regardless of the outage, the best preparation is to ensure personnel are trained to follow these steps:

Step 1 Switch to alternate equipment if necessary.

Step 2 Notify the appropriate personnel per company policy.

Step 3 Complete any required documentation.

2.3.2 System Communications Failures

Control center personnel must recognize an abnormal condition, investigate immediately, and react according to company policy. The condition must be verified, and the cause(s) should be determined. If the failure involves a major pipeline component such as a remote gate valve or block valve, control center personnel must try to solve the problem as quickly as possible to prevent an incident. In some instances, control center personnel might be able to bypass the problem component until field personnel can perform appropriate maintenance. In other cases, they may be required to shut down.

If one of the pumps at a station that is running suddenly indicates a stop status with no apparent changes to the station pressures, control center personnel should notify appropriate field personnel, and investigate as required per company policy. If a control center loses its monitoring capabilities for a remotely unmanned station, but SCADA is showing normal operating pressures and flow rates upstream and downstream of the station, control center personnel should not make multiple unit switches upstream or downstream of the facility before it is confirmed that a problem actually exists. Appropriate reactions include restoring communications with an alternate system, notifying repair personnel, and having the pipeline facility manned.

The control center must notify field personnel and/or systems support about system communications failures so that they can isolate, investigate, and address the problem as needed. If field personnel can operate the component manually, the problem can be alleviated until repairs can be made. If the component cannot be manually operated, the pipeline may have to be shut down. Control center personnel must monitor pressure remotely or shut down. If pressure is remaining within limits or the component is not one that will affect the operation directly, appropriate personnel must decide if the pipeline can safely remain in operation. If the pipeline remains in operation, the condition must be monitored closely.

After normal conditions have been restored, control center personnel must continue to monitor the condition to make certain that the procedures have been successful. If SCADA communication is lost, control center personnel must follow their company's policies and procedures, usually covered in the business continuity plan. These procedures may include switching to a backup system that may be in another location.

2.4.0 Reacting to Power Interruptions

The proper reaction to a power failure depends on which part(s) and how much of the pipeline is affected. Regardless of how the outage affects the pipeline facility, the appropriate personnel should always notify the power company of the outage as soon as possible. If only a single component or station is involved, control center and/or field personnel may be able to adjust line pressure and flow to keep the pipeline in operation. For example, if a power interruption occurs while a valve is in operation, field personnel may have to operate the valve manually. However, if a station is needed to keep pressure within nnormal operating limits, control center personnel may have to isolate a pipeline segment. If a shutdown is not required, the gas or liquid shipment will continue to arrive, and the schedule must be followed and the tasks completed manually.

Power interruptions can happen in the control center itself. If backup power isn't available, the control center personnel should contact field personnel to go to their control stations to maintain pipeline operations.

Each company has specific procedures for responding properly to a power interruption. Always follow company policy and make appropriate notifications per company policy.

2.5.0 Reacting to a Fire, Explosion, or Natural Disaster

The pipeline operator must have procedures in place so that personnel know what to do in case of a fire, explosion, or natural disaster that affects a pipeline. Always react according to company policy, which includes the policies regarding shutting down the pipeline.

When any of these events occur, the safety of personnel is always top priority. Personnel must ensure that any injured individuals receive medical attention as soon as possible. Consider any alarm as an actual alarm and react appropriately and decisively until it has been verified as false. Control center personnel must investigate the cause of any alarms or other notifications and confirm the nature of the problem in order to react appropriately. Personnel must make appropriate notifications and complete the required documentation.

In emergency situations, including evacuation if needed, follow the company emergency procedures manual as appropriate. Personnel must be adequately trained and familiar with emergency procedures in advance.

2.5.1 Emergency Shutdowns

Some events occurring on the pipeline that present immediate hazards require an emergency shutdown. Such an event might be caused by human error, mechanical failure, electrical failure, or an accident on the pipeline. It might also be brought about by threats from weather or civil disobedience. Regardless of the reason, when an emergency shutdown is required the line must be shut down quickly and safely. Each company has a standard operating procedure for emergency shutdowns.

In the case of a fire, explosion, or release, time for discussion may be very limited. Once the need is confirmed, decisive action must be taken. The pipeline must be shut down by control center personnel who will shut down the appropriate pumps. Control center personnel may isolate or authorize the isolation of the system by closing block valves as necessary.

In emergency conditions such as a possible release, the pipeline operator has full responsibility and authority to make shutdown decisions and take appropriate action. These actions are governed by the company release policy above all other considerations.

2.5.2 Evacuations

The safety of pipeline personnel and the public is the first priority when any AOC occurs. If there is any threat of injury or death to any persons, an evacuation of pipeline personnel must occur. Events that could result in the need for an evacuation are fire, explosion, release of hydrocarbon or other hazardous vapors, seismic activity, or other natural disasters.

Companies have standard procedures in place to guide evacuation of personnel. By law, emergency response plans must be in place and evaluated annually. Facility layouts with evacuation maps are required to be posted in every area or department. Safe areas are designated for all personnel. When the command to evacuate is given, trained personnel may act as safety coordinators. Their job is to conduct a roll call to confirm that all personnel are safe. To assist in this process, personnel must report to their designated safe areas and identify themselves to the designated safety coordinator. If any personnel cannot be accounted for, authorized personnel must be informed so that missing individuals can be located and rescued, if necessary.

2.6.0 Reacting to the Unexpected Release of Hazardous Gas or Liquid

When a hazardous gas or liquid is released from a pipeline unexpectedly, personal safety should be the first consideration. Beyond that, the proper reaction to the unexpected release of a product depends on the characteristics of the product and the location of the release. Some gases and liquids can explode under certain conditions, and some are toxic to people and the environment. To identify the hazards associated with the chemical being released, consult the material safety data sheets (MSDSs) / safety data sheets (SDSs).

Once notified by field personnel of a release, the source of the release must be determined quickly. Control center personnel may need to reduce pressure and shut down the pipeline. If power cannot safely be cut off at the switchgear, control center personnel must call the power company and have power cut off from outside the pipeline facility. Releases in outdoor areas include the additional concern for the safety of the general public, as well as the safety of the environment. In all cases, follow company policies and procedures, make the appropriate notifications, and complete the required documentation.

Personnel must take special precautions when dealing with the release of products that can become toxic. Examples of toxic products include carbon dioxide (CO_2), hydrogen sulfide (H_2S), propane, anhydrous ammonia, and natural gas.

Carbon dioxide is a colorless, odorless, gaseous compound used as a propellant in many domestic applications, such as fire extinguishing equipment, and is a by-product of combustion. Because it is heavier than air, it can collect in low-lying areas where it may displace oxygen; exposure can result in suffocation and death. A leak detector can detect a hazardous atmosphere, but without it, personnel can be overcome without realizing the problem. Some pipeline operations pump carbon dioxide commercially, and in such operations the hazards are increased. To enter an area where a carbon dioxide release is suspected, personnel must use the proper gas detection meter, calibrated for the detection of carbon dioxide, and wear the appropriate PPE, such as a **self-contained breathing apparatus (SCBA)** (see *Figure 6*).

65102-13_F06.EPS

Figure 6 Self-contained breathing apparatus.

Hydrogen sulfide (H_2S) is a colorless gas that often smells like rotten eggs. However, in fairly low concentrations, it can deaden the sense of smell. Exposure to a higher concentration can cause death. In addition, it is flammable and may cause a flash fire. When a gas detection unit shows that H_2S is present, personnel must wear the appropriate PPE or immediately leave the area.

Propane is a common natural gas that is colorless and heavier than air, so it can collect in low-lying areas where it may displace oxygen. Exposure can cause death. At high concentrations, it is flammable.

Anhydrous ammonia is a liquefied gas that has a sharp odor. Exposure in high concentration can cause death. Even exposure in low concentrations can cause severe reactions from mild irritation to severe burns. The reaction depends on the length of exposure and concentration amounts. If anhydrous ammonia is released, call emergency personnel.

Naturally occurring radioactive materials (NORMs) may be present when oil and gas is extracted from the earth. These NORMs are then introduced into the product streams and transported in the pipeline. Natural gas liquids (NGLs), along with processing associated with these products, may trap NORMs in the form of sludge, scale, and particulates. Filters, valves, scrapers, and even internal surfaces of pipelines may contain high levels of NORMs. Precautions should be taken to protect personnel by monitoring for NORMs in accordance with all local, state, federal, and company requirements. If high levels are detected, personnel should distance and/or shield themselves from the source and notify the appropriate personnel.

WARNING!

Exposure to hydrogen sulfide (H_2S), carbon dioxide (CO_2), and propane can be fatal. H_2S is flammable and may cause a flash fire. When a gas detection unit shows that H_2S is present, wear the appropriate PPE or immediately leave the area. CO_2 and propane are colorless, odorless gases that are heavier than air, so they can collect in low-lying areas where they may displace oxygen.

2.7.0 Reacting to Unexplained Flow Rate/Pressure Changes

Control center personnel must respond to a flow rate/pressure change alarm immediately to prevent serious damage to the pipeline. The sooner a response is initiated, the more quickly the problem will be solved and the more likely a serious condition will be averted. Control center personnel should check pressure and flow rates upstream and downstream for comparison. If the time frame involved in the change is very short, a release might be indicated and the pipeline shut down per company policy.

2.7.1 Unexplained Flow Rate Changes

Control center personnel must determine if there has been an intentional change in pumping status or compressor status on the pipeline that was not scheduled. A valve might be the cause of the problem. If the main-line flow has dropped, a pump unit or compressor unit may have shut down upstream, and control center personnel can adjust pumping or compressor units upstream in an attempt to compensate. It is the responsibility of control center personnel to initiate corrective action.

If a pipeline block valve has been closed without authorization or if personnel notice a high flow rate at the origin station, personnel should follow established troubleshooting procedures, perform a shutdown as necessary, and make all proper notifications.

If a pipeline is running, and the origin meter flow rate increases without justification, possible causes for this include SCADA malfunction, a pipeline rupture downstream of the origin meter, or maintenance personnel interfering with data transmission prior to notification. Again, control center personnel should follow established troubleshooting procedures, perform a shutdown as necessary, and make all proper notifications.

If there is not a release, it may be possible to safely work around the AOC. Control center personnel should try to adjust the pipeline pressure and flow to achieve a steady-state operation. During line upsets or valve swings, control center personnel must avoid frequent or unnecessary line upsets because data indicating a release is very difficult to identify during these conditions. If the pressure and flow cannot be corrected within a reasonable length of time, control center personnel should shut the pipeline down as quickly as possible without jeopardizing safety. Company policy determines exact procedures that must be followed. Appropriate personnel should be notified of the shutdown as soon as possible.

2.7.2 Unexplained Pressure Changes

Unexplained pressure changes can include an absence of pressure, a sudden increase in pressure, or a sudden decrease in pressure. Deviations in pressure may indicate an unusual condition on the pipeline. Occurrences such as unintended valve closures may cause AOCs that could endanger the pipeline. The condition must be recognized, and the reaction must be quick and decisive.

If a high-pressure alarm is received from a compressor or pumping unit, personnel should investigate the alarm and determine the appropriate action. This could include shutting down the unit or line as needed.

If the condition is a valve closure or a sudden decrease in discharge pressure followed by a sudden increase in discharge pressure, control center personnel must troubleshoot to determine the cause of the closure. If there is not a high-pressure condition downstream and a valve closure is identified between the two stations, then a command to open the valve must be sent by control center personnel.

If the valve cannot be opened automatically or if the condition occurs at a location with personnel available, opening the valve manually may be possible. If the valve does not open, the pipeline may have to be shut down due to increasing pressure. Company policy and procedure must be followed.

If control center personnel conclude that there may be time to make adjustments before the pipeline reaches MOP/MAOP, immediate action must be taken. Personnel must notify all operations upstream to reduce or stop pipeline injections. Personnel may reduce strategic pumping units upstream or initiate other means to relieve the pressure. Appropriate personnel may be consulted if there is time. If the condition can be corrected by such efforts, the pipeline operation may be continued. Control center personnel must monitor the pressure readings closely until satisfied that the pipeline has stabilized.

If there is no time to accomplish the adjustments or if it appears that the efforts are having no effect, control center personnel must perform a shutdown according to company policy and procedure.

At a steady-state, multipump station pipeline, if there is an increased pressure reduction between two stations, a flow increase upstream of the pressure loss, and all analog data indicates the high probability of a release, control center personnel should shut down the line with the lowest possible pressure on the segment with the suspected release. If personnel notice a main-line block valve status has changed from opened to closed with an increase in discharge pressure at the upstream pump station, personnel should shut down the pipeline until the correct valve position is determined.

Additional Resources

It is assumed that you will learn to perform these procedures under the direct supervision of your crew leader, foreman, or instructor. For independent study, refer to the recommended Additional Resources at the back of this module.

2.0.0 Section Review

1. If damaged pipeline is reported to the control center, the proper reaction always includes shutting down the pipeline.
   a. True
   b. False

2. If a safety device activates, personnel should focus *first* on _____.
   a. protecting the pipeline
   b. ensuring the safety of personnel
   c. notifying PHMSA
   d. verifying that the alarm is not false

3. Field personnel may need to operate equipment in the field manually if there is a systems failure.
   a. True
   b. False

4. When a power outage occurs, the appropriate pipeline personnel should contact _____.
   a. the One-Call notification system
   b. the power company
   c. OSHA
   d. local authorities

5. When reacting to a fire, explosion, or natural disaster, the first priority is always to protect _____.
   a. the pipeline facility
   b. the environment
   c. people
   d. the pipeline product

6. One reason that the unexpected release of carbon dioxide can be dangerous is because it is _____.
   a. odorless
   b. a compound
   c. lighter than air
   d. white in color

7. A pipeline rupture downstream of the origin meter could be the cause of _____.
   a. a power outage
   b. an unexplained communication failure
   c. an unexplained flow rate change
   d. damaged pipe

SUMMARY

Pipeline operators have developed effective routines and policies that enable personnel to respond quickly, safely, and effectively to AOCs. Usually an AOC is a routine matter that can be handled safely by pipeline personnel. Sometimes, however, an AOC can result in an emergency situation that may involve working with outside agencies.

In this module, you learned about the different types of AOCs that can occur on the pipeline and the proper reactions to each type of AOC. During your career, you may have to respond to some or all of the AOCs outlined in this module. When responding to these AOCs, follow company policies and procedures. Protecting people, property, and the environment should be of the utmost concern.

Review Questions

1. By definition, an AOC is a condition that may exceed _____.
   a. maximum operating pressure
   b. design limits
   c. safety laws
   d. size limits

2. By definition, a pipeline facility includes the right-of-way.
   a. True
   b. False

3. Which of the following is an example of pipeline system damage?
   a. Gouges or dents in the pipeline
   b. Transformer failure
   c. Decreased pressure
   d. Activation of a pressure-relief valve

4. Pressure-relief systems, tank-level alarms, and main-line pressure shutdown alarms are examples of _____.
   a. prevention devices
   b. warning devices
   c. hazardous atmosphere alarms
   d. fire alarms

5. The loss of SCADA is an example of which type of AOC?
   a. Hydraulic failure
   b. Mechanical failure
   c. Communications failure
   d. Power failure

6. To keep the pipeline running in case of a power failure, some companies use an _____.
   a. operations override power source
   b. uninterruptible power system
   c. interface power system
   d. indirect power system

7. The failure of a valve or a transmitter may be indicated by _____.
   a. ongoing pressure increases or decreases
   b. sudden pressure increases or decreases
   c. high-level pressure increases or decreases
   d. maximum pressure increases or decreases

8. When pipeline support damage is discovered, notification is especially important if the pipe _____.
   a. is in danger of excessive movement
   b. has changed color
   c. is dented
   d. is corroded

9. If a relief device alarm is activated, control center personnel must immediately _____.
   a. shut down or call 911
   b. reduce pressure or shut down
   c. contact field personnel
   d. complete the required reports

10. At a station that is running, if a pump suddenly indicates a stop status with no apparent changes to the station pressures, control center personnel should _____.
    a. notify appropriate field personnel
    b. reduce pressure
    c. shut down the SCADA system
    d. activate safety devices

11. If SCADA fails, it may be necessary for equipment to be operated manually by _____.
    a. PHMSA
    b. field personnel
    c. control center personnel
    d. local authorities

12. When a power interruption occurs, personnel may be able to keep the pipeline in operation by adjusting _____.
    a. MOP/MAOP
    b. line pressure and pipe location
    c. line pressure and flow
    d. product temperature

13. Regardless of the nature of an emergency, if there is a threat of injury or death, pipeline personnel must be _____.
    a. contacted
    b. evacuated
    c. warned
    d. involved

14. The top priority is always the safety of _____.
    a. personnel
    b. the pipeline
    c. the environment
    d. equipment
15. Emergency shutdowns are generally required when _____.
    a. equipment is reported on the ROW
    b. natural disasters are anticipated
    c. there is an immediate hazard to persons
    d. pipeline damage is reported by third parties
16. Product releases in outdoor areas involve an extra concern for _____.
    a. safety of third-party equipment
    b. safety of pipeline equipment
    c. safety of the pipeline
    d. safety of the environment
17. Control center personnel should rely solely on reports of unexpected releases by field personnel or third parties to determine the proper reaction.
    a. True
    b. False
18. AOCs are not necessarily considered safety-related conditions.
    a. True
    b. False
19. All of the following are examples of toxic products, *except* _____.
    a. CO_2
    b. anhydrous ammonia
    c. propane
    d. H_2O
20. What type of device is preset to signal an alarm in the control center when the temperature reaches a critical level?
    a. RTD
    b. seal failure alarm
    c. LEL
    d. high-pressure alarm

Trade Terms Quiz

Fill in the blank with the correct term that you learned from your study of this module.

1. A computer-based system or systems used by a controller in a control room that collects and displays information about a pipeline facility and may have the ability to send commands back to the pipeline facility is called ________.
2. The ________ is the area along the pipeline which has been legally permitted to be used by the pipeline operator to construct, operate, and maintain the pipeline.
3. Activity carried out by and/or equipment belonging to outside parties on the pipeline ROW is referred to as ________.
4. A(n) ________ is a device worn by an individual to provide breathable air in an atmosphere that is hazardous.
5. ________ is the separation of the protective coating from the pipeline.
6. A(n) ________ is a backup system that provides AC power to the systems it serves.
7. The maximum pressure at which a gas pipeline or segment of a gas pipeline may be normally operated is the ________.
8. A(n) ________ is used to relieve surges in a hazardous liquid pipeline system or receive and store hazardous liquid transported by a pipeline for reinjection and continues transportation by pipeline.
9. A device used on pumps that reads the temperature of the pump case and the product in the pipeline that will sound an alarm when the temperature reaches a critical level and shut down the pump automatically if needed is a(n) ________.
10. A(n) ________ duplicates the operations of a system so that a complete backup system is available in the event of systems failure.
11. A(n) ________ is identified by the operator and may indicate (1) a malfunction of a component or (2) a deviation from normal operations that may indicate a condition exceeding design limits or result in a hazard to persons, property, or the environment.
12. The pressure produced by a change in the velocity of the moving stream that can result from the closure of a valve is called ________.
13. The maximum pressure at which a liquid pipeline or segment of a liquid pipeline may be operated is the ________.
14. The ________ is a centralized communication system established by utilities, government agencies, and operators of underground facilities that provides one telephone number for contractors and the general public to call to notify of their intent to use equipment on or near a pipeline.
15. A hazardous liquid that has a vapor pressure exceeding 40 pounds per square inch absolute at 100°F is classified as a(n) ________.
16. The government agency that is responsible for developing and enforcing regulations for pipeline operations in the United States is the ________.
17. The ________ is the company that operates the pipeline.

18. Materials with radioactive elements that can be extracted during oil and gas extraction from the earth and can get into product streams and become transported in the pipeline are called _______.

19. An organization that adopts standards for a variety of products, services, and processes in the United States to ensure consistency is the _______.

20. The lowest concentration of a gas or a vapor in the atmosphere capable of producing fire in the presence of an ignition source is the _______.

21. Technology that provides monitoring tools used by control center personnel to detect temperature, flow, and pressure irregularities in a pipeline is _______.

22. New and existing pipe on the pipeline itself, the right-of-way (ROW), and any equipment, facility, or building used in the transportation of a liquid or gas are defined as part of a(n) _______.

Trade Terms

Abnormal operating condition (AOC)
American National Standards Institute (ANSI)
Breakout tank
Computational pipeline monitoring (CPM)
Disbondment
Encroachment
Highly volatile liquid (HVL)
Lower explosive limit (LEL)
Maximum allowable operating pressure (MAOP)
Maximum operating pressure (MOP)
Naturally occurring radioactive materials (NORMs)
One-Call notification system
Pipeline and Hazardous Materials Safety Administration (PHMSA)
Pipeline facility
Pipeline operator
Redundant system
Resistance temperature device (RTD)
Right-of-way (ROW)
Self-contained breathing apparatus (SCBA)
Supervisory control and data acquisition (SCADA)
Surge pressure
Uninterruptible power system (UPS)

Trade Terms Introduced in This Module

Abnormal operating condition (AOC): A condition identified by the operator that may indicate (1) a malfunction of a component or (2) a deviation from normal operations that may indicate a condition exceeding design limits or result in a hazard to persons, property, or the environment.

American National Standards Institute (ANSI): An organization that adopts standards for a variety of products, services, and processes in the United States to ensure consistency.

Breakout tank: A tank used to (a) relieve surges in a hazardous liquid pipeline system or (b) receive and store hazardous liquid transported by a pipeline for reinjection and continued transportation by pipeline.

Computational pipeline monitoring (CPM): Technology that provides monitoring tools used by control center personnel to detect temperature, flow, and pressure irregularities in a pipeline.

Disbondment: Separation of the protective coating from the pipeline.

Encroachment: Activity carried out by and/or equipment belonging to outside parties on the pipeline right-of-way (ROW). A particular encroachment might be authorized or not, so each situation requires specific handling.

Highly volatile liquid (HVL): a hazardous liquid which will form a vapor cloud when released to the atmosphere and which has a vapor pressure exceeding 276 kPa (40 psia) at 37.8°C (100°F).

Lower explosive limit (LEL): The lowest concentration of a gas or a vapor in the atmosphere capable of producing fire in the presence of an ignition source. Different products have different LELs.

Maximum allowable operating pressure (MAOP): The maximum pressure at which a pipeline or segment of a pipeline may be operated under this part (liquid).

Maximum operating pressure (MOP): The maximum pressure at which a pipeline or segment of a pipeline may be normally operated under this part (gas).

Naturally occurring radioactive materials (NORMs): Materials with radioactive elements that can be extracted during oil and gas extraction from the earth and can get into product streams and become transported in the pipeline.

One-Call notification system: A centralized communication system established by utilities, government agencies, and operators of underground facilities. This system provides one telephone number for contractors and the general public to call to notify of their intent to use equipment on or near a pipeline. It is also in place for any individual to call if pipeline damage is discovered. The One-Call system is generally set up on a state-by-state basis.

Pipeline and Hazardous Materials Safety Administration (PHMSA): An agency of the US Department of Transportation responsible for developing and enforcing regulations for pipeline operations in the United States.

Pipeline facility: New and existing pipe; rights-of-way; and any equipment, facility, or building used in the transportation of hazardous liquids or carbon dioxide, or new and existing pipelines; rights-of-way; and any equipment, facility, or building used in the transportation of gas or in the treatment of gas during the course of transportation.

Pipeline operator: The company that operates the pipeline, as opposed to an individual person who works for the company that operates the pipeline.

Redundant system: A system that duplicates the operations of an entire system so that, in the event of a failure, a backup system is available.

Resistance temperature device (RTD): A device used on pumps that reads the temperature of the pump case and the product in the pipeline. It will sound an alarm when the temperature reaches a critical level and shut down the pump automatically if needed.

Right-of-way (ROW): The area along the pipeline which has been legally permitted to be used by the pipeline operator to construct, operate, and maintain the pipeline. The ROW extends the entire length of the pipeline.

Self-contained breathing apparatus (SCBA): A device worn to provide breathable air for an individual in an area where the atmosphere is hazardous.

Supervisory control and data acquisition (SCADA): A computer-based system or systems used by a controller in a control room that collects and displays information about a pipeline facility and may have the ability to send commands back to the pipeline facility.

Surge pressure: Pressure produced by a change in the velocity of the moving stream that results from shutting down a pump station or pumping unit, closure of a valve, or any other change in the velocity of the moving stream.

Uninterruptible power system (UPS): A system that provides continuity of power when the normal power supply is interrupted. A UPS provides AC power to the system it serves.

Appendix A

Weather and Natural Disaster Information Sources

Emergency Managers Weather Information. US Department of Commerce, National Oceanic and Atmospheric Administration, National Weather Service, 1325 East-West Highway, Silver Spring, MD 20910. **www.nws.noaa.gov**

National Earthquake Information Center, US Geological Survey. Box 25046, DFC, MS 967 Denver, CO 80225. **www.earthquake.usgs.gov**

National Hurricane Center Tropical Prediction Center. 11691 SW 17th Street, Miami, FL 33165-2149. **www.nhc.noaa.gov**

National Weather Service and National Weather Service Weather Radio, National Oceanic and Atmospheric Administration (NOAA). 1325 East-West Highway, Silver Spring, MD 20910. **www.weather.gov**

Appendix B

Organizations of Interest to the Pipeline Industry

American Gas Association (AGA). 400 N. Capitol Street, NW, Washington, DC 20001. **www.aga.org**

American Petroleum Institute (API). 1220 L Street, NW, Washington, DC 20005. **www.api.org**

Interstate Natural Gas Association of America (INGAA). 20 F Street, NW, Suite 450, Washington, DC 20001. **www.ingaa.org**

National Institute for Occupational Safety and Health (NIOSH). 395 E Street, SW, Suite 9200, Patriots Plaza Building, Washington, DC 20201. **www.cdc.gov/niosh**

National Response Center (NRC). c/o United States Coast Guard (CG-5335 - Stop 7581), 2100 2nd Street, SW, Washington, DC 20593-0001. **www.nrc.uscg.mil**

Occupational Safety and Health Administration (OSHA). 200 Constitution Avenue, NW, Washington, DC 20210. **www.osha.gov**

Pipeline Hazardous Material Safety Administration (PHMSA). East Building, 2nd Floor, 1200 New Jersey Ave., SE, Washington, DC 20590. **www.phmsa.dot.gov**

Additional Resources

The following recommended resources can provide additional helpful information related to the requirements for this covered task.

Code of Federal Regulations 49, Parts 192 and 195.

OSHA *Occupational Safety and Health Standards* 1910, Subpart L, Standard 1910.157.

Figure Credits

Courtesy of Sunoco Logistics, Module Opener, Figure 1

U.S. Geological Survey/photo by Don Becker, Figure 2

PORV image courtesy of Farris Engineering, Figure 4

Spectrex Inc., Figure 5

Courtesy of MSA, Figure 6

Section Review Answer Key

Answer	Section Reference	Objective
Section One		
1. b	1.1.0	1a
2. a	1.2.0	1b
3. d	1.3.2	1d
4. b	1.4.0	1d
5. b	1.5.1	1e
6. a	1.6.0	1f
7. b	1.7.0	1g
Section Two		
1. b	2.1.1	2a
2. b	2.2.0	2b
3. a	2.3.2	2d
4. b	2.4.0	2e
5. c	2.5.0	2f
6. a	2.6.0	2h
7. c	2.7.1	2i

NCCER CURRICULA — USER UPDATE

NCCER makes every effort to keep its textbooks up-to-date and free of technical errors. We appreciate your help in this process. If you find an error, a typographical mistake, or an inaccuracy in NCCER's curricula, please fill out this form (or a photocopy), or complete the online form at **www.nccer.org/olf**. Be sure to include the exact module ID number, page number, a detailed description, and your recommended correction. Your input will be brought to the attention of the Authoring Team. Thank you for your assistance.

Instructors – If you have an idea for improving this textbook, or have found that additional materials were necessary to teach this module effectively, please let us know so that we may present your suggestions to the Authoring Team.

NCCER Product Development and Revision

13614 Progress Blvd., Alachua, FL 32615

Email: curriculum@nccer.org

Online: www.nccer.org/olf

❑ Trainee Guide ❑ Lesson Plans ❑ Exam ❑ PowerPoints Other ______________

Craft / Level: Copyright Date:

Module ID Number / Title:

Section Number(s):

Description:

Recommended Correction:

Your Name:

Address:

Email: Phone:

AOCFG-17

Abnormal Operating Conditions – Field and Gas

Overview

Pipeline operators and contractors have programs, procedures, training, safety equipment, and warning devices for practically every facet of pipeline operations. Situations that are outside of normal operating conditions may arise at any time, and field personnel must know what to do to help keep pipeline operations safe for everyone. In some cases, an abnormal operating condition (AOC) is a routine matter that can be quickly and safely addressed by personnel. In other cases, an AOC can result in an emergency situation. In this module, personnel will learn how to recognize, properly react to, and properly report AOCs that may occur during pipeline operations.

AOCFG-17
ABNORMAL OPERATING CONDITIONS – FIELD AND GAS

Objectives

When you have completed this module, you will be able to do the following:

1. Explain how to recognize abnormal operating conditions for both gas and liquid pipelines.
   a. Explain how to recognize abnormal pipeline facility conditions.
   b. Explain how to recognize the activation of a safety device.
   c. Explain how to recognize a communications failure and a control system failure.
   d. Explain how to recognize power interruptions.
   e. Explain how to recognize a fire, explosion, and natural disaster occurring with the pipeline.
   f. Explain how to recognize the unexpected release of hazardous liquid or gas.
   g. Explain how to recognize unexplained pressure/flow rate changes.
2. Describe the proper reaction to abnormal operating conditions for both gas and liquid pipelines.
   a. Describe the proper reaction to abnormal pipeline facility conditions.
   b. Describe the proper reaction to the activation of a safety device.
   c. Describe the proper reaction to a communications failure and a control system failure.
   d. Describe the proper reaction to power interruptions.
   e. Describe the proper reaction to a fire, explosion, and natural disaster occurring with the pipeline.
   f. Describe the proper reaction to the unexpected release of hazardous liquid or gas.
   g. Describe the proper reaction to unexplained pressure/flow rate changes.

Performance Tasks

This is a knowledge-based module; there are no performance tasks.

Trade Terms

Aboveground span
Abnormal operating condition (AOC)
American National Standards Institute (ANSI)
Breakout tank
Disbondment
Encroachment
Flame ionization detection unit
Highly volatile liquid (HVL)
Incipient fire
Intrinsically safe/explosionproof
Lower explosive limit (LEL)
Maximum allowable operating pressure (MAOP)
Maximum operating pressure (MOP)
Naturally occurring radioactive materials (NORMs)
One-Call notification system
Pipeline and Hazardous Materials Safety Administration (PHMSA)
Pipeline facility
Pipeline operator
Redundant system
Right-of-way (ROW)
Self-contained breathing apparatus (SCBA)
Supervisory control and data acquisition (SCADA)
Surge pressure
Uninterruptible power system (UPS)

Industry-Recognized Credentials

If you're training through an NCCER-accredited sponsor, you may be eligible for credentials from NCCER's Registry. The ID number for this module is AOCFG-17. Note that this module may have been used in other NCCER curricula and may apply to other level completions. Contact NCCER's Registry at 888.622.3720 or go to **www.nccer.org** for more information.

Contents

Topics to be presented in this module include:

1.0.0 Recognizing Abnormal Operating Conditions 1
1.1.0 Recognizing Abnormal Pipeline Facility Conditions 2
1.1.1 Pipeline System Damage 4
1.1.2 Damaged Pipeline Supports 5
1.2.0 Recognizing the Activation of a Safety Device 5
1.2.1 Activation of a Pressure-Relief System 6
1.2.2 Unintended Valve Closure or Shutdown 6
1.3.0 Recognizing a Communications Failure 6
1.3.1 Failure of Personnel Communication Tools 6
1.3.2 System Communications Failures 7
1.4.0 Recognizing Power Interruptions 7
1.5.0 Recognizing Fire, Explosion, or Natural Disaster 8
1.5.1 Fire 8
1.5.2 Explosion 9
1.5.3 Natural Disaster 9
1.6.0 Recognizing the Unexpected Release of Hazardous Gas or Liquid 9
1.7.0 Recognizing Unexplained Flow Rate / Pressure Changes 10
2.0.0 Reacting to Abnormal Operating Conditions 12
2.1.0 Reacting to Abnormal Facility Conditions 12
2.1.1 Pipeline System Damage 12
2.1.2 Damaged Pipeline Supports 13
2.2.0 Reacting to the Activation of a Safety Device 13
2.3.0 Reacting to a Communications Failure 13
2.3.1 Failure of Personnel Communication Tools 13
2.3.2 System Communications Failures 14
2.4.0 Reacting to Power Interruptions 14
2.5.0 Reacting to a Fire, Explosion, or Natural Disaster 14
2.5.1 Fire 14
2.5.2 Explosion 16
2.5.3 Natural Disaster 16
2.5.4 Emergency Shutdowns 16
2.5.5 Evacuations 16
2.6.0 Reacting to the Unexpected Release of Hazardous Gas or Liquid .. 17
2.6.1 Releases in Enclosed Areas 18
2.6.2 Releases in Outdoor Areas 18
2.6.3 Toxic Products 18
2.7.0 Reacting to Unexplained Flow Rate / Pressure Changes 19
Appendix A Weather and Natural Disaster Information Sources 28
Appendix B Organizations of Interest to the Pipeline Industry 29

Figures and Tables

Figure 1 Dead wildlife may be a sign of an unexpected release 5
Figure 2 Aboveground span of pipeline on pipe supports 6
Figure 3 Relief valve 7
Figure 4 Flame detector 8
Figure 5 Self-contained breathing apparatus 18

Table 1A Abnormal Operating Conditions: Examples and General Reactions 3
Table 1B Abnormal Operating Conditions: Examples and General Reactions 4
Table 2 Recognizing Warning Signs of Hazardous Gases or Liquids 5
Table 3 Fire Class Ratings 15

SECTION ONE

1.0.0 RECOGNIZING ABNORMAL OPERATING CONDITIONS

Objective

Explain how to recognize abnormal operating conditions for both gas and liquid pipelines.

a. Explain how to recognize abnormal pipeline facility conditions.
b. Explain how to recognize the activation of a safety device.
c. Explain how to recognize a communications failure and a control system failure.
d. Explain how to recognize power interruptions.
e. Explain how to recognize a fire, explosion, and natural disaster occuring with the pipeline.
f. Explain how to recognize the unexpected release of hazardous liquid or gas.
g. Explain how to recognize unexplained pressure/flow rate changes.

Trade Terms

Aboveground span: A length of pipeline that is entirely aboveground and supported by pipe supports.

Abnormal operating condition (AOC): A condition identified by the operator that may indicate (a) a malfunction of a component or (b) a deviation from normal operations that may indicate a condition exceeding design limits or result in a hazard to persons, property, or the environment.

American National Standards Institute (ANSI): An organization that adopts standards for a variety of products, services, and processes in the United States to ensure consistency.

Breakout tank: A tank used to (a) relieve surges in a hazardous liquid pipeline system or (b) receive and store hazardous liquid transported by a pipeline for reinjection and continued transportation by pipeline.

Disbondment: Separation of the protective coating from the pipeline.

Encroachment: Activity carried out by, and/or equipment belonging to outside parties on the pipeline right-of-way (ROW). A particular encroachment might be authorized or not, so each situation requires specific handling.

Flame ionization detection unit: A device that detects some hazardous gases in the atmosphere.

Highly volatile liquid (HVL): A hazardous liquid which will form a vapor cloud when released to the atmosphere and which has a vapor pressure exceeding 276 kPa (40 psia) at 37.8°C (100°F).

Intrinsically safe/explosion-proof: Describes electrical equipment that is designed specifically for use in explosive or flammable locations so that a spark cannot ignite the atmosphere.

Lower explosive limit (LEL): The lowest concentration of a gas or a vapor in the atmosphere capable of producing fire in the presence of an ignition source. Different products have different LELs.

Maximum allowable operating pressure (MAOP): The maximum pressure at which a pipeline or segment of a pipeline may be operated (gas).

Maximum operating pressure (MOP): The maximum pressure at which a pipeline or segment of a pipeline may be normally operated (liquids).

One-Call notification system: A centralized communication system established by utilities, government agencies, and operators of underground facilities. This system provides one telephone number for contractors and the general public to call to notify of their intent to use equipment on or near a pipeline. It is also in place for any individual to call if pipeline damage is discovered. The One-Call system is generally set up on a state-by-state basis.

Pipeline and Hazardous Materials Safety Administration (PHMSA): An agency of the US Department of Transportation responsible for developing and enforcing regulations for pipeline operations in the United States.

Pipeline facility: New and existing pipe, rights-of-way, and any equipment, facility, or building used in the transportation of hazardous liquids or carbon dioxide, or new and existing pipelines, rights-of-way, and any equipment, facility, or building used in the transportation of gas or in the treatment of gas during the course of transportation.

Pipeline operator: The company that operates the pipeline, as opposed to an individual person who works for the company that operates the pipeline.

Redundant system: A system that duplicates the operations of a system so that, in the event of a failure, a backup system is available.

Right-of-way (ROW): The area along the pipeline which has been legally permitted to be used by the pipeline operator to construct, operate, and maintain the pipeline. The ROW extends the entire length of the pipeline.

Supervisory control and data acquisition (SCADA): A computer-based system or systems used by a controller in a control room that collects and displays information about a pipeline facility and may have the ability to send commands back to the pipeline facility.

Surge pressure: Pressure produced by a change in velocity of the moving stream that results from shutting down a pump station or pumping unit, closure of a valve, or any other blockage of the moving stream.

Uninterruptible power system (UPS): A system that provides continuity of power when the normal power supply is interrupted. A UPS provides AC power to the system it serves.

An **abnormal operating condition (AOC)** is defined under federal regulations as a condition identified by the **pipeline operator** that may indicate (1) a malfunction of a component or (2) a deviation from normal operations that may indicate a condition exceeding design limits or result in a hazard to persons, property, or the environment. AOCs are defined as such by federal pipeline safety law 49 *United States Code (U.S.C.)* 60102 and pipeline safety regulations 49 *Code of Federal Regulations (CFR)* 192 (for transportation of gas) and 49 *CFR* 195 (for transportation of liquids). The federal agency that regulates pipeline maintenance is the **Pipeline and Hazardous Materials Safety Administration (PHMSA)** in Washington, DC.

Personnel in the field are responsible for being aware of pipeline surroundings at all times so they can recognize AOCs when they occur. Types of AOCs, along with examples and proper reactions, are listed in *Table 1.* Note that the reactions are general.

Some pipeline equipment is fitted with warning devices to alert personnel to problems. Sometimes field workers will be notified of AOCs by control center personnel. In other cases, field workers must use their skills and training to observe the physical signs of AOCs. Field personnel play a critical role as the "eyes, ears, and nose" of the pipeline operator. They must be able to recognize the signs of AOCs and report them appropriately. They must stay in close communication with control center personnel who may need to take the proper action remotely. In the most serious cases, the control center may need to shut down the pipeline.

One of the most important responsibilities of field personnel is to be able to recognize the unexpected release of hazardous gases or liquids. In some cases, early recognition can prevent releases from turning into fires and explosions. Dead vegetation, dead wildlife (*Figure 1*), and an oily sheen on water are just a few of the signs that a release has occurred. *Table 2* illustrates the ways that field personnel must use their senses to be aware of the signs of releases.

1.1.0 Recognizing Abnormal Pipeline Facility Conditions

Any problem with the physical structure of the pipeline is defined as an abnormal **pipeline facility** condition. A pipeline facility is defined as new and existing pipe, **right-of-way (ROW)**, and any equipment, facility, or building used in the transportation of hazardous liquids or carbon dioxide, or new and existing pipelines, rights-of-way, and any equipment, facility, or building used in the transportation of gas or in the treatment of gas during the course of transportation. Examples of components of a pipeline facility include compressors (for gas pipelines) and pump stations (for liquid pipelines).

The following are examples of abnormal pipeline facility conditions:

- Exposed pipeline that should be covered
- Pipeline system damage
- Missing or damaged pipeline supports
- Corrosion
- Coating separating from the pipeline (**disbondment**)
- Unauthorized activity and/or equipment on the pipeline right-of-way (unauthorized **encroachment**)
- Missing signs or markers
- Sunken floating tank roof
- Incorrect data or records caused by equipment malfunction
- Incorrect **American National Standards Institute (ANSI)** pressure rating of a component or material type
- Damage caused by fire or explosion
- Malfunctions due to human error

Work crews, line patrols, or other field personnel are often the first to discover abnormal pipeline facility conditions. However, a report may come in as a result of a call made to the **One-Call notification system** or from a person calling the pipeline operator directly. When a report comes in, personnel must obtain complete information so that they can quickly find the problem and address it. Many pipeline operators use a form to capture the following information:

- Caller's name and telephone number
- Date and time of the call

Table 1A Abnormal Operating Conditions: Examples and General Reactions (1 of 2)

Abnormal Operating Condition	Examples	Appropriate Reaction
Abnormal pipeline facility condition	• Exposed pipe • Evidence of corrosion • Incorrect ANSI pressure rating of a component or material type • Disbondment (coatings separating from the pipe) • Missing signs or markers • Unauthorized activity and/or equipment on the right-of-way (unauthorized encroachment) • Incorrect data or records caused by equipment malfunction • Mechanical/hydraulic system failures • Fire and explosion • Pipeline system damage, such as: – Missing or damaged pipeline supports – Damaged or inoperable pipeline components (pipe, valves, pumps, compressors, meters, tank gauges) due to external force – Gouges, dents, pitting, wrinkles, buckles, or cracks – Cracked or corroded flange or bolts – Sunken floating tank roof – Damaged or malfunctioning pontoon – Defective tank roof seal – Damaged tank shell – Damaged pipe due to natural disaster	Follow standard operating procedures and make notifications per company policy. Isolate and/or shut down the pipeline segment or facility if necessary. Complete repairs as necessary. Complete the required reporting.
Activation of a safety device	• Emergency or abnormal shutdown alarm • High-pressure shutdown alarm • Activation of a pressure relief valve • Activation of a smoke or fire detector • Activation of a hazardous atmosphere alarm • Activation of an overfill protection device • High sump-level shutdown alarm • High tank-level shutdown alarm • High pump-case pressure alarm • Seal failure alarm	Follow standard operating procedures and make notifications per company policy. Isolate and/or shut down the pipeline segment or facility if necessary. Complete the required reporting.
Communications failures	Personnel communication tools: • Lost use of radio communications • Lost use of landline telephone • Lost use of cell phone System communications: • Facility outages • SCADA communication outage • System programming errors/incorrect command logic • Loss of satellite communications • Loss of microwave signals • Stale data (communications going from the control center to the field, but not being received by the control center from the field)	Switch to alternate equipment or backup systems. Follow standard operating procedures and make notifications per company policy. Complete the required reporting.

Table 1B Abnormal Operating Conditions: Examples and General Reactions (2 of 2)

Power interruptions	• Transformer failure • Downed power lines • Broken leads or wires • Incorrect wire termination • Station or terminal power problems (due to motor overload or a tripped breaker, for example) • Weather conditions, such as an ice storm	Follow standard operating procedures. Switch to an alternate control system or power source if necessary. Notify the power company of the outage. Make additional notifications per company policy. Complete the required reporting.
Fire, explosion, or natural disaster	• Electrical fire • Ignition of a hazardous vapor • Unintentional detonation • Fire, explosion, or pipeline movement caused by natural disaster, such as: – Lightning strike – Flooding – Landslide – Earthquake –Tornado	Follow company emergency action plans. Call 911 if appropriate. Isolate and/or shut down the pipeline segment or facility if necessary. Limit access to the area as much as possible. Make notifications per company policy. Complete the required reporting.
Unexpected release of a hazardous gas or liquid	Signs include: • Vapor clouds or frost balls • Smell of hydrocarbon release • Dead vegetation/contaminated soil • Dead wildlife • Rainbow sheen on water Hazardous products include: • Hydrogen sulfide • Carbon dioxide • Propane • Natural gas • Refined fuels/products Note that the controlled transfer, sampling, venting, draining, and handling of hazardous products are not considered AOCs.	Follow company emergency action plans. Call 911 if appropriate. Isolate and/or shut down the pipeline segment or facility if necessary. Make notifications per company policy. Complete the required reporting.
Unexplained flow rate/ pressure change	Any change from the expected flow or pressure, such as: • Increased pressure • Decreased pressure • No pressure • High-pressure limit • Low-pressure limit • High flow • Low flow • No flow	Follow standard operating procedures and make notifications per company policy. Isolate and/or shut down the pipeline segment or facility if necessary. Complete the required reporting.

- Location of the problem (obtain the latitude/ longitude of the location from a global positioning system [GPS] device or a phone, if possible)
- Nearby landmarks such as highways, streams, mile markers, or any other structure that can help repair crews pinpoint the location
- The nature of the damage and any conditions observed at the site

1.1.1 Pipeline System Damage

Accidents that can damage the pipeline and its components occasionally happen. Personnel working on the right-of-way may strike the pipeline without realizing it. An excavator operator may accidentally hit the line. Although the locations of pipelines are clearly marked with signs or markers, construction work near the pipeline may hide markings. Ships' anchors may hit offshore

67107-13_F01.EPS

Figure 1 Dead wildlife may be a sign of an unexpected release.

pipelines. In some parts of the country, hunters have accidentally shot pipelines. Intentional damage due to vandalism, terrorist activity, or other criminal activity may also occur.

Pipeline damage may be discovered by field personnel, control center personnel, or a third party. Field personnel may notice physical signs of damage while out in the field. The control center may receive the indication of pipeline damage on the control center console as an unexplained change in operating conditions.

Physical signs of damage to the pipeline and its components may include gouges, cracks, pitting, dents, buckles, and inoperable valves, pumps, or tank gauges. Flanges or bolts may be cracked or show significant corrosion. Tank shells or pipeline supports may show damage. A sunken roof or a damaged pontoon can also indicate damage. Less obvious signs of damage could include a rainbow sheen on the water near submerged pipelines, which may indicate a release. The release of a highly volatile liquid (HVL) may be discovered by observing a frost ball or vapor cloud along the ROW of a liquid pipeline.

Table 2 Recognizing Warning Signs of Hazardous Gases or Liquids

Sense	Examples of Warning Signs
Sight	• Dead, discolored vegetation, often near healthy vegetation • Dead wildlife • Dry or frozen spots in soil during warm weather • Oily sheen on water/"rainbow" sheen • Vapor fog • Blowing dirt around the pipeline • Bubbles in standing water
Sound	Hissing, blowing, or roaring sound
Smell	Gaseous, skunk-like, or "rotten egg" odor is sometimes recognizable. Never rely on sense of smell only; a release can be completely odorless.

1.1.2 Damaged Pipeline Supports

Damage to pipeline supports can be a serious problem and can lead to a shutdown. In other cases, pipeline operations may continue while repairs are made. Personnel may recognize pipeline support damage while performing maintenance or inspection tasks in the field. Examples of damage include evidence of corrosion, visible cracking, sagging, or buckling of a support, or bolts or other components of the support that have broken or become loose. Damage to supports may be caused by corrosion, settling of soil, erosion due to wind or water, landslides, washouts, or earthquakes.

Any aboveground span must adhere to specific regulations. These spans refer to lengths of pipeline that are entirely aboveground and supported by pipe supports (see *Figure 2*). 49 *CFR* Part 195.110 requires that pipeline systems be designed to handle anticipated external loads from such conditions as earthquakes, vibration, thermal expansion, and contraction. In addition, the pipe and other components must be supported in such a way that the support does not cause excess stress to the pipe wall. 49 *CFR* Part 195.254 states each component of the pipeline that is aboveground must be protected from the forces exerted by the anticipated loads.

1.2.0 Recognizing the Activation of a Safety Device

Safety devices are a part of the pipeline system and are designed to sense and alert personnel to potential danger and/or hazardous conditions.

67107-13_F02.EPS

Figure 2 Aboveground span of pipeline on pipe supports.

Hazardous conditions can include changes in pressure that may exceed pipeline design limits as well as the release of harmful gases. The purpose of a safety device is to prevent injury or accidents. Safety devices include but are not limited to pressure relief systems, pressure limiting devices, high pump-case-pressure alarms, seal failure alarms, overfill protection devices, tank-level alarms, main-line pressure shutdown alarms, and hazardous atmosphere alarms.

When a safety device activates, it may sound an alarm. The device may ring loudly, beep continuously, or flash a light. A sensor on the device may also send a signal to the control center, where it appears as an alert on the control center console. Never ignore a safety alarm and never turn off a safety alarm until the problem has been checked and cleared by authorized personnel. Notify the control center when an alarm sounds, to maintain communication about appropriate steps to take.

1.2.1 Activation of a Pressure Relief System

The activation of relief devices along the pipeline system is a time-sensitive occurrence. Relief valves reduce pressure by discharging product. Depending on the product, it may be discharged into the atmosphere, tanks, flares, or sumps. Relief valves (see *Figure 3*) can be installed on pipeline or **breakout tanks**. The purpose of a pressure relief device is to protect the pipeline or pipeline facility from overpressure. Pressure relief valves are usually self-actuating at preset limits determined by the operator. However, according to DOT regulations, the pressure in a pipeline cannot exceed 110 percent of the operating pressure, so all preset limits must be within acceptable parameters.

1.2.2 Unintended Valve Closure or Shutdown

The unintended closure of a main-line block valve, remote block valve, manual block valve, pump-station suction or discharge valve, and certain terminal valves (manifold/tank) may cause a pressure change which could activate a pressure-relief system. Breakout tanks can be used to temporarily absorb **surge pressure** from the main-line and manifold pressure-relief valves to lower pressure on the line. According to 49 *CFR* 195, relief venting must be provided for each breakout tank. Pressure relief devices must be provided for each low-pressure and high-pressure breakout tank. There are different specifications for low-pressure and high-pressure tanks.

The unintended closure of the valves could also result in a shutdown of the pipeline. In these cases, personnel should inspect the equipment to determine why closure occurred, and reset, open, or close the valve when it is safe to do so. Such occurrences may not cause an AOC, but if the event causes the pipeline to exceed design limits in any way, its classification changes to abnormal.

1.3.0 Recognizing a Communications Failure

Communications refers to both the tools that personnel use to communicate with each other and with communication systems. Communications failures include the failure of communications among personnel, or the failure of communications between the control center and equipment on the pipeline.

1.3.1 Failure of Personnel Communication Tools

Pipeline personnel use landline telephones, cell phones, and radios to communicate with one another on the job. If using a cell phone near the pipeline facility, personnel must make sure it is **intrinsically safe/explosion-proof**. Personnel must keep their equipment in good working order and test their equipment regularly to ensure that it is working properly.

A breakdown in personnel communications equipment is easy to recognize. Personnel cannot hear messages or make phone calls. Some equipment may beep or flash a liquid crystal display to alert users to a problem. These tools are usually reliable, but breakdowns can occur. If radio communication goes down, personnel must address the problem immediately. A communications breakdown can also occur due to weather-related conditions and location.

67107-13_F03.EPS

Figure 3 Relief valve.

1.3.2 System Communications Failures

System communications refers to the computer equipment located in the control center and on some pipeline equipment. Personnel in the control center use the **supervisory control and data acquisition (SCADA)** system to manage pipeline operations. They receive signals from and send signals to pipeline equipment that may be many miles away. The SCADA system allows them to monitor pressure, volume, and temperature and keep products flowing safely and on schedule. If the system fails, control center personnel lose the ability to communicate with equipment on the pipeline. Some systems are set up to switch to a **redundant system**—a backup that comes on line if the main system fails.

Some failures of communications systems show up as warnings or messages on the control center console. A system failure may include any of the following events:

- Random facility outages
- Partial or complete facility outages
- Loss of the SCADA system
- Loss of satellite communications
- Loss of microwave signals
- Stale data
- Loss of signal point within the system

Mechanical and hydraulic failures are related to system failures. For example, the SCADA system sends a signal to a piece of equipment, but the equipment does not respond to the signal. In some cases, the problem is a result of a programming or command logic error. In other cases, equipment may not work because of a mechanical, electrical, or hydraulic problem. Examples of monitoring device failures include, but are not limited to, frozen valves, electrical component burnouts, and compressor breakdowns. A valve may have a mechanical failure that prevents it from being operated remotely by the control system. Sometimes electrical components or connections burn out, preventing the equipment from responding to a command by the system to start up or shut down.

1.4.0 Recognizing Power Interruptions

Electric power is essential to pipeline operations. Nearly every part of the pipeline depends on electricity in some way. A loss of power impacts the abilities of the control center or pipeline facility operations to safely control and monitor the pipeline facility. Power interruptions that result from bad weather or a problem at the local utility company, such as transformer failure or downed power lines, are not uncommon.

Some pipeline operators have an **uninterruptible power system (UPS)** that turns on if electrical power is off for more than a few seconds. UPS systems supply emergency alternating current (AC) power to pipeline components to keep the pipeline operating until power is restored. Instead of electricity, these systems run on engines fueled with gasoline, diesel, or natural gas. When power fails, an AC power failure alarm may be sent to the control center. To ensure that the backup system has turned on, control center personnel check to see whether they are receiving updated information from the pipeline.

If power is lost, electronic indicators will turn off. Some compressors and remote control valves may not operate. The SCADA system may switch off and pumps may shut down. In addition, lights, heat, and air conditioning may fail. An area-wide power outage may cause a loss of telephone communications. Local power outages may shut down only a portion of the pipeline system. Power outages can also cause pumps at the receiving terminal to stop operating, causing a gas or liquid shipment to overflow in tanks or be unable to be shipped further without pumps.

Some equipment is protected from power failure by batteries that supply temporary low voltage power. These batteries run on direct current (DC).

The following types of equipment can operate on DC power:

- Fire detection equipment and alarms
- Fire suppression systems
- Emergency lighting systems
- SCADA
- Alarm panels
- Control modules
- Process control panels
- Specialized compressor units

> **NOTE**
> Equipment plugged into electric sockets uses AC power. AC power reverses its direction at regular intervals. Batteries supply DC power. DC power flows in one direction.

1.5.0 Recognizing Fire, Explosion, or Natural Disaster

Fire, explosion, or natural disaster can occur at any time. Some events may be anticipated, but many cannot. Field personnel must be trained to recognize these events.

1.5.1 Fire

Personnel must take steps to minimize the likelihood of fire and explosions in a pipeline facility. Two examples include using intrinsically safe/explosion-proof electronic equipment and checking atmospheric conditions before starting any hot work. Always follow company procedures related to the prevention of fire and explosion. Detection devices installed along the pipeline and in pipeline facility buildings alert personnel in case of a fire. These devices send an alarm when their sensors detect heat or flame. Such equipment includes thermal detectors, flame detectors, and smoke detectors.

> **NOTE**
> Before performing any work that may produce an ignition source, verify that a hazard analysis has been performed. Examples of such work include welding or cutting, opening electrical boxes, or operating non-explosionproof tools in hazardous classified areas.

A thermal detector is a heat-sensitive unit. It is fitted with electrical contacts that stay open as long as conditions are normal. When the unit senses rapidly rising temperatures, the contacts start to close. When the contacts connect, the unit may send an alarm, which may be audible, to the control center and the fire control panel in the workstation. If a thermal detector activates, personnel should evacuate the area and, investigate the cause of the alarm (if appropriately trained according to company procedures).

A flame detector (*Figure 4*) is a light-sensitive unit. It is fitted with an ultraviolet detector that turns on an alarm when the unit senses light from a flame. The unit is sensitive enough to detect a flame even in the earliest stages of a fire. Some flame detectors are connected to a firefighting system that may dump fire suppression material into the area. Personnel can also dump this material over a fire manually by using pull switches.

A smoke detector can be a light-sensitive unit or an ionization detector, or it can be a combination of both types. Ionization detectors are more sensitive to fires in the early smoldering stage. Light detectors are more sensitive to the flaming stage of fires.

An ionization smoke detector detects ionization in air caused by smoke. If any smoke particles pass into the chamber of the detector, the ions will attach to the particles, and so become less able to carry the current. An electronic circuit detects the current drop and sounds the alarm. The light detector operates by detecting light beams. In the absence of smoke, the light passes in front of the detector in a straight line. When smoke enters the chamber across the path of the light beam, light is scattered by smoke particles, directing it at the sensor that triggers the alarm.

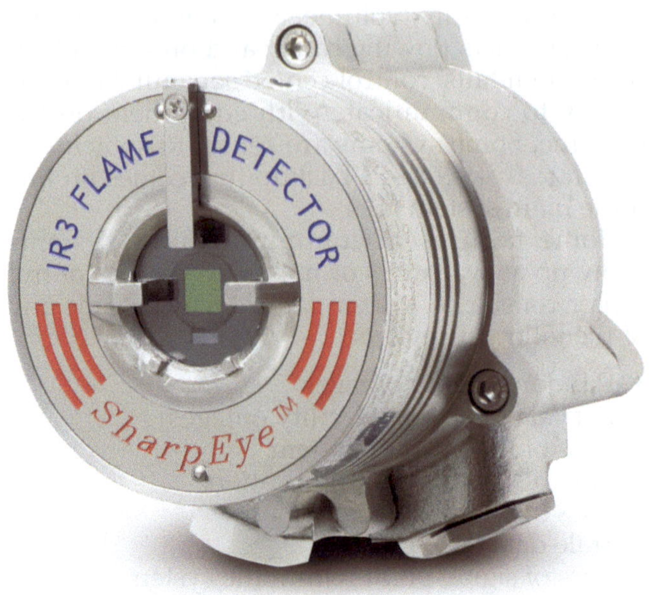

67107-13_F04.EPS

Figure 4 Flame detector.

CAUTION

Before using any fire suppression material, personnel must know how to handle it. Be sure to read the material safety data sheet (MSDS)/safety data sheet (SDS) for information about how to use the material safely.

Monitoring devices that detect a lower explosive limit (LEL) produce alarms to indicate that there is a risk of explosion. The LEL is the lowest concentration of a gas or a vapor in the atmosphere capable of producing fire in the presence of an ignition source. Different products have different LELs. The monitoring device must be set to detect the particular LEL in an enclosed area. If an LEL alarm activates while welding is taking place, the first action is to stop the welding immediately. Always check atmospheric conditions before reentering an area after an LEL alarm has activated.

1.5.2 Explosion

Pipeline operators install a number of warning devices to warn personnel about conditions that could cause explosions. However, operators cannot control every circumstance, such as intentional damage, that could lead to an explosion on the pipeline. An explosion on a pipeline registers in the control center as an unexplained change in pressure. An explosion may occur in any enclosed space where flammable vapors can accumulate to dangerous levels. Examples of such spaces include deep pipeline trenches and pipeline facilities.

A natural gas explosion may occur when natural gas leaks into an enclosed space. The odorant added to the gas sometimes alerts people to the problem. However, gas can sometimes accumulate undetected. For example, the building may be vacant or the odors may be too far away from people inside the building. In addition, products may not have an odorant added during transport. Be aware that a spark from electrical equipment can cause the gas to explode.

WARNING!

Personnel should never rely on their sense of smell to detect a natural gas leak. Products do not always have a detectable odor.

For an explosion to take place in an enclosed space, there must be a source of ignition and fuel. The vapors creating the fuel for an explosion can be of various types. Each type of flammable vapor or gas has a different lower and upper explosive limit (LEL/UEL). For example, when natural gas escapes into the air, its molecules generally dissipate or scatter because natural gas is lighter than air. However, in enclosed areas natural gas can accumulate, and if there is an ignition source an explosion may occur.

1.5.3 Natural Disasters

Natural disasters include earthquakes, hurricanes, tornadoes, landslides, wildfires, and flooding. These events may cause soil movement, which may result in forces sufficient to cause pipeline failure or product release. Even heavy rains resulting in saturated soil conditions may cause soil movement in some areas.

Pipeline operators generally have designated personnel to monitor weather conditions that can develop into a natural disaster. Control center personnel monitor for earthquakes with seismic indicators. In addition, they may obtain real-time information on the internet. Up-to-date earthquake information is available from the National Earthquake Information Center, which is located in Denver, Colorado. Personnel may track weather alerts and warnings through the National Weather Service website. In case of a power failure, personnel can continue to receive information through the National Weather Service's National Weather Radio or other electronic devices. A list of weather and natural disaster information sources is included in *Appendix A*.

Operators should also have field personnel designated to patrol the ROW to monitor for soil movement and to ensure the accessibility of pipeline components, such as valves, that are needed in case of emergency response actions such as shutdown, isolation, and containment in case of a pipeline failure.

1.6.0 Recognizing the Unexpected Release of Hazardous Gas or Liquid

Pipeline releases are often noticed when someone sees the physical signs. Depending on the product, the release may result in a pool of liquid on the ground, a frost ball, vapor cloud, mist in the air, dead or discolored vegetation, or a rainbow sheen on water. Other signs include dead fish in waterways or other dead wildlife. Gas leaks may also cause a whistling or hissing sound. Natural gas leaks may be identified by a loud, roaring noise. Note that the controlled transfer, sampling, venting, draining, and handling of hazardous products are not AOCs.

Personnel may use a combustible gas indicator or a flame ionization detection unit to detect hazardous gases such as hydrogen sulfide (H_2S), propane, natural gas, and in some cases, carbon dioxide (CO_2). An unexpected release may lead to a hazardous atmosphere, an explosion, or other serious condition. Additional PPE may be required with an unexpected release. Always follow company procedures when dealing with any release.

Personnel must not rely on their sense of smell to detect a hazardous gas for the following reasons:

- Some hazardous gases are odorless. Products may not have an odorant added to alert workers to the presence of gas.
- A low concentration of some hazardous gases can deaden the sense of smell.
- A person's sense of smell may be deadened by a medical problem.

1.7.0 Recognizing Unexplained Flow Rate / Pressure Changes

Flow rate and pressure are important indicators of both normal and abnormal operating conditions. Minor changes in pressure may occur normally in the pipeline. However, significant deviations from the expected flow or pressure are considered AOCs. Unexplained pressure changes include absences of, increases in, and decreases in pressure. The maximum internal pressure permitted during the operation of a gas pipeline is called the **maximum allowable operating pressure (MAOP)** and is defined in 49 *CFR* 192. The maximum internal pressure permitted during the operation of a liquid pipeline is called the **maximum operating pressure (MOP)** and is defined in 49 *CFR* 195.

Pressure-indicating devices located near mainline valves, compressor stations, and custody transfer stations monitor pressure and send readings to the control center. When an unexpected change in pressure occurs, an alarm may alert control center personnel, who then notify field personnel. Sudden pressure increases or decreases might indicate a problem such as the failure of a valve or a transmitter. Personnel must be familiar enough with the pipeline system, equipment, product characteristics, and environmental factors to quickly identify any important changes in pressure and flow.

Internal Combustion Engines

Pipeline facilities/systems that process or store flammable liquids or gases are particularly vulnerable to the risk of fire and explosion. But did you know that the internal combustion engines that are typically used in pipeline facilities can ignite flammable liquids and gases? Internal combustion engines include vehicles of any kind (trucks, cranes, excavators, fire engines, and ambulances) as well as stationary or portable engines such as compressors, generators, pumps, and welders.

Be sure that you know and understand your company's specific safety policies about internal combustion engines to ensure your own safety, the safety of other personnel, and the safety of the entire pipeline facility. To prevent internal combustion engines from becoming an ignition source, pipeline operators are responsible for implementing safety measures such as:

- Ensuring that materials and equipment are stored and used in accord with applicable Occupational Safety and Health Administration (OSHA) standards
- Ensuring that permanently mounted internal combustion engines are not stored where flammable vapors or gases could exist
- Establishing and marking safe boundaries and traffic routes for vehicles and portable equipment through hazardous areas
- Installing flammable gas and vapor detectors in areas where hazards could exist
- Providing training to personnel about the signs and hazards of flammable vapors and gas, the safe handling of internal combustion engines, and all relevant emergency procedures.

Additional Resources

It is assumed that you will learn to perform these procedures under the direct supervision of your crew leader, foreman, or instructor. For independent study, refer to the recommended Additional Resources at the back of this module.

1.0.0 Section Review

1. Which of the following is an abnormal pipeline facility condition?
   a. Aboveground pipe on pipe supports
   b. Missing signs or markers
   c. Valves operating properly
   d. Authorized encroachment

2. Safety devices include _____.
   a. hazardous atmosphere alarms
   b. tank shells
   c. SCADA
   d. compressors

3. SCADA serves as a _____.
   a. regulatory agency
   b. notification system for all utilities
   c. device worn to provide breathable air where the atmosphere is hazardous
   d. command system that assists control center personnel

4. An uninterruptible power system (UPS) supplies _____.
   a. temperature readings
   b. emergency backup power
   c. nuclear power
   d. pressure readings

5. A thermal detector, unlike a flame detector, senses _____.
   a. rapidly rising temperatures
   b. ultraviolet light
   c. flames
   d. pipeline pressure

6. Which of the following is a reason that personnel should *not* rely on their sense of smell to detect a hazardous gas release?
   a. Hazardous gases always have an odor.
   b. Low concentrations of hazardous gases heighten the sense of smell.
   c. A person's sense of smell may be deadened by a medical problem.
   d. Personnel can hear a gas release.

7. The maximum operating pressure of a gas pipeline is called _____.
   a. MOP
   b. MAOP
   c. surge pressure
   d. compression

SECTION TWO

2.0.0 REACTING TO ABNORMAL OPERATING CONDITIONS

Objective

Describe the proper reaction to abnormal operating conditions for both gas and liquid pipelines.

a. Describe the proper reaction to abnormal pipeline facility conditions.
b. Describe the proper reaction to the activation of a safety device.
c. Describe the proper reaction to a communications failure and a control system failure.
d. Describe the proper reaction to power interruptions.
e. Describe the proper reaction to a fire, explosion, and natural disaster occuring with the pipeline.
f. Describe the proper reaction to the unexpected release of hazardous liquid or gas.
g. Describe the proper reaction to unexplained pressure / flow rate changes.

Trade Terms

Incipient fire: A fire in its beginning stage.

Naturally occurring radioactive materials (NORMs): Materials with radioactive elements that can be extracted during oil and gas extraction from the earth and can get into product streams and become transported in the pipeline.

Self-contained breathing apparatus (SCBA): A device worn to provide breathable air for an individual in an area where the atmosphere is hazardous.

Time is of the essence when AOCs occur and field personnel are expected to take decisive action. Personnel must always follow established company-specific policies and procedures. As regulation 49 *CFR* 195.402 states, each pipeline operator must ensure its personnel maintain a thorough knowledge of its operations, maintenance, and emergency manuals. Personnel must make use of their training, skills, and good judgment to maintain safe operation until solutions are found.

2.1.0 Reacting to Abnormal Facility Conditions

Some abnormal pipeline facility conditions may require immediate action. Others may require scheduling repairs for a later date. If field personnel recognize an abnormal pipeline facility condition, they must notify the appropriate personnel so that repairs can be made safely and in a timely manner. If notified of a condition by a third-party call, it is important that personnel record information received from the third party. They must also keep in mind that a caller's description may not be entirely accurate. The problem may be different than described. In rare cases, a call might not be valid. After a report is confirmed, field personnel should generally follow these steps:

Step 1 Investigate and evaluate the report to determine the extent of the damage, if any.

Step 2 Carry out company safety procedures to protect people, property, and the environment if necessary.

Step 3 Notify the appropriate personnel so that proper notifications or repairs can be made.

2.1.1 Pipeline System Damage

When field personnel notice or accidently cause pipeline damage, they investigate the extent of the damage, note the location on the pipeline, and isolate the damage, as applicable. They must also contact the control center, according to company procedures. If an immediate hazard exists, contact 911 emergency personnel.

NOTE

Before excavation or other digging near a pipeline begins, responsible personnel must contact the One-Call notification system.

Some incidents may be serious enough to require an emergency shutdown of the pipeline. One example of a serious incident is someone losing control of a vehicle on the ROW, extensively gouging or denting a large section of pipeline or damaging a valve. Whenever pipeline damage results in a release and/or explosion, field personnel must notify the control center as soon as possible. In some cases, control center personnel may be able to drain a portion of the pipeline into a delivery system to reduce the amount of product released from the pipeline.

Once the pipeline has resumed normal operations, the pipeline operator conducts an internal investigation. This investigation focuses on evaluating control systems and mechanical equipment as well as personnel performance.

Each company has specific procedures for reacting to pipeline damage. These procedures explain how and when to contact the control center, emergency personnel, and local, state, and federal law enforcement agencies. Always follow company policy, make appropriate notifications per company policy, and complete the required reporting.

2.1.2 Damaged Pipeline Supports

The proper reaction to the discovery of damaged pipeline supports varies depending on the severity of the damage. If the damage to the support is critical or indicative of a possible release, the individual making the discovery should contact the control center or other appropriate personnel immediately. Notification is especially important if the pipe is in danger of excessive movement. The appropriate personnel must then follow company policy precisely to investigate the condition and dispatch qualified personnel to the scene. Until the arrival of qualified personnel, all other personnel should stay a safe distance away from the site, evacuate the area if necessary, and make the appropriate notifications as required by company policy.

In some cases, the pipeline operator may need to isolate the pipeline segment. Whether or not the segment needs to be isolated, a damage report must be submitted so that the repair work can be scheduled. The decision to continue or halt operations is a judgment call made by management and qualified inspection personnel. If the pipeline segment must be isolated, affected personnel must be notified.

Field personnel must know and follow their company's procedures. Those procedures are especially important when a condition requires a decision about whether to isolate a segment of the pipeline facility. Make notifications and complete the reporting as required by company policy.

2.2.0 Reacting to the Activation of a Safety Device

Personnel must respond to safety device alarms quickly and calmly. To react to an alarm, personnel should follow these steps:

Step 1 Notify affected personnel.

Step 2 Carry out company warning procedures to protect people, property, and the environment.

Step 3 Troubleshoot the problem and make necessary repairs if qualified to do so.

Step 4 Reset and test the warning device alarm.

Step 5 Make the appropriate notifications and complete the required reporting.

It is important to ensure that the alarm is not a false alarm. In case of a false alarm, the control center should notify affected personnel and have the alarm equipment checked. Generally, no further action is required when an alarm is false.

When a hazardous atmosphere alarm sounds, the proper reaction is somewhat different. In this case, personnel wearing the appropriate personal protective equipment (PPE) must follow these steps:

Step 1 Evacuate the area and don the appropriate PPE if available.

Step 2 Make the appropriate notifications per company policy.

During an emergency, personnel must focus on three key areas: ensuring personnel safety, reducing the damage to equipment, and protecting the environment. The general rule is to put people first. The goal is also to prevent small problems from turning into larger incidents. All reactions must be made safely. Personnel investigating the alarm must wear the appropriate PPE. If told by authorized personnel to evacuate an area, they must leave immediately.

If reacting to a seal failure alarm, personnel should investigate the alarm, check the station sump level if it is equipped with one, and notify the appropriate personnel.

If a pressure-relief system activates, an audible alarm may activate. Immediate action must be taken to reduce pressure. A pressure reduction is usually achieved by shutting down the upstream pumping station, closing a block valve, or shutting down the pipeline system. Field personnel may need to inspect the relief system, test it, and recalibrate it as necessary.

2.3.0 Reacting to a Communications Failure

The proper reaction to a communications failure depends on whether the failure involves a communications tool that personnel use or the failure of system communications.

2.3.1 Failure of Personnel Communication Tools

When communication tools fail, personnel should first check to see if the problem is a simple one that can be quickly and easily fixed. Immediately correct any communications failure, if possible. For example, some telephones have a volume control that may accidentally be turned off. The telephone may seem to be dead, but a simple adjustment to the volume corrects the problem. Other simple solutions include replacing a battery or recharging a unit.

Some communications problems are more serious. Weather or problems at the local utility company can knock out communications completely. Regardless of the outage, the best preparation is to ensure personnel are trained to follow these steps:

Step 1 Switch to alternate equipment if necessary.

Step 2 Notify the appropriate personnel per company policy.

Step 3 Complete any required reporting or documentation.

If communications are interrupted for long periods, it may not be possible to ask for instructions from a remote location. Training before an emergency situation occurs is the best way to ensure that field personnel know how to react properly. However, field personnel must do everything that they can to re-establish communications as soon as possible.

2.3.2 System Communications Failures

The control center will generally notify field personnel about system communications failures so that they can isolate, investigate, and react according to company policy. The proper reactions for field personnel may include some or all of the following activities:

- Switch to alternate or backup systems
- Isolate a pipeline segment if necessary
- Determine the scope of the problem
- Take corrective actions to restore the system
- Estimate how long the main system will be down
- Notify affected personnel about how long repairs will take
- Complete the required reporting or documentation

If the problem is with the SCADA system, or equipment cannot otherwise be operated from the control center, field personnel may have to operate equipment manually. The pipeline may be able to continue to operate while repairs are made. Operations are dependent on company procedures.

2.4.0 Reacting to Power Interruptions

The proper reaction to a power failure depends on which part(s) and how much of the pipeline is affected. Regardless of how the outage affects the pipeline facility, the appropriate personnel should always notify the power company of the outage as soon as possible. If only a single component or station is involved, control center and/or field personnel may be able to adjust line pressure and flow to keep the pipeline in operation. For example, if a power interruption occurs while a valve is in operation, field personnel may have to operate the valve manually. However, if a station is needed to keep pressure within normal operating limits, for example, control center personnel may have to isolate a pipeline segment. If a shutdown is not required, the gas or liquid shipment will continue to arrive, and the schedule must be followed and the tasks completed manually.

Each company has specific procedures for responding to a power interruption. Always follow company policy, make appropriate notifications per company policy, and complete the required documentation.

2.5.0 Reacting to a Fire, Explosion, or Natural Disaster

The pipeline operator must have a training program in place so that personnel know what to do in case of a fire, explosion, or natural disaster that affects a pipeline. Many companies conduct regular drills so that personnel can practice their responses to such events. Drills help ensure that personnel are prepared to handle emergencies safely and calmly. Each operator has its own procedures for emergencies. In general, trained personnel should know the following information:

- The location of fire suppression equipment and how to operate it; certain fire suppression equipment and PPE may require additional training
- The process for shutting down equipment manually if communication with the control center is lost
- The company's evacuation plan
- The procedures for contacting local emergency response agencies, if necessary

Depending on the situation, pipeline personnel may be able to safely handle an emergency without calling for outside help. Follow company procedures for contacting local emergency response agencies.

2.5.1 Fire

When a fire occurs, the safety of personnel is always top priority, so personnel should evacuate the area. Protection of property is always second.

Personnel must ensure that any injured personnel receive medical attention as soon as possible. Consider any alarm as an actual alarm and react appropriately and decisively.

Personnel must also eliminate the source of the fire, but only as authorized. For example, only qualified personnel are authorized to turn off valves. Company policy dictates procedures for emergency actions and must be followed. Procedures are site-specific and may allow for some flexibility in responding. However, personnel should not need to consult a manual in case of an alarm. They must be adequately trained and familiar with emergency procedures in advance. It may even be necessary to shut down the pipeline if required to control the emergency. Always notify the appropriate personnel and emergency agencies and complete any required documentation or reporting.

According to 49 *CFR* 192 and 195, operators shall establish and conduct a continuing training program to instruct emergency personnel in the proper use of firefighting procedures and equipment, in the identification of hazardous conditions likely to cause emergencies, in the steps necessary to control accidental releases, and in the characteristics of hazardous liquids or carbon dioxide being transported.

The proper reaction differs according to the type of fire in progress. If the fire is small or just beginning (**incipient fire**), personnel should use the available fire extinguishing equipment to control the situation. According to *OSHA 1910.157*, companies must provide designated personnel with training on fire extinguishing equipment as well as the hazards involved with incipient firefighting.

Use the proper type of fire extinguisher for each class of fire. It is important to know the extinguisher's fire class rating before using it. The fire class rating is printed at the bottom of the label on a fire extinguisher. There are four fire class ratings (see *Table 3*). Most extinguishers can be used for multiple types of fires.

If the fire is larger than the immediate workforce can contain, personnel should try to mitigate the situation, notify appropriate personnel and emergency agencies, and complete the required documentation and/or reporting. For example, during an electrical storm, if lightning strikes a full tank at a facility and the roof seal catches on fire, personnel should follow company policy and contact the fire department to contain the fire.

Once the fire is out, the air is cleared, and any injured personnel have received medical attention, the following actions are taken:

Table 3 Fire Class Ratings

Fire Class	Materials and Proper Fire Extinguisher
A	Involves ordinary combustibles, such as wood or paper. This type of fire is suppressed by cooling the fire. Class A extinguishers contain water. A Class A fire extinguisher can never be used on a Class B, C, or D fire.
B	Involves grease, flammable liquids, and gases. Class B extinguishers contain carbon dioxide or another material that smothers the fire by removing oxygen.
C	Involves electrical equipment. Class C extinguishers are designed to protect the firefighter from electric shock. This type of extinguisher smothers the fire.
D	Involves metals. Class D extinguishers contain a powder that either forms a crust around the burning metal or gives off gases that prevent oxygen from reaching the fire. Some metals may keep burning even after they have been coated with powder from a Class D extinguisher. The best way to fight these fires is to keep using the extinguisher so that the fire does not spread.

- Alarms are reset and tested.
- Equipment damage is assessed and reported to the appropriate personnel.
- Work orders are issued and required permits are obtained for repairs.
- Personnel performance is evaluated and corrective action is determined if necessary.
- Injuries or deaths are reported to the appropriate regulatory agencies if required.

The evaluation of personnel performance during a fire may reveal problems in personnel response. Management generally recommends retraining so that personnel can improve their responses to a fire.

A false alarm or an expected alarm due to maintenance activities requires no further action unless the control center determines that the system needs to be investigated. If an alarm has been verified as a false alarm, the appropriate personnel follow these steps:

Step 1 Notify affected personnel.

Step 2 Test alarms to ensure that they are operating properly.

Step 3 Reset alarms or ensure that they have automatically reset.

2.5.2 Explosion

Each operator has specific procedures for reacting to an explosion. In addition, the correct order of the steps to take may vary from operator to operator. A response to an explosion must be appropriate, decisive, and calm. If the main-line valves do not close automatically, designated personnel must initiate actions to ensure that the valves are closed. In general, the proper reactions to an explosion are:

- Evacuate and notify appropriate personnel
- Evaluate and assess the situation.
- Carry out company safety procedures to protect people, property, and the environment.
- Call local emergency response agencies if necessary.
- Confirm the location of all pipeline personnel near the explosion location.
- Confirm the location of any members of the public near the explosion location.
- Call local, state, or federal law enforcement if the cause indicates criminal or terrorist activity.
- Make the appropriate notifications and complete the required documentation and/or reporting.
- Conduct an incident investigation.

2.5.3 Natural Disaster

Much of the pipeline's ability to withstand extreme weather that can result in natural disaster is already present in its design and installation.

Each company has specific procedures for dealing with natural disasters. These procedures reflect local conditions. For example, hurricanes occur more frequently in Florida or the Gulf states. Therefore, companies located in these areas include hurricane response in their procedures.

The following are examples of proper reactions to weather-related problems:

- Monitor weather conditions such as tropical storms that may become hurricanes
- Carry out company safety procedures for protecting people, property, and the environment
- Adjust work schedules as needed to respond to emergency conditions
- Follow company procedures for responding to fire, explosion, or damage resulting from natural disasters
- Set up a command post if conditions warrant such a step
- Work with local emergency response agencies if required
- Conduct investigations when appropriate

2.5.4 Emergency Shutdowns

Events occurring on the pipeline that present immediate hazard to persons or property require an emergency shutdown. An event might be caused by human error, mechanical failure, electrical failure, or an accident on the pipeline. It might also be brought about by threats from weather or civil disobedience. Regardless of the reason, the line must be shut down quickly and safely. Each company has a standard operating procedure for emergency shutdowns.

In the case of a fire, explosion, or release, time for discussion may be very limited. Once the need is confirmed, decisive action must be taken. The pipeline must be isolated by control center personnel by safely shutting down the originating station and continuing downstream in a safe manner until all pumps are offline. Control center personnel will then close block valves to complete the shutdown and notify other personnel as needed. In case of a pump shutdown from low pressure, there may be some time to discuss the decision with appropriate management if the line has enough flexibility to withstand the stress.

Keep in mind that threats from natural disaster or civil disobedience may come without warning. If field personnel witness these kinds of threats to the pipeline, they should contact their supervisor or control center immediately. Affected facilities may need to be isolated. Once facilities are isolated, pipeline shutdown may follow.

In emergency conditions such as a possible release, the pipeline operator has full responsibility and authority to make shutdown decisions and take appropriate action. These actions are governed by the company release policy above all other considerations.

2.5.5 Evacuations

The safety of pipeline personnel and the public is the first priority when any AOC occurs. If there is any threat of injury or death to any persons, an evacuation of pipeline personnel must occur. Events that could result in the need for an evacuation are fire, explosion, release of hydrocarbon or other hazardous vapors, seismic activity, or other natural disasters.

Companies have standard procedures in place to guide the evacuation of personnel. By law, emergency response plans must be in place and evaluated annually. Facility layouts with evacuation maps are required to be posted in every area or department. Safe areas are designated for all personnel. When the command to evacuate is given, trained personnel may act as safety coordinators. Their job is to conduct a roll call to confirm that all personnel are safe. To assist in this process, personnel must report to their designated safe areas and identify themselves to the designated safety coordinator. If any personnel cannot be accounted for, authorized personnel must locate and rescue them if necessary.

Some emergencies may require an evacuation of the public located near the pipeline. The pipeline operator may work with the local authorities to provide details about the incident, determine whether a general evacuation is required, provide contact information for people residing or working near the pipeline, and provide personnel to help staff a command center if requested.

NOTE

Local emergency planning committees (LEPCs) exist in every jurisdiction in the United States. Law enforcement agencies, firefighting companies, local government agencies, hospitals, companies, and citizen's groups make up the memberships of LEPCs. Committee members meet on a regular basis to plan for emergencies. Planning for a general evacuation is just one example of the projects handled by an LEPC.

2.6.0 Reacting to the Unexpected Release of Hazardous Gas or Liquid

When a hazardous gas or liquid is released from a pipeline unexpectedly, personal safety should be the first consideration. Beyond that, the proper reaction to the unexpected release of a product depends on the characteristics of the product and the location of the release. Some gases and liquids can explode under certain conditions, and some are toxic to people and the environment.

NOTE

To identify the hazards associated with the chemical being released, consult the material safety data sheets (MSDS)/safety data sheets (SDS).

Regardless of location or the type of product, when a product is released personnel *must* take the following steps:

Step 1 Evacuate the area.

Step 2 Identify the source of the release and eliminate potential ignition sources.

Step 3 Vent the area if possible.

Step 4 Prevent anyone who is not wearing the appropriate PPE from entering the area (personnel must be trained to use PPE).

Step 5 Call for emergency rescue and medical help, if required.

Step 6 Contact the control center; appropriate personnel may need to isolate the pipeline.

Step 7 Have the power company turn off the power, if appropriate.

WARNING

When witnessing a release and there is a downed person in the area of the release, call 911 and evacuate the area. Do not attempt to rescue personnel.

NOTE

OSHA stipulates that employers must ensure that contract personnel are trained in the work practices necessary to perform their job safely.

Of equal importance are actions personnel must avoid. When reacting to a release, personnel *must not*:

- Open switchgear doors.
- Operate electrical switches or equipment in the facility, including lights, doors, louvers, or vents.
- Operate personal electrical equipment (including cell or landline telephones) unless the equipment is classified as intrinsically safe/explosion-proof.

Intrinsically safe/explosion-proof electrical equipment is designed so that a spark cannot ignite the atmosphere. It is made to meet the following specifications:

- Outside air cannot penetrate the casing.
- The casing must be strong enough to contain an explosion inside the equipment without igniting the surrounding atmosphere.
- The interior must be filled with nonflammable powder or inert gas to prevent ignition.

This equipment is determined to be safe by the National Electrical Manufacturers Association (NEMA), and is certified according to the standards of the *National Electric Code*®. It must also meet the requirements of Underwriters Laboratories (UL). Equipment that is intrinsically safe/explosion-proof should always be labeled as such.

WARNING

Electronic equipment must not be used in the presence of hazardous gases/vapors unless it is classified as intrinsically safe/explosion-proof. Failure to use intrinsically safe/explosion-proof equipment in a hazardous atmosphere may result in an explosion.

2.6.1 Releases in Enclosed Areas

If a product release occurs in an enclosed area, the source of the release must be determined quickly. It is also important that personnel move to a safe location as quickly as possible and then notify the appropriate personnel, who can operate equipment remotely. If trained, personnel may close the cutoff valve to mitigate the situation.

2.6.2 Releases in Outdoor Areas

Any release must be approached by personnel from the upwind side. Releases in outdoor areas include the additional concern for the safety of the general public, as well as the safety of the environment. Personnel should evacuate the area, go upwind, and take additional appropriate action per company policy. If an odor of hydrocarbon is detected even a few hundred feet from the source, contact control center personnel to shut down the pipeline. Report the incident and any action taken, and complete the required documentation.

2.6.3 Toxic Products

Personnel must take special precautions when dealing with the release of products that can become toxic. Examples include carbon dioxide (CO_2), hydrogen sulfide (H_2S), propane, anhydrous ammonia, and natural gas.

Carbon dioxide is a colorless, odorless, gaseous compound used as a propellant in many domestic applications, such as fire extinguishing equipment, and is a byproduct of combustion. Because it is heavier than air, it can collect in low-lying areas where it may displace oxygen; exposure can result in suffocation and death. A leak detector can detect a hazardous atmosphere, but without it, personnel can be overcome without realizing the problem. Some pipeline operations pump carbon dioxide commercially, and in such operations, the hazards are increased. To enter an area where a carbon dioxide release is suspected, personnel must use the proper gas detection meter, calibrated for the detection of carbon dioxide, and wear the appropriate PPE, such as **self-contained breathing apparatus (SCBA)** (see *Figure 5*).

67107-13_F05.EPS

Figure 5 Self-contained breathing apparatus.

WARNING

Carbon dioxide (CO_2) is a colorless, odorless gas that is heavier than air, so it can collect in low-lying areas where it may displace oxygen. Exposure can be fatal.

Hydrogen sulfide (H_2S) is a colorless gas that smells like rotten eggs. However, in fairly low concentrations, it can deaden the sense of smell. Exposure to high concentrations can be fatal.

WARNING

Exposure to high concentrations of hydrogen sulfide (H_2S) can be fatal. Even in low concentrations, it can cause headaches or eye irritation. In addition, it is flammable and may cause a flash fire. When a gas detection unit shows that H_2S is present, wear the appropriate PPE or immediately leave the area.

Anhydrous ammonia is a liquefied gas that has a sharp odor. Exposure in high concentration can cause death. Even exposure in low concentrations can cause severe reactions from mild irritation to severe burns. The reaction depends on the length of exposure and concentration amounts. If anhydrous ammonia is released, call emergency personnel.

> **WARNING**
>
> Exposure to high concentrations of anhydrous ammonia can cause death. Even in low concentrations, it can cause severe reactions. If detected, wear the appropriate PPE or immediately leave the area. If a release occurs, call emergency personnel.

Propane is a common natural gas that is colorless and heavier than air, so it can collect in low-lying areas where it may displace oxygen. Exposure can cause death. At high concentrations, it is flammable.

> **WARNING**
>
> Exposure to propane gas can cause death. At high concentrations, it is flammable. If detected, wear the appropriate PPE or immediately leave the area.

Naturally occurring radioactive materials (NORMs) may be present when oil and gas is extracted from the earth. These NORMs are then introduced into the product streams and transported in the pipeline. Natural gas liquids (NGLs), along with the processing associated with these products, may trap NORMs in the form of sludge, scale, and particulates. Filters, valves, scrapers, and even internal surfaces of pipelines may contain high levels of NORMs. Precautions should be taken to protect personnel by monitoring for NORMs in accordance with all local, state, federal, and company requirements. If high levels are detected, personnel should distance and/or shield themselves from the source and notify the appropriate personnel.

Only appropriately trained personnel may enter an area where a hazardous gas has been detected. These personnel must be trained and qualified to use SCBA. Other personnel must stay a safe distance away until the deadly atmosphere is cleared. Warning signs must be posted at access points so that personnel do not enter until the area has been inspected and tested.

2.7.0 Reacting to Unexplained Flow Rate/ Pressure Changes

If a change in pressure or flow rate occurs that is different from what is expected, alarms may be initiated to alert personnel to the problem. Personnel must act quickly to address the problem. Qualified personnel will investigate and evaluate the problem. Some problems may require field personnel to manually adjust components, such as valves and transmitters, or make repairs. More serious problems may result in the need to isolate a segment of the pipeline.

Additional Resources

It is assumed that you will learn to perform these procedures under the direct supervision of your crew leader, foreman, or instructor. For independent study, refer to the recommended Additional Resources at the back of this module.

2.0.0 Section Review

1. If a damaged pipeline support is discovered and the damage appears to be critical, the individual who made the discovery should _____.
   a. begin repairs immediately
   b. stay close to the site
   c. report it to the appropriate personnel
   d. assume it is safe

2. If a hazardous atmosphere alarm sounds, field personnel should first _____.
   a. call 911
   b. evacuate the area
   c. notify control center personnel
   d. verify that the alarm is not false

3. Field personnel may need to operate equipment in the field manually if there is a system failure.
   a. True
   b. False

4. When a power outage occurs, the appropriate pipeline personnel should contact _____.
   a. the One-Call notification system
   b. the power company
   c. OSHA
   d. local authorities

5. Fire extinguishing equipment can be used on an incipient fire.
   a. True
   b. False

6. Electronic equipment must not be used in the presence of hazardous gases unless it is classified as _____.
   a. OSHA approved
   b. intrinsically safe/explosion-proof
   c. NRC approved
   d. mechanically safe/disruption proof

7. When an unexplained flow rate or pressure change occurs, field personnel may need to _____.
   a. notify the One-Call system
   b. contact the power company
   c. make repairs
   d. remove pipe supports

SUMMARY

Pipeline operators have developed effective procedures and policies that enable personnel to respond quickly, safely, and effectively to AOCs. Usually an AOC is a routine matter that can be handled safely by pipeline personnel. Sometimes, however, an AOC can result in an emergency situation that may involve working with outside agencies.

In this module, you learned about the different types of AOCs that can occur on the pipeline and the proper reactions to each type of AOC. During your career, you may have to respond to some or all of the AOCs outlined in this module. When responding to these AOCs, you must follow your company's policies and procedures. Protecting people, property, and the environment should be your utmost concern.

Review Questions

1. By definition, an AOC is a condition that may exceed _____.
   a. company operating budget
   b. design limits
   c. safety laws
   d. size limits

2. The federal agency that regulates pipeline maintenance is _____.
   a. OSHA
   b. GAO
   c. EPA
   d. PHMSA

3. By definition, a pipeline facility includes the right-of-way.
   a. True
   b. False

4. Which of the following is *not* always an AOC?
   a. Exposed pipe
   b. Reduced pressure
   c. Lost radio communication
   d. Unexpected release of a hazardous gas

5. Pressure relief systems, tank-level alarms, and main-line pressure shutdown alarms are examples of _____.
   a. prevention devices
   b. safety devices
   c. hazardous atmosphere alarms
   d. fire alarms

6. The loss of SCADA is an example of which type of AOC?
   a. Hydraulic failure
   b. Mechanical failure
   c. Communications failure
   d. Power failure

7. In the event of a power failure, pipeline companies may be able to keep the pipeline running by using an _____.
   a. operations override power source
   b. uninterruptible power system
   c. interface power system
   d. indirect power system

8. In the control center, one indication of a pipeline explosion could be a(n) _____.
   a. change in product temperature
   b. valve malfunction
   c. power outage
   d. unexplained change in pressure

9. Which of the following is *not* true of hazardous gas?
   a. Odorants are often added.
   b. High concentrations can deaden a person's sense of smell.
   c. Some gases are odorless.
   d. Low concentrations can deaden a person's sense of smell.

10. Which of the following best defines an unexplained change in flow rate or pressure?
    a. Reduced pressure only
    b. No flow only
    c. High flow rate
    d. Any change from the intended flow or pressure

11. When pipeline support damage is discovered, notification is especially important if the pipe _____.
    a. is in danger of excessive movement
    b. has changed color
    c. is metal
    d. is corroded

12. When a safety device sounds an alarm, field personnel must _____.
    a. turn it off
    b. never ignore it
    c. reset it
    d. contact the local emergency agency

13. If personnel communication tools fail, personnel should first _____.
    a. check for problems that have simple solutions
    b. contact the control center
    c. call 911
    d. complete any required reporting

14. If SCADA fails, equipment may have to operated manually by _____.
    a. PHMSA
    b. field personnel
    c. control center personnel
    d. local authorities

15. When a power interruption occurs, personnel may be able to keep the pipeline in operation by adjusting the _____.
    a. MAOP
    b. MOP
    c. line pressure and flow
    d. product temperature

16. A false alarm may trigger an investigation of the system, which is determined by _____.
    a. the control center
    b. field personnel
    c. the job site supervisor
    d. SCADA

17. Before using a fire extinguisher, it is important to know its _____.
    a. fire class rating
    b. OSHA rating
    c. combustibility
    d. capacity

18. The pipeline operator may work with local authorities to facilitate an evacuation of the area surrounding a pipeline.
    a. True
    b. False

19. An area where a hazardous gas has been detected may only be entered by personnel trained in rescue operations and qualified to use a _____.
    a. combustible gas indicator
    b. flame ionization indicator
    c. self-contained breathing apparatus
    d. fire suppression kit

20. All of the following are examples of a pipeline facility *except* _____.
    a. compressors
    b. pump stations
    c. control center
    d. rights-of-way

Trade Terms Quiz

Fill in the blank with the correct term that you learned from your study of this module.

1. The computer-based system(s) used by a controller in a control room that collects information about a pipeline facility and may have the ability to send commands back to the pipeline facility is called________.
2. The ________ is the area along the pipeline which has been legally permitted to be used by the pipeline to construct, operate, and maintain the pipeline.
3. A fire in its beginning stage is called a(n) ________.
4. ________ is the separation of the protective coating from the pipeline.
5. A(n) ________ is a system that provides continuity of power when the normal power supply is interrupted.
6. The maximum pressure at which a pipeline or segment of a pipeline may be operated under this part (gas) is called ________.
7. A(n) ________ is used to relieve surges in a hazardous liquid pipeline system or to receive and store hazardous liquid transported by a pipeline for injection and continued transportation by pipeline.
8. The lowest concentration of a gas or a vapor in the atmosphere capable of producing fire in the presence of an ignition source is the ________.
9. A ________ duplicates the operations of a system so that, in the event of a failure, a backup system is available.
10. A(n) ________ indicates a malfunction of a component or a deviation from normal operations that may indicate a condition exceeding design limits or result in a hazard to persons, property, or the environment.
11. A device that detects some hazardous gases in the atmosphere is a(n) ________.
12. The pressure produced by a change in the velocity of the moving stream that results from shutting down a pump station or pumping unit, closure of a valve, or any other blockage of the moving stream is called ________.
13. ________ is the maximum pressure at which a pipeline or segment of a pipeline may be normally operated under this part (liquids).
14. ________ is a centralized communication system established by utilities, government agencies, and operators of underground facilities.
15. A hazardous liquid which will form a vapor cloud when released to the atmosphere and which has a vapor pressure exceeding 276 kPa (40 psia) at 37.8°C (100°F) is classified as a(n) ________.
16. The government agency that is responsible for developing and enforcing regulations for pipeline operations in the United States is the ________.
17. Electrical equipment that is ________ is designed specifically for use in explosive or flammable locations so that a spark cannot ignite the atmosphere.
18. An organization that adopts standards for a variety of products, services, and processes in the United States to ensure consistency is the ________.
19. A ________ is a device worn by an individual to provide breathable air in an atmosphere that is hazardous.
20. The ________ is the company that operates the pipeline, as opposed to an individual person who works for the company that operates the pipeline.

21. New and existing pipe on the pipeline itself, the right-of-way (ROW), and any equipment, facility, or building used in the transportation of a liquid or gas are defined as part of a(n) _______.

22. Activity carried out by, and/or equipment belonging to outside parties on the pipeline ROW is referred to as _______.

23. A length of pipeline that is entirely aboveground and supported by pipe supports is called a(n) _______.

24. Materials with radioactive elements that can be extracted during oil and gas extraction from the earth and can get into product streams and become transported in the pipeline are called _______.

Trade Terms

Aboveground span
Abnormal operating condition (AOC)
American National Standards Institute (ANSI)
Breakout tank
Disbondment
Encroachment
Flame ionization detection unit
Highly volatile liquid (HVL)
Incipient fire
Intrinsically safe/explosion-proof
Lower explosive limit (LEL)
Maximum allowable operating pressure (MAOP)
Maximum operating pressure (MOP)
Naturally occurring radioactive materials (NORMs)
One-Call notification system
Pipeline and Hazardous Materials Safety Administration (PHMSA)
Pipeline facility
Pipeline operator
Redundant system
Right-of-way (ROW)
Self-contained breathing apparatus (SCBA)
Supervisory control and data acquisition (SCADA)
Surge pressure
Uninterruptible power system (UPS)

Trade Terms Introduced in This Module

Aboveground span: Length of pipeline that is entirely aboveground and supported by pipe supports.

Abnormal operating condition (AOC): A condition within a pipeline system, identified by the operator, that may indicate (1) a malfunction of a component or (2) a deviation from normal operations that may indicate a condition exceeding design limits or result in a hazard to persons, property, or the environment.

American National Standards Institute (ANSI): An organization that adopts standards for a variety of products, services, and processes in the United States to ensure consistency.

Breakout tank: a tank used to (a) relieve surges in a hazardous liquid pipeline system or (b) receive and store hazardous liquid transported by a pipeline for reinjection and continued transportation by pipeline.

Disbondment: Separation of the protective coating from the pipeline.

Encroachment: Activity carried out by, and/or equipment belonging to outside parties on the pipeline right-of-way (ROW). A particular encroachment might be authorized or not, so each situation requires specific handling.

Flame ionization detection unit: A device that detects some hazardous gases in the atmosphere.

Highly volatile liquid (HVL): A hazardous liquid which will form a vapor cloud when released to the atmosphere and which has a vapor pressure exceeding 276 kPa (40 psia) at 37.8°C (100°F).

Incipient fire: A fire in its beginning stage.

Intrinsically safe/explosion-proof: Describes electrical equipment that is designed specifically for use in explosive or flammable locations so that a spark cannot ignite the atmosphere.

Lower explosive limit (LEL): The lowest concentration of a gas or a vapor in the atmosphere capable of producing fire in the presence of an ignition source. Different products have different LELs.

Maximum allowable operating pressure (MAOP): The maximum pressure at which a pipeline or segment of a pipeline may be operated under this part (gas).

Maximum operating pressure (MOP): The maximum pressure at which a pipeline or segment of a pipeline may be normally operated under this part (liquids).

Naturally occurring radioactive materials (NORMs): Materials with radioactive elements that can be extracted during oil and gas extraction from the earth and can get into product streams and become transported in the pipeline.

One-Call notification system: A centralized communication system established by utilities, government agencies, and operators of underground facilities. This system provides one telephone number for contractors and the general public to call to notify of their intent to use equipment on or near a pipeline. It is also in place for any individual to call if pipeline damage is discovered. The One-Call system is generally set up on a state-by-state basis.

Pipeline and Hazardous Materials Safety Administration (PHMSA): An agency of the US Department of Transportation responsible for developing and enforcing regulations for pipeline operations in the United States.

Pipeline facility: New and existing pipe, rights-of-way and any equipment, facility, or building used in the transportation of hazardous liquids or carbon dioxide, or new and existing pipelines, rights-of-way, and any equipment, facility, or building used in the transportation of gas or in the treatment of gas during the course of transportation.

Pipeline operator: The company that operates the pipeline, as opposed to an individual person who works for the company that operates the pipeline.

Redundant system: A system that duplicates the operations of a system so that, in the event of a failure, a backup system is available.

Right-of-way (ROW): The area along the pipeline which has been legally permitted to be used by the pipeline operator to construct, operate, and maintain the pipeline. The ROW extends the entire length of the pipeline.

Self-contained breathing apparatus (SCBA): A device worn to provide breathable air for an individual in an area where the atmosphere is hazardous.

Supervisory control and data acquisition (SCADA): a computer-based system or systems used by a controller in a control room that collects and displays information about a pipeline facility and may have the ability to send commands back to the pipeline facility.

Surge pressure: Pressure produced by a change in velocity of the moving stream that results from shutting down a pump station or pumping unit, closure of a valve, or any other blockage of the moving stream.

Uninterruptible power system (UPS): A system that provides continuity of power when the normal power supply is interrupted. A UPS provides AC power to the system it serves.

Additional Resources

The following recommended resources can provide additional helpful information related to the requirements for this covered task.

American National Standard Institute (ANSI), 1899 L Street, NW, 11th Floor, Washington, DC 20036. **www.ansi.org**

American Petroleum Institute (API), 1220 L Street, NW, Washington, DC 20005. **www.api.org**

American Society for Mechanical Engineers (ASME), Two Park Avenue, New York, NY 10016-5990. **www.asme.org**

American Society for Testing and Materials (ASTM) Specification D396.

Association of Oil Pipe Lines (AOPL), 1808 Eye Street, NW, Suite 300, Washington, DC 20006. **www.aopl.org**

Code of Federal Regulations 49, Parts 192 and 195. **www.ecfr.gov**

International Society of Automation (ISA), 67 TW Alexander Drive, Research Triangle Park, NC 27709. **www.isa.org**

National Association of Corrosion Engineers (NACE), 1440 South Creek Drive, Houston, TX 77084. **www.nace.org**

National Electrical Code®. **www.nfpa.org**

Occupational Safety and Health Administration (OSHA), 200 Constitution Avenue, NW, Washington, DC 20210. **www.osha.gov**

Pipeline and Hazardous Material Safety Administration (PHMSA), East Building, 2nd Floor, 1200 New Jersey Avenue, SE, Washington, DC 20590. **www.phmsa.dot.gov**

US Bureau of Labor Statistics, Postal Square Building, 2 Massachusetts Avenue, NE, Washington, DC 20212. **www.bls.gov**

US Energy Information Administration, 100 Independence Avenue, SW, Washington DC 20585. **www.eia.gov**

Figure Credits

U.S. Geological Survey/photo by Don Becker, Module Opener

Energy Information Administration, Office of Oil & Gas, Natural Gas Division, Gas Transportation Information System, 2009, Figure 1

Photograph No. 5998083 (Photographer Eric Vance), September 2006; Records of the Environmental Protection Agency, 1944–2006, Record Group 412; National Archives and Records Administration–College Park, Figure 9

Cameron Strother, Southland Safety, Figure 15

Photo courtesy of STV, Figure 34

Courtesy of Liquid Controls, LLC, Figure 35

Photographs and illustrations provided courtesy of Daniel Measurement and Control, Inc. Daniel Senior and Simplex are registered trademarks of Daniel Measurement and Control, Figure 36

Courtesy of Sunoco Logistics, Figure 39

Appendix A

Weather and Natural Disaster Information Sources

Emergency Managers Weather Information. US Department of Commerce, National Oceanic and Atmospheric Administration, National Weather Service, 1325 East-West Highway, Silver Spring, MD 20910.**www.nws.noaa.gov/emwin/index.htm**

National Earthquake Information Center, US Geological Survey. Box 25046, DFC, MS 967 Denver, CO 80225. **www.earthquake.usgs.gov/regional/neic**

National Hurricane Center Tropical Prediction Center. 11691 SW 17th Street, Miami, FL 33165-2149. **www.nhc.noaa.gov**

National Weather Service and National Weather Service Weather Radio, National Oceanic and Atmospheric Administration (NOAA). 1325 East-West Highway, Silver Spring, MD 20910. **www.weather.gov**

Appendix B

Organizations of Interest to the Pipeline Industry

American Gas Association (AGA). 400 N. Capitol Street, NW, Washington, DC 20001. **www.aga.org**

American Petroleum Institute (API). 1220 L Street, NW, Washington, DC 20005. **www.aga.org**

Interstate Natural Gas Association of America (INGAA). 20 F Street, NW, Suite 450, Washington, DC 20001. **www.ingaa.org**

National Institute for Occupational Safety and Health (NIOSH). 395 E Street, SW, Suite 9200, Patriots Plaza Building, Washington, DC 20201. **www.cdc.gov/niosh**

National Response Center (NRC). c/o United States Coast Guard (CG-5335 - Stop 7581), 2100 2nd Street, SW, Washington, DC 20593-0001. **www.nrc.uscg.mil**

Occupational Safety and Health Administration (OSHA). 200 Constitution Avenue, NW, Washington, DC 20210. **www.osha.gov**

Pipeline Hazardous Material Safety Administration (PHMSA). East Building, 2nd Floor, 1200 New Jersey Ave., SE, Washington, DC 20590. **www.phmsa.dot.gov**

Section Review Answer Key

Answer	Section Reference	Objective
Section One		
1. b	1.1.0	1a
2. a	1.2.0	1b
3. d	1.3.2	1c
4. b	1.4.0	1d
5. a	1.5.1	1e
6. c	1.6.0	1f
7. b	1.7.0	1g
Section Two		
1. c	2.1.2	2a
2. b	2.2.0	2b
3. a	2.3.2	2c
4. b	2.4.0	2d
5. a	2.5.1	2e
6. b	2.6.0	2f
7. c	2.7.0	2g

NCCER CURRICULA — USER UPDATE

NCCER makes every effort to keep its textbooks up-to-date and free of technical errors. We appreciate your help in this process. If you find an error, a typographical mistake, or an inaccuracy in NCCER's curricula, please fill out this form (or a photocopy), or complete the online form at **www.nccer.org/olf**. Be sure to include the exact module ID number, page number, a detailed description, and your recommended correction. Your input will be brought to the attention of the Authoring Team. Thank you for your assistance.

Instructors – If you have an idea for improving this textbook, or have found that additional materials were necessary to teach this module effectively, please let us know so that we may present your suggestions to the Authoring Team.

NCCER Product Development and Revision

13614 Progress Blvd., Alachua, FL 32615

Email: curriculum@nccer.org
Online: www.nccer.org/olf

❑ Trainee Guide ❑ Lesson Plans ❑ Exam ❑ PowerPoints Other ______________________

Craft / Level: Copyright Date:

Module ID Number / Title:

Section Number(s):

Description:

Recommended Correction:

Your Name:

Address:

Email: Phone:

CT25_1-17

Inspect, Test, and Calibrate Pressure Switches

Overview

Pipelines and modern process systems rely on the accurate and precise measurement of process variables to operate as safely, efficiently, and reliably as possible. Installing the appropriate instruments for measuring process variables is only the first step. Technicians must regularly inspect, test, and properly calibrate these instruments to keep them functioning correctly. This module describes types of pipeline pressure switches and explains how to inspect, test, and calibrate them.

Trainees with successful module completions may be eligible for credentialing through the NCCER Registry. To learn more, go to **www.nccer.org** or contact us at 1.888.622.3720. Our website has information on the latest product releases and training, as well as online versions of our *Cornerstone* magazine and Pearson's product catalog.

Your feedback is welcome. You may email your comments to **curriculum@nccer.org**, send general comments and inquiries to **info@nccer.org**, or fill in the User Update form at the back of this module.

This information is general in nature and intended for training purposes only. Actual performance of activities described in this manual requires compliance with all applicable operating, service, maintenance, and safety procedures under the direction of qualified personnel. References in this manual to patented or proprietary devices do not constitute a recommendation of their use.

CT25_1-17

Inspect, Test, and Calibrate Pressure Switches

Objectives

When you have completed this module, you will be able to do the following:

1. Identify and describe pressure switches and the issues associated with testing them.
   a. Identify and describe pressure switches.
   b. Describe the issues associated with testing pressure switches.
2. Describe the procedures used to inspect, test, and calibrate pressure switches (CT25_1-17).
   a. Identify the prerequisites, competencies, and abnormal operating conditions (AOCs) associated with inspecting, testing, and calibrating pressure switches.
   b. Describe the procedures used to inspect, test, and calibrate pressure switches.

Performance Task

Under the supervision of the instructor, you should be able to do the following:

1. Inspect, test, and calibrate pressure switches (CT25_1-17).
   - Identify potential abnormal operating conditions (AOCs) that may occur during performance of this CT, and know the appropriate actions to take in response to them.
   - Utilize the appropriate personal protective equipment according to relevant company procedures.
   - Verify certification that test equipment has been calibrated prior to performing calibrations.
   - Verify device set point for switches prior to performing calibration.
   - Determine correct test medium and range to be used for testing.
   - Notify control center and/or affected personnel before work begins.
   - Isolate the pressure switch from the process system and inspect all connection for leakage.
   - Apply test medium pressure and verify device setpoint or range "as found" and document the results.
   - Re-apply test medium for desired setpoint or range and adjust if necessary according to device manufacturer's specifications to establish desired setpoint. Document final setpoint setting(s) "as left".
   - Notify control center and/or affected personnel after completion of work.
   - Complete appropriate documentation as required by operator's procedures.

Trade Terms

Abnormal operating conditions (AOCs)
Alarm
Bench condition testing (BCT)
Bourdon tube
Dynamic condition testing (DCT)
End device
Intrinsically safe
Pressure switch deadband
Pressure switch setpoint
Process system
Shutdown device
Static condition testing (SCT)
Test medium
Trace pressure switch

Industry Recognized Credentials

If you are training through an NCCER-accredited sponsor, you may be eligible for credentials from NCCER's Registry. The ID number for this module is CT25_1-17. Note that this module may have been used in other NCCER curricula and may apply to other level completions. Contact NCCER's Registry at 888.622.3720 or go to **www.nccer.org** for more information.

Contents

1.0.0 Pressure Switches ... 1
1.1.0 Characteristics of Pressure Switches ... 1
1.1.1 Pressure Switch Operations ... 1
1.2.0 Testing Pressure Switches ... 2
1.2.1 Protection Parameters ... 2
1.2.2 Calibrating Pressure Switches in Explosive Atmospheres ... 4
2.0.0 Inspect, Test, and Calibrate Pressure Switches ... 5
2.1.0 Prerequisites, Competencies, and Abnormal Operating Conditions ... 5
2.1.1 Prerequisites and Competencies ... 6
2.1.2 Abnormal Operating Conditions ... 6
2.2.0 Procedure ... 6

Figures

Figure 1 Typical pressure switch ... 1
Figure 2 Trace pressure switch designs ... 3
Figure 3 Intrinsically safe pressure calibrator ... 4

Section One

1.0.0 Pressure Switches

Objective

Identify and describe pressure switches and the issues associated with testing them.

a. Identify and describe pressure switches.
b. Describe the issues associated with testing pressure switches.

Trade Terms

Alarm: A status event from the control system, notifying personnel that a process parameter has moved out of its normal range; alarms can be audible, visual, or both.

Bourdon tube: A mechanical pressure-measuring instrument that uses for its sensing element a curved or twisted metal tube that is flattened in cross section and closed. It is also called a Bourdon pressure gauge.

End device: A device that must operate to provide protection for a pipeline. For example, a pressure switch activates and, through the control sequences, causes disconnection of power to a motor. The motor is the end device.

Intrinsically safe: A test instrument that is safe to use in an explosive atmosphere because it does not generate sufficient electrical or thermal energy to ignite in it.

Pressure switch deadband: The pressure value at which a tripped pressure switch resets. Also known as the zero pressure.

Pressure switch setpoint: The pressure value at which a pressure switch trips.

Process system: A facility within which devices for measuring process variables such as temperature, pressure, level, and flow are installed.

Shutdown device: A device that protects a pipeline in response to an unacceptable process condition.

Trace pressure switch: A pressure switch in which the sensing element moves or traces with every change in pressure.

Accurate and precise measurement of process variables is essential to the proper operation and control of any process system. In turn, the installation, calibration, and repair of measuring devices are primary responsibilities of pipeline E&I technicians. To perform quality work, technicians must have a thorough understanding of process variables and the instruments used to measure and control processes.

1.1.0 Characteristics of Pressure Switches

The four process variables most frequently measured in instrument and control systems are flow, pressure, level, and temperature. Each of these variables affects the operation of pipelines. Various meters and other devices measure the variables. These measurements then trigger switches and transmitters to automatically control the process. This section addresses the qualities and characteristics of pipeline pressure switches.

1.1.1 Pressure Switch Operations

Pressure switches are used to control the pressure of liquids and gases by starting and stopping an end device such as a motor at the proper time. Air compressors, for example, are started directly or indirectly on a call for more air by a pressure switch.

> **CAUTION**
>
> When selecting pressure switches, pay particular attention to the pressure rating of the switch and the operating pressure of the system. Never exceed the switch ratings.

Pressure switches use mechanical motion from pressure changes to operate one or more sets of contacts. *Figure 1* shows a typical pressure switch that uses a bellows as the pressure-sensing element.

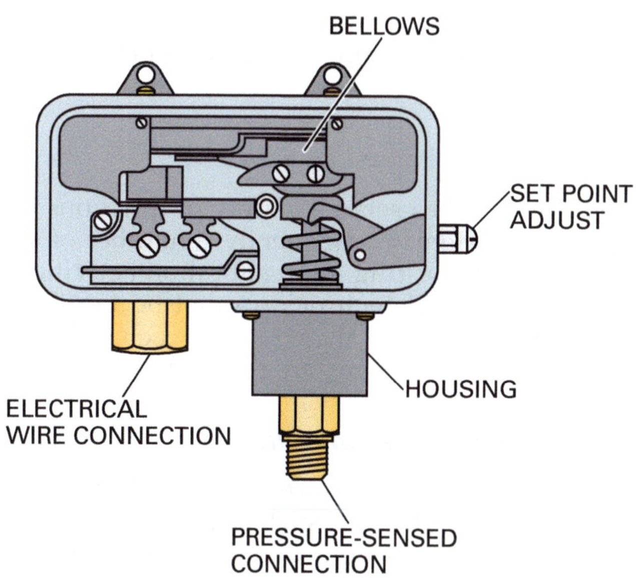

Figure 1 Typical pressure switch.

Other common sensing elements are the diaphragm and the **Bourdon tube**. Most pressure switches have an adjustment screw to adjust the setpoint. Pressure switches often come with two sets of contacts: one normally open (NO) set and one normally closed (NC) set. When the pressure reaches the setpoint, one set of contacts will open and the other will close.

A **trace pressure switch** is one in which the sensing element moves with, or traces, every change in pressure. *Figure* 2 shows some trace pressure switch designs.

When selecting a pressure switch for a particular application, take each of the following factors into consideration:

- *Adjustable range* – This is the span of pressures within which the pressure-sensing element can be set to actuate the switch contacts. For example, a pressure switch may have a range of 20 to 100 psi (138 to 689 kPa).
- *Adjustable differential pressure* – This is the span of pressure between the **pressure switch setpoint**, which actuates the contacts, and the **pressure switch deadband**, which resets the contacts to their normal position. As an example, a pressure switch may have an adjustable range between 20 and 100 psi (138 and 689 kPa) with an adjustable differential pressure of only 5 to 15 psi (34 to 103 kPa).
- *Rating of the switch and contacts* – The rating includes the type of switch used, such as single-pole single-throw (SPST), single-pole double-throw (SPDT), etc. It also specifies the amount of current that the switch can carry and the maximum voltage that the contacts can safely switch.
- *Accuracy* – The accuracy of the switch refers to the ability of the switch to repeatedly actuate at the setpoint. This value is typically stated as a percent of the maximum operating pressure of the switch.

1.2.0 Testing Pressure Switches

Pressure switches serve a variety of functions in a pipeline system. Consequently, a number of issues surround testing and calibration activities. The pipeline E&I technician must be familiar with these issues and always assess each individual situation to see if any are relevant.

1.2.1 Protection Parameters

Protection parameters are developed and applied to control devices, such as process controllers or PLCs, to prevent damage from various conditions that could cause process upsets and/or dangerous operating conditions. Some examples of such conditions are: abnormal operating pressures, flow rate deviations, high or low temperature extremes, high or low vibration, and excessively high or low tank levels. Protection parameters enable a control device to initiate actions, such as an **alarm**, and to act as a **shutdown device** to protect process systems and their components, and to protect the personnel who operate the systems.

The following steps outline the general procedures used to verify and set protection parameters for a control device:

Step 1 Obtain verification of the target value(s) for the device setpoint or operating parameters that are to be implemented or changed. Follow a management of change (MOC) procedure, if company policy requires technicians to do so.

> **NOTE**
> Changes in control device parameter settings can be made for various reasons. For example, to protect refined product from tank contamination, the parameters for API gravity limits may be changed to correspond with changes in product gravity that resulted from refinery crude oil slate changes.

> **NOTE**
> Changes to parameter settings in timers, valve sequences, tank statuses, alarm sequences and statuses, or many other ladder logic rungs and elements may be temporary or they may be permanent.

Step 2 Ensure proper isolation of the affected section from the process system. Follow appropriate lockout/tagout procedures, as required.

Step 3 Verify data within the current program and document the data "as found".

Step 4 Implement changes to the setpoints.

Step 5 Remove isolation from the affected section of the process system.

Step 6 Verify program acceptance through software or dynamic testing methods. Be sure to save program changes. Back up the changes on a different medium, as well.

Step 7 Document system alterations "as left".

DESIGN PRINCIPLE	CONSTRUCTION (SENSING ELEMENT)	ADVANTAGES	DISADVANTAGES
BELLOWS	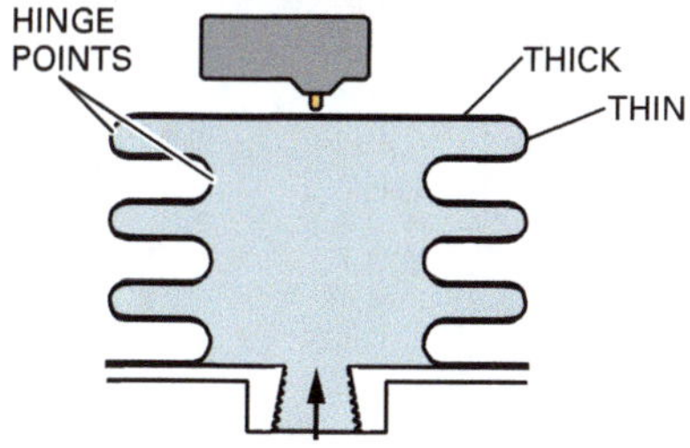	• Available materials suitable for a wide range of pressures and media • Inexpensive manufacturing costs • Small diameter • Narrow deadband	• Sensitive to ambient temperature changes • Critical to vibration • Unpredictable service life • Settings drift during use
METAL DIAPHRAGM	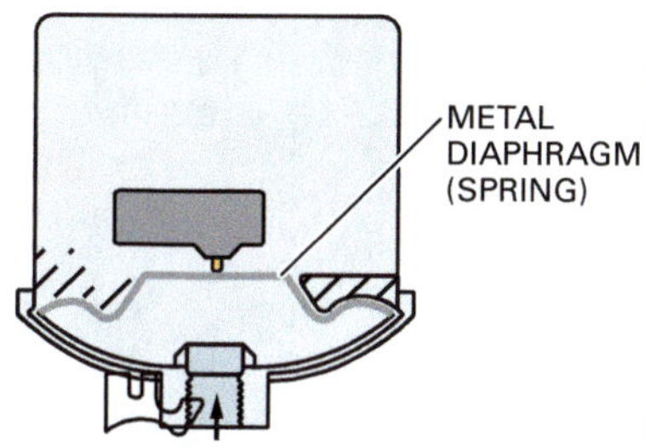	• Narrow deadbands • Different diaphragm materials yield wide media capability • Extremely low pressure range capability • Highly accurate within upper $\frac{2}{3}$ of range	• Costly to produce • Settings drift with metal fatigue • Shorter service life of tracing element • Critical to shock • Critical to vibration
LIMP DIAPHRAGM	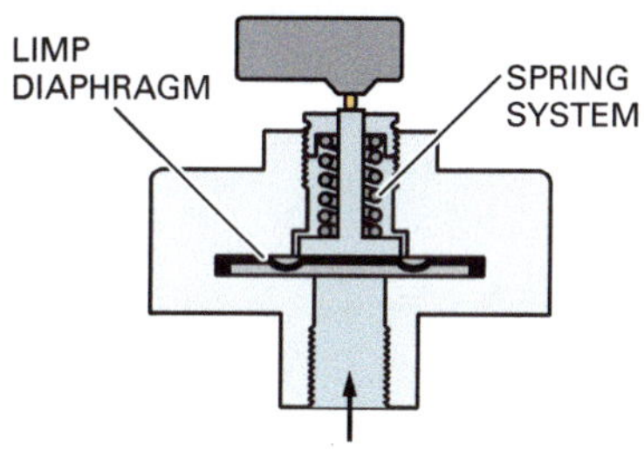	• Different diaphragm materials yield wide media capability • Low pressure range capabilities • Relatively narrow deadbands • Highly accurate within upper $\frac{2}{3}$ of range	• Some diaphragms highly susceptible to damage • Spring system critical to shock and vibration • Temperature sensitivity of some diaphragm materials • Possible aging of synthetic diaphragms, which results in unpredictable service life
BOURDON TUBE	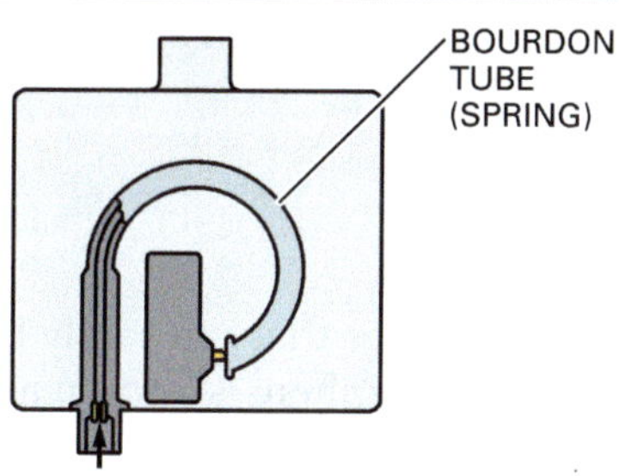	• Highly versatile pressure range capability • Highly sensitive to pressure changes • Narrow deadband for high pressures • Highly accurate within upper $\frac{2}{3}$ of range	• Tracing-type element (reduced service life) • High cost of quality tubes • Critical to ambient temperature changes due to tube length • Critical to vibration due to length • Highly critical to surge pressure or pump
PISTON (UNSEALED)	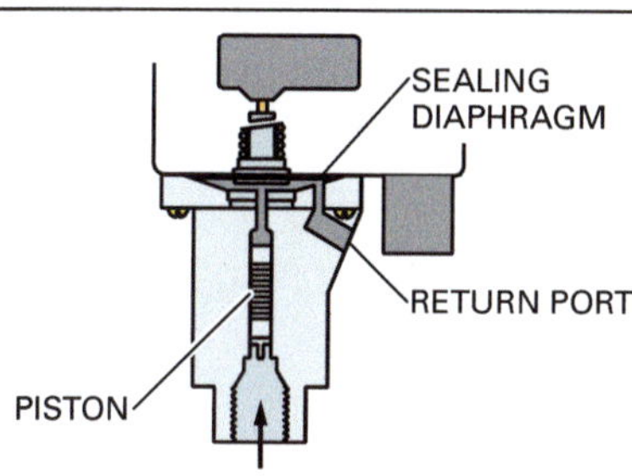	• Broad pressure range capability • Not critical to vibration • Not critical to surgepressure or pump ripple • No O-ring seal means no breakaway force required on first actuation • Narrower deadband than the sealed piston switch	• High installation cost due to drain line • Highly critical to dirt, including very fine particles
PISTON (SEALED)	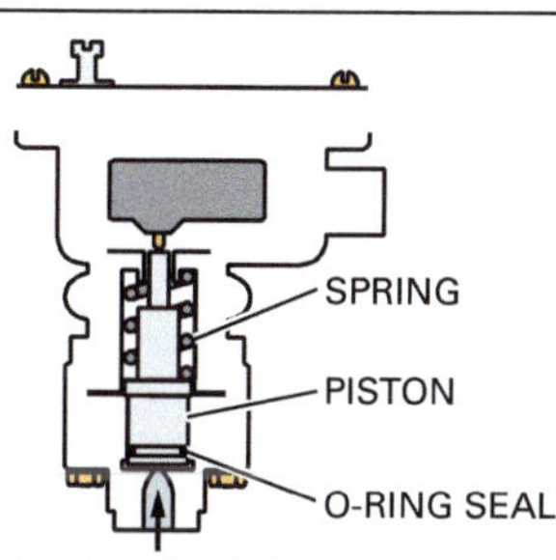	• Not critical to vibration • Not critical to dirt • Not critical to surge pressure or pump ripple	• Seal friction (stiction) means higher initial breakaway force for first actuation • Seal friction causes wider deadband

Figure 2 Trace pressure switch designs.

1.2.2 Calibrating Pressure Switches in Explosive Atmospheres

When calibrating pressure switches in potentially explosive environments, use a pressure calibrator that is certified as being **intrinsically safe** (*Figure 3*). Such devices are designed to be unable to release sufficient thermal or electrical energy to cause ignition of flammable material (such as gas, dust, or particulates).

In addition to using an intrinsically safe pressure calibrator, adhere to safe work practices for potentially explosive environments, including, but not limited to, the following guidelines:

- Follow the appropriate lockout/tagout procedures.
- Secure the work area to prevent anyone from entering the area during the calibration.
- Safely purge or vent the system to remove any remaining gases.
- Use a gas detector to make sure the atmosphere is safe before starting the calibration.

Figure 3 Intrinsically safe pressure calibrator.

Additional Resources

It is assumed that you will learn to perform these procedures under the direct supervision of your crew leader, foreman, or instructor. For independent study, refer to the recommended Additional Resources at the back of this module.

1.0.0 Section Review

1. The pressure at which a pressure switch resets after being tripped is the _____.
   a. pressure switch reset value
   b. pressure switch deadband
   c. Bourdon clearance point
   d. diaphragm de-flex value

2. A calibrator that can safely be used in an explosive atmosphere is known as being _____.
   a. intrinsically safe
   b. volatile sealed
   c. essentially flameproof
   d. pneumatically isolated

Section Two

2.0.0 Inspect, Test, and Calibrate Pressure Switches

Objective

Describe the procedures used to inspect, test, and calibrate pressure switches (CT25_1-17).

a. Identify the prerequisites, competencies, and abnormal operating conditions (AOCs) associated with inspecting, testing, and calibrating pressure switches.
b. Describe the procedures used to inspect, test, and calibrate pressure switches.

Performance Task

1. Inspect, test, and calibrate pressure switches (CT25_1-17).
   - Identify potential abnormal operating conditions (AOCs) that may occur during performance of this CT, and know the appropriate actions to take in response to them.
   - Utilize the appropriate personal protective equipment according to relevant company procedures.
   - Verify certification that test equipment has been calibrated prior to performing calibrations.
   - Verify device set point for switches prior to performing calibration.
   - Determine correct test medium and range to be used for testing.
   - Notify control center and/or affected personnel before work begins.
   - Isolate the pressure switch from the process system and inspect all connection for leakage.
   - Apply test medium pressure and verify device setpoint or range "as found" and document the results.
   - Re-apply test medium for desired setpoint or range and adjust if necessary according to device manufacturer's specifications to establish desired setpoint. Document final setpoint setting(s) "as left".
   - Notify control center and/or affected personnel after completion of work.
   - Complete appropriate documentation as required by operator's procedures.

Trade Terms

Abnormal operating conditions (AOCs): Conditions within a pipeline system, identified by the operator, that may indicate (1) a malfunction of a component or (2) a deviation from normal operations that may indicate a condition exceeding design limits or result in a hazard to persons, property, or the environment.

Bench condition testing (BCT): Any test procedure that must be performed while the device or equipment is completely removed from the process system. It is usually a safer and more precise means of calibrating and testing; however, it is also more expensive and time-consuming.

Dynamic condition testing (DCT): Any test procedure that must be performed while the device or equipment being tested remains under normal operating conditions with associated energy sources active. It requires extreme caution on the part of pipeline E&I technicians. DCT must sometimes be used to determine the correct data for a given situation or as a diagnostic procedure.

Static condition testing (SCT): When possible, the device to be tested should be completely isolated from the process system. The term static condition testing applies when a device remains in the process system but has no electrical, mechanical, or process connection that may affect the process system.

Test medium: The material used in the system being tested that allows reliable and repeatable test measurements.

Like many other instruments, pressure switches require regular inspection, testing, and calibration to keep them functioning properly. Manufacturers usually specify inspection, testing, and calibration procedures for their instruments. To achieve accuracy and repeatability from the instruments, follow the procedures that are recommended by the manufacturer. The general procedures presented in this section should be applicable in most cases.

2.1.0 Prerequisites, Competencies, and Abnormal Operating Conditions

Personnel must possess a variety of skills in order to successfully inspect, test, and calibrate pressure switches. The various operations require a series of steps, which technicians must be competent to perform. Personnel must also be aware of the particular abnormal operating conditions (AOCs) that may occur during the procedure and know the proper responses to them.

2.1.1 Prerequisites and Competencies

Pipeline E&I technicians must possess the following basic knowledge and skills prior to inspecting, testing, and calibrating pressure switches:

- They must be skilled with test equipment and know how to operate the following types of equipment:
  - Pressure gauges of all types
  - Multifunction calibrators
  - Multimeters
  - Digital pressure modules
- They must be skilled with test media and the equipment used to manipulate it, including the following:
  - Hand pumps
  - Pressure regulators
  - Process connections
  - Air and nitrogen
  - Glycol
- They must be able to work safely with electrical systems.
- They must be familiar with the control system and its instrumentation, including alarm, control, indication, and recording functions.

Personnel must also be familiar with the company safety, evacuation, and response procedures, as well as the proper use of personal protective equipment (PPE).

2.1.2 Abnormal Operating Conditions

Personnel inspecting, testing, and calibrating pressure switches must be able to identify potential abnormal operating conditions (AOCs) and know the proper response to each, should they occur. The following are some potential AOCs and the appropriate action the E&I technician should take in response to them:

- *The device does not respond when pressure is in the target range or the device does not respond consistently* – Alert appropriate personnel to take the requisite corrective action.
- *During testing, a safety or control device activates unintentionally, causing a shutdown or undesired valve movement resulting in control loss or an overpressure condition* – Take appropriate action to restore normal conditions and then alert appropriate personnel as required by company procedure.
- *An isolation valve fails to isolate the process pressure due to incomplete closure or other malfunction* – Halt the test and return the valve to normal in order to prevent a loss of control or other negative side effect. The technician must also alert appropriate personnel as required by company procedure.
- *A hazardous gas or liquid is accidentally released* – Halt the test, place equipment in a secure state, and alert appropriate personnel, such as the process operator, to take the requisite corrective action. Applicable emergency procedures should be followed.
- *Contact with an electrically energized component, resulting in shock* – Call 911.

2.2.0 Procedure

The following general guidelines present procedures for the inspection, testing, and calibration of pressure switches:

Step 1 Determine the device's number and nameplate data in order to verify that it is the correct unit to calibrate/test, and that any data coming from the unit will be properly identified as such.

NOTE

If the nameplate is missing or illegible, it must be replaced according to company policy and procedures.

Step 2 Before beginning the calibration process, check that the test equipment itself has been certified, calibrated, and verified. Some technicians will not use test equipment until it has been tested on a known source, even if its calibration is current and properly documented.

Step 3 Check the appropriate company documentation to determine and verify the pressure switch setpoint values.

Step 4 Calibration requires simulation of pipeline pressure in order to activate the switch. Identify and obtain the correct **test medium** (air, nitrogen, etc.) for this procedure.

WARNING!

Safety or control devices can activate due to the incorrect application of a test medium, resulting in loss of control or regulation and/or overpressure condition on the device. If this occurs, technicians should respond by correcting the application of the test medium and contacting the control center for further instructions.

Step 5 Before beginning the process, notify the control center, local personnel, and any other personnel who might be affected by the calibration/test procedure.

> **NOTE**
> The control center may have to provide feedback about the pressure switch alarm response during the process. It may also have to override automatic shutdowns or safety systems to protect the process from interruption. These events are often known as Critical Interlock and Trips (CIATs). They usually require higher authorization to override. Pipeline E&I technicians should be aware of these rules and know what they're allowed to do.

Step 6 Visually inspect the device and any associated equipment. Check for mechanical condition, corrosion, leakage, electrical connection problems, or problems with associated electronics. Also confirm that the device is appropriate for the intended application.

> **WARNING!**
> Always exercise extreme caution when working with energized circuits to avoid contact and electrical shock. If shock occurs, call 911.

Step 7 Determine the appropriate test procedure to be used.

- **Bench condition testing (BCT)** is performed with the device completely removed from the process.
- **Dynamic condition testing (DCT)** is performed under normal operating conditions with the associated energy sources active.
- **Static condition testing (SCT)** is performed with the device completely isolated from the process system; there is no electrical, mechanical, or process connection that may affect the process system.

Step 8 Using appropriate measures, isolate the pressure switch from the process. Following the proper procedure, disconnect the switch from the process connections and attach the test equipment in its place. Confirm that there are no leaks, as these would invalidate the test/calibration run.

Step 9 Apply pressure to the switch to determine its current setpoint value. If necessary, repeat this step as many times as necessary to determine the switch's repeatability. Based on the results, decide if the switch is tripping at the correct setpoint.

Step 10 Document the results of Step 9 to provide historical data and the "as found" status of the device. Follow company procedure.

Step 11 Based on your conclusions from Step 9, calibrate the device if necessary, following the manufacturer's guidelines and company procedure. Repeat this step as often as required to establish a repeatable calibration.

Step 12 Document the results of Step 11 to provide historical data and the "as left" status of the device. Follow company procedure.

Step 13 Disconnect the test equipment and restore the unit's normal connections, confirming that they are leak-free.

Step 14 Notify the control center, local personnel, and any affected personnel that the unit has been returned to normal operation.

Step 15 Complete all required documentation.

> **NOTE**
> Documenting data for compliance, historical review, and trending is an important element of valve maintenance and should not be neglected.

Additional Resources

It is assumed that you will learn to perform these procedures under the direct supervision of your crew leader, foreman, or instructor. For independent study, refer to the recommended Additional Resources at the back of this module.

2.0.0 Section Review

1. A pipeline E&I technician needs to know how to work with test media and the associated equipment, including _____.
   a. thermocouple heat blocks
   b. a nitrogen strainer
   c. a hand pump
   d. a certified valve aligner

2. Before connecting the test equipment to the pressure switch, the technician must _____.
   a. calibrate the meter pressure module
   b. disable the alarm
   c. set SCADA to "Ignore" mode
   d. isolate the switch from the process

SUMMARY

Keeping pressure switches working reliably is key to efficient and safe pipeline operation. Pressure switches function as normal operation control devices, switching pumps and compressors on and off as needed. But they also act as safety devices, detecting overpressure situations and triggering alarms and initiating shutdowns. Both functions are important, so proper switch calibration is critical. This module provided a basic overview of the procedures for inspecting, testing, and calibrating pressure switches.

Review Questions

1. Common sensing elements in pressure switches include the diaphragm, Bourdon tube, and _____.
   a. lever
   b. cam
   c. stem
   d. bellows

2. The pressure span between the pressure switch setpoint and the pressure switch deadband is the _____.
   a. adjustable pressure range
   b. adjustable differential pressure
   c. pressure window
   d. spanning envelope

3. Settings programmed into a controller or PLC designed to prevent damage and process upsets are called _____.
   a. protection parameters
   b. shutdown thresholds
   c. lockout lines
   d. monitor alerts

4. A common test medium used for calibrating a pressure switch is _____.
   a. heliox
   b. argon
   c. hydrazine
   d. nitrogen

5. If a control device activates unintentionally during testing, the first response should be to _____.
   a. call the local 911
   b. alert appropriate personnel
   c. restore normal conditions
   d. follow emergency procedures

6. In order to know that the data being read is coming from the correct pressure switch, the pipeline E&I technician must first _____.
   a. talk to the control room
   b. perform a fieldbus scan
   c. ping the transmitter with a handheld communicator
   d. determine the device number and nameplate data

Trade Terms Quiz

Fill in the blank with the correct term that you learned from your study of this module.

1. A device that protects a pipeline in response to an unacceptable process condition is a(n) ________.
2. A test procedure that must be performed as the device remains under normal operating conditions with associated energy sources active is called ________.
3. A test instrument that is safe to use in an explosive atmosphere is considered ________ because it does not generate sufficient electrical or thermal energy to ignite in it.
4. The pressure value at which a pressure switch trips is called a(n) ________.
5. A test procedure in which the device is completely isolated from the process system with no electrical, mechanical, or process connection is known as ________.
6. Conditions within a pipeline system that may indicate (1) a malfunction of a component or (2) a condition exceeding design limits or resulting in a hazard to persons, property, or the environment are called ________.
7. A mechanical pressure-measuring instrument that uses for its sensing element a curved or twisted metal tube that is flattened in cross section and closed is called a(n) ________.
8. The material used in the system being tested that allows reliable and repeatable test measurements is the ________.
9. The pressure value at which a tripped pressure switch resets is called the ________.
10. An audible and/or visual status event from the control system, notifying personnel that a process parameter has moved out of its normal range is called a(n) ________.
11. A device that must operate to provide protection for a pipeline is the ________.
12. A pressure switch in which the sensing element moves with every change in pressure is called a(n) ________.
13. A test procedure that must be performed with the device completely removed from the process system is called ________.
14. A facility within which devices for measuring process variables such as temperature, pressure, level, and flow are installed is called a(n) ________.

Trade Terms

Abnormal operating conditions (AOCs)
Alarm
Bench condition testing (BCT)
Bourdon tube
Dynamic condition testing (DCT)
End device
Intrinsically safe
Pressure switch deadband
Pressure switch setpoint
Process system
Shutdown device
Static condition testing (SCT)
Test medium
Trace pressure switch

Trade Terms Introduced in This Module

Abnormal operating conditions (AOCs): Conditions within a pipeline system, identified by the operator, that may indicate (1) a malfunction of a component or (2) a deviation from normal operations that may indicate a condition exceeding design limits or result in a hazard to persons, property, or the environment.

Alarm: A status event from the control system, notifying personnel that a process parameter has moved out of its normal range; alarms can be audible, visual, or both.

Bench condition testing (BCT): Any test procedure that must be performed while the device or equipment is completely removed from the process system. It is usually a safer and more precise means of calibrating and testing; however, it is also more expensive and time-consuming.

Bourdon tube: A mechanical pressure-measuring instrument that uses for its sensing element a curved or twisted metal tube that is flattened in cross section and closed. It is also called a Bourdon pressure gauge.

Dynamic condition testing (DCT): Any test procedure that must be performed while the device or equipment being tested remains under normal operating conditions with associated energy sources active. It requires extreme caution on the part of pipeline E&I technicians. DCT must sometimes be used to determine the correct data for a given situation or as a diagnostic procedure.

End device: A device that must operate to provide protection for a pipeline. For example, a pressure switch activates and, through the control sequences, causes disconnection of power to a motor. The motor is the end device.

Intrinsically safe: A test instrument that is safe to use in an explosive atmosphere because it does not generate sufficient electrical or thermal energy to ignite in it.

Pressure switch deadband: The pressure value at which a tripped pressure switch resets. Also known as the zero pressure.

Pressure switch setpoint: The pressure value at which a pressure switch trips.

Process system: A facility within which devices for measuring process variables such as temperature, pressure, level, and flow are installed.

Shutdown device: A device that protects a pipeline in response to an unacceptable process condition.

Static condition testing (SCT): When possible, the device to be tested should be completely isolated from the process system. The term static condition testing applies when a device remains in the process system but has no electrical, mechanical, or process connection that may affect the process system.

Test medium: The material used in the system being tested that allows reliable and repeatable test measurements.

Trace pressure switch: A pressure switch in which the sensing element moves or traces with every change in pressure.

Additional Resources

The following recommended resources can provide additional helpful information related to the requirements for this covered task:

API Recommended Practice 1161, Third Edition, © 2014 American Petroleum Institute. API Publishing Services, 1220 L Street, NW, Washington, DC 20005. Distributed by Thompson Reuters (Scientific) LLC, **www.techstreet.com**.

NCCER Module AOCCC-17, *Abnormal Operating Conditions Control Center*

NCCER Module AOCFG-17, *Abnormal Operating Conditions Field and Gas*

NCCER Module 60102-02, *Pipeline E&I Safety*

NCCER Module 64202-02, *Use of Meters and Test Equipment*

Instrument Engineers' Handbook, Volume 1: Process Measurement and Analysis. Béla G. Lipták (editor). Fourth Edition. 2003. Boca Raton, FL: CRC Press.

Instrument Engineers' Handbook, Volume 2: Process Control and Optimization. Béla G. Lípták (editor). Fourth Edition. 2005. Boca Raton, FL: CRC Press.

Instrumentation. Franklyn W. Kirk, Thomas A. Weedon, and Philip Kirk. Fifth Edition. 2010. Orland Park, IL: American Technical Publishers.

Instrumentation Reference Book. Walt Boyes (editor). Fourth Edition. 2009. Oxford, UK: Butterworth-Heinemann.

Measurement and Control Basics. Thomas A. Hughes. Fifth Edition. 2014. Research Triangle Park, NC: The International Society of Automation.

Purdy's Instrument Handbook. Ralph Dewey. 2015 Edition. Deer Park, TX: Glen Enterprises.

The Condensed Handbook of Measurement and Control. N.E. Battikha. Third Edition, 2006. Research Triangle Park, NC: The International Society of Automation.

The following websites offer resources for products and training:

Dwyer Instruments, Inc., **www.dwyer-inst.com**

Emerson Electric Co., **www.emerson.com**

The International Society of Automation, **http://www.isa.org** www.isa.org

Figure Credits

Bigelow-Liptak of Canada, Module Opener

Fluke Corporation, Figure 3

Section Review Answer Key

Answer	Section Reference	Objective
Section One		
1. b	1.1.1	1a
2. a	1.2.2	1b
Section Two		
1. c	2.1.1	2a
2. d	2.2.0	2b

NCCER CURRICULA — USER UPDATE

NCCER makes every effort to keep its textbooks up-to-date and free of technical errors. We appreciate your help in this process. If you find an error, a typographical mistake, or an inaccuracy in NCCER's curricula, please fill out this form (or a photocopy), or complete the online form at **www.nccer.org/olf**. Be sure to include the exact module ID number, page number, a detailed description, and your recommended correction. Your input will be brought to the attention of the Authoring Team. Thank you for your assistance.

Instructors – If you have an idea for improving this textbook, or have found that additional materials were necessary to teach this module effectively, please let us know so that we may present your suggestions to the Authoring Team.

NCCER Product Development and Revision

13614 Progress Blvd., Alachua, FL 32615

Email: curriculum@nccer.org
Online: www.nccer.org/olf

❑ Trainee Guide ❑ Lesson Plans ❑ Exam ❑ PowerPoints Other ______________________

Craft / Level: Copyright Date:

Module ID Number / Title:

Section Number(s):

Description:

Recommended Correction:

Your Name:

Address:

Email: Phone:

CT25_2-17

Inspect, Test, and Calibrate Pressure Transmitters

Overview

Pipelines and modern process systems rely on the accurate and precise measurement of process variables to operate safely and in the most efficient and reliable way possible. Installing the appropriate instruments for measuring process variables is only the first step. Technicians must regularly inspect, test, and properly calibrate the instruments to keep them functioning correctly. This module reviews types of pipeline pressure transmitters and surveys how to inspect, test, and calibrate them.

Trainees with successful module completions may be eligible for credentialing through the NCCER Registry. To learn more, go to **www.nccer.org** or contact us at 1.888.622.3720. Our website has information on the latest product releases and training, as well as online versions of our *Cornerstone* magazine and Pearson's product catalog.

Your feedback is welcome. You may email your comments to **curriculum@nccer.org**, send general comments and inquiries to **info@nccer.org**, or fill in the User Update form at the back of this module.

This information is general in nature and intended for training purposes only. Actual performance of activities described in this manual requires compliance with all applicable operating, service, maintenance, and safety procedures under the direction of qualified personnel. References in this manual to patented or proprietary devices do not constitute a recommendation of their use.

CT25_2-17

Inspect, Test, and Calibrate Pressure Transmitters

Objectives

When you have completed this module, you will be able to do the following:

1. Identify and describe pressure transmitters and the issues associated with testing them.
   a. Identify and describe pressure transmitters.
   b. Describe the issues associated with testing pressure transmitters.
2. Describe the procedures used to inspect, test, and calibrate pressure transmitters (CT25_2-17).
   a. Identify the prerequisites, competencies, and abnormal operating conditions (AOCs) associated with inspecting, testing, and calibrating pressure transmitters.
   b. Describe the procedures used to inspect, test, and calibrate pressure transmitters.

Performance Task

Under the supervision of the instructor, you should be able to do the following:

1. Inspect, test, and calibrate pressure transmitters (CT25_2-17).
   - Identify potential abnormal operating conditions (AOCs) that may occur during performance of this CT, and know the appropriate actions to take in response to them.
   - Utilize the appropriate personal protective equipment according to relevant company procedures.
   - Verify certification that test equipment has been calibrated prior to performing calibrations.
   - Verify range of transmitter prior to performing calibration.
   - Determine correct test medium and the range to be used for testing.
   - Notify control center and/or affected personnel before work begins.
   - Isolate the pressure transmitter from the process system and inspect all connections for leakage.
   - Apply test medium pressure and verify device setpoint or range "as found". Verify pressure transmitters at zero, mid, and full scale calibration points. Document results "as found."
   - Re-apply test medium for zero, mid-scale and full scale. Adjust if necessary according to device manufacturer's specifications to establish desired calibration.
   - Verify activation of digital control output and/or operation of the end device at the established setpoint. Document final setpoint settings "as left."
   - Notify control center and/or affected personnel after completion of work.
   - Complete appropriate documentation as required by operator's procedures.

Trade Terms

Abnormal operating conditions (AOCs)
Alarm
Bourdon tube
Force balance transmitter
High-static DP transmitter
Intrinsically safe
Motion balance transmitter
Process system
Range
Shutdown device
Test Medium
Variable capacitance cell

Industry Recognized Credentials

If you are training through an NCCER-accredited sponsor, you may be eligible for credentials from NCCER's Registry. The ID number for this module is CT25_2-17. Note that this module may have been used in other NCCER curricula and may apply to other level completions. Contact NCCER's Registry at 888.622.3720 or go to **www.nccer.org** for more information.

Contents

1.0.0 Pressure Transmitters 1
  1.1.0 Types of Pipeline Transmitters 1
    1.1.1 Pneumatic Transmitters 1
    1.1.2 Electronic Transmitters 6
    1.1.3 Optical Transmitters 7
  1.2.0 Issues Associated with Pressure Transmitter Testing 8
    1.2.1 Protection Parameters 8
    1.2.2 Calibrating Pressure Transmitters in Explosive Atmospheres 9
2.0.0 Inspecting, Testing, and Calibrating Pressure Transmitters 11
  2.1.0 Prerequisites, Competencies, and Abnormal Operating Conditions 11
    2.1.1 Prerequisites and Competencies 11
    2.1.2 Abnormal Operating Conditions 12
  2.2.0 Procedure 12

Figures

Figure 1 Force balance differential-pressure transmitter 2
Figure 2 Measuring section 2
Figure 3 Force bar section 3
Figure 4 Diaphragm effect 3
Figure 5 Balancing section 4
Figure 6 Pneumatic input/output section 4
Figure 7 Pneumatic relay 4
Figure 8 Motion balance transmitter 5
Figure 9 Common measuring elements 5
Figure 10 Link and flapper nozzle assembly 5
Figure 11 Bellows and relay assembly 6
Figure 12 High-static DP transmitter 6
Figure 13 Electronic force balance DP transmitter 6
Figure 14 Sensor assembly 7
Figure 15 Output section 7
Figure 16 Intrinsically safe pressure calibrator 9

SECTION ONE

1.0.0 PRESSURE TRANSMITTERS

Objective

Identify and describe pressure transmitters and the issues associated with testing them.

a. Identify and describe pressure transmitters.
b. Describe the issues associated with testing pressure transmitters.

Trade Terms

Alarm: A status event from the control system, notifying personnel that a process parameter has moved out of its normal range; alarms can be audible, visual, or both.

Bourdon tube: A mechanical pressure-measuring instrument that uses for its sensing element a curved or twisted metal tube that is flattened in cross section and closed. It is also called a Bourdon pressure gauge.

Force balance transmitter: A type of pneumatic transmitter that is operated through differential pressure. It is also called a DP cell.

High-static DP transmitter: A type of differential-pressure transmitter that is often used in very high pressure applications.

Intrinsically safe: A test instrument that is safe to use in an explosive atmosphere because it does not generate sufficient electrical or thermal energy to ignite it.

Motion balance transmitter: A type of pneumatic transmitter that converts the input motion of basic measuring elements into a pneumatic output that is directly proportional to the measured variable.

Process system: A facility within which devices for measuring process variables such as temperature, pressure, level, and flow are installed.

Range: The values that an instrument can measure, bounded by lower and upper limits.

Shutdown device: A device that protects a pipeline in response to an unacceptable process condition.

Variable capacitance cell: An electronic transmitter system that uses an electronic capacitance bridge circuit as its output device.

Accurate and precise measurement of process variables is essential to the proper operation and control of any process system. In turn, the installation, calibration, and repair of measuring devices are primary responsibilities of pipeline E&I technicians. To perform quality work, technicians must have a thorough understanding of process variables and the instruments used to measure and control processes.

1.1.0 Types of Pipeline Transmitters

Transmitters were originally very specific in function. The device prepared a signal to be sent from one location to another over some distance. Now the meaning of the term transmitter has grown to include its original function as well as the functions of a transducer/converter and an amplifier. A transmitter might also include the function of a detector, because modern transmitters typically provide an output in a form that is ready for direct transmission to a remote location.

Transmitters can be pneumatic, electronic, or optical. They usually consist of a sensor and an output device. The sensor detects a signal from a primary element that is measuring a process variable. The output device converts the signal to a transmittable form that is proportional to the measurement. This signal can then be used to indicate, record, or control the process variable. Transmitters are used on pipelines to transmit the following information:

- Pressure measurements
- Flow measurements
- Level measurements
- Temperature measurements

1.1.1 Pneumatic Transmitters

Pneumatic transmitters fall into one of two main operating types, a force balance transmitter or a motion balance transmitter. These units differ in their basic operating principles.

The most common type of pneumatic transmitter is the force balance differential-pressure transmitter, also called a *DP cell* (*Figure 1*). The DP cell usually operates on a regulated 20 psi (138 kPa) air supply. The output signal of the DP is usually 3 to 15 psi (21 to 103 kPa).

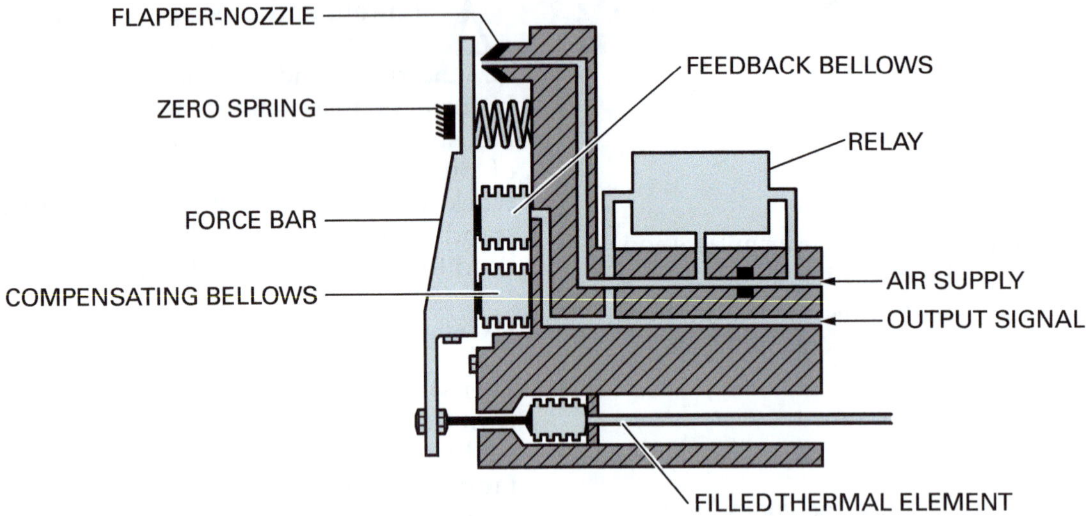

Figure 1 Force balance differential-pressure transmitter.

A pneumatic DP cell can be divided into four basic sections:

- Process measuring section
- Force bar section
- Balancing section
- Pneumatic input/output section

The process measuring section consists of a transmitter body and a diaphragm capsule. The diaphragm capsule splits the transmitter body into a high-pressure chamber and a low-pressure chamber. Fluid is piped to the high- and low-pressure sides, and pressure acts on the sealed metallic diaphragm to cause a small movement of the diaphragm link. The measuring section is shown in *Figure 2*.

The force bar section of a pneumatic DP cell (*Figure 3*) is connected to the diaphragm link. A force bar seal prevents process fluid from leaking out of the transmitter body. The top of the force bar is connected to the pneumatic flapper nozzle.

Any difference in pressure in the body causes a movement of the diaphragm link, which is transferred to the force bar at the link connection. An increase in the high-pressure side causes the base of the force bar to move toward the low-pressure side. This action draws the flapper closer to the nozzle. An increase in pressure on the low-pressure side moves the flapper away from the nozzle. This effect is illustrated in *Figure 4*.

The balancing section of a DP cell (*Figure 5*) consists of a feedback bellows, range rod, range wheel, and zero adjustment. Movement of the flapper nozzle by the force bar causes changes in the amount of air supplied to the feedback bellows. The bellows air supply is charged until the bellows exerts enough pressure on the range rod to balance the original movement of the force bar. The term *force balance* comes from these counterbalancing actions.

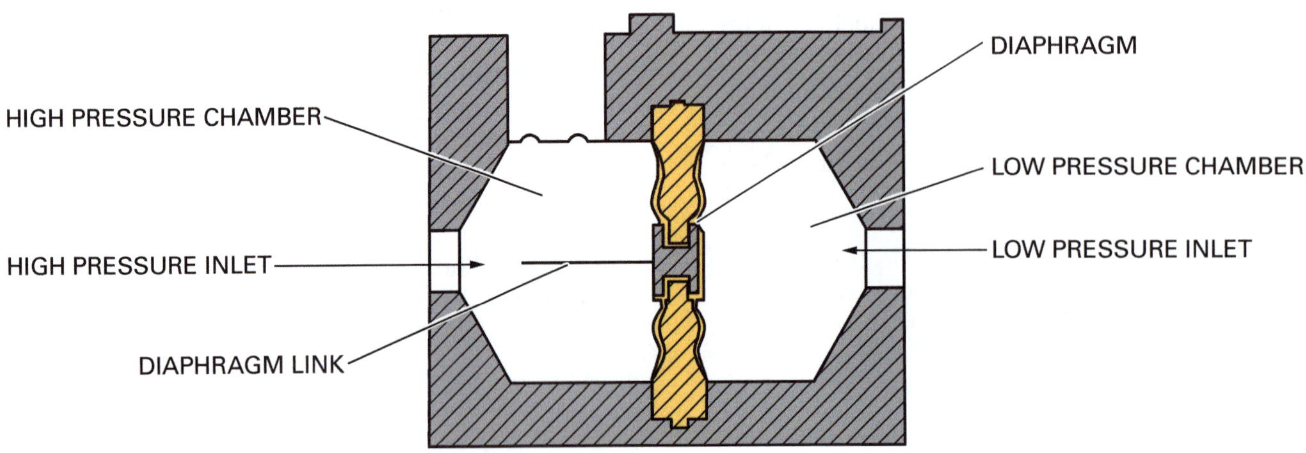

Figure 2 Measuring section.

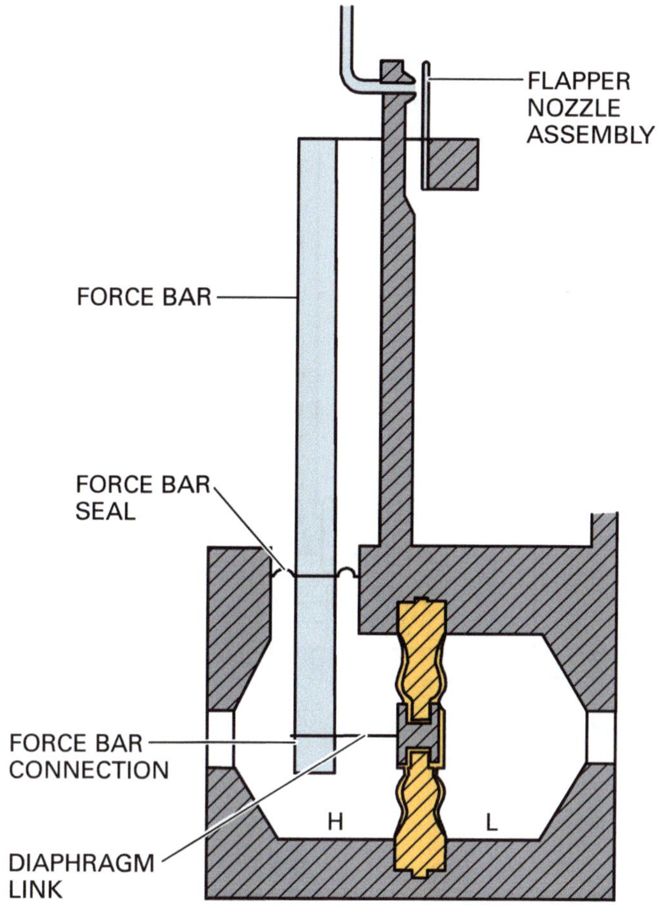

Figure 3 Force bar section.

The zero adjustment establishes zero, usually 3 psi (21 kPa), by allowing a manual setting of the flapper nozzle position. The range wheel is the balancing point for the motion of the range rod. By adjusting the range wheel up or down on the range rod, the total measuring range of the instrument can be changed.

The pneumatic input/output section of a DP cell (*Figure 6*) consists of a regulated 20 psi (138 kPa) air supply, a pneumatic relay, a nozzle, and an output connection. The pneumatic relay operates from a 20 psi (138 kPa) air supply. A fixed restrictor within the relay body provides a constant supply of air to the nozzle. The nozzle pressure acts on a diaphragm inside the relay. The diaphragm acts on a ball check valve that supplies air to the feedback bellows and output signal. A pneumatic relay is shown in*Figure 7*.

In a pneumatic DP cell, an increase in process pressure on the high-pressure side of the capsule causes the flapper to cover the nozzle. This causes air pressure to build up on the nozzle supply side of the diaphragm. The downward movement of the diaphragm causes the ball check valve to open and increase the air pressure to the feedback bellows and output signal. This continues until the pressure of the bellows against the range rod counterbalances the original movement of the force bar. If the flapper moves away from

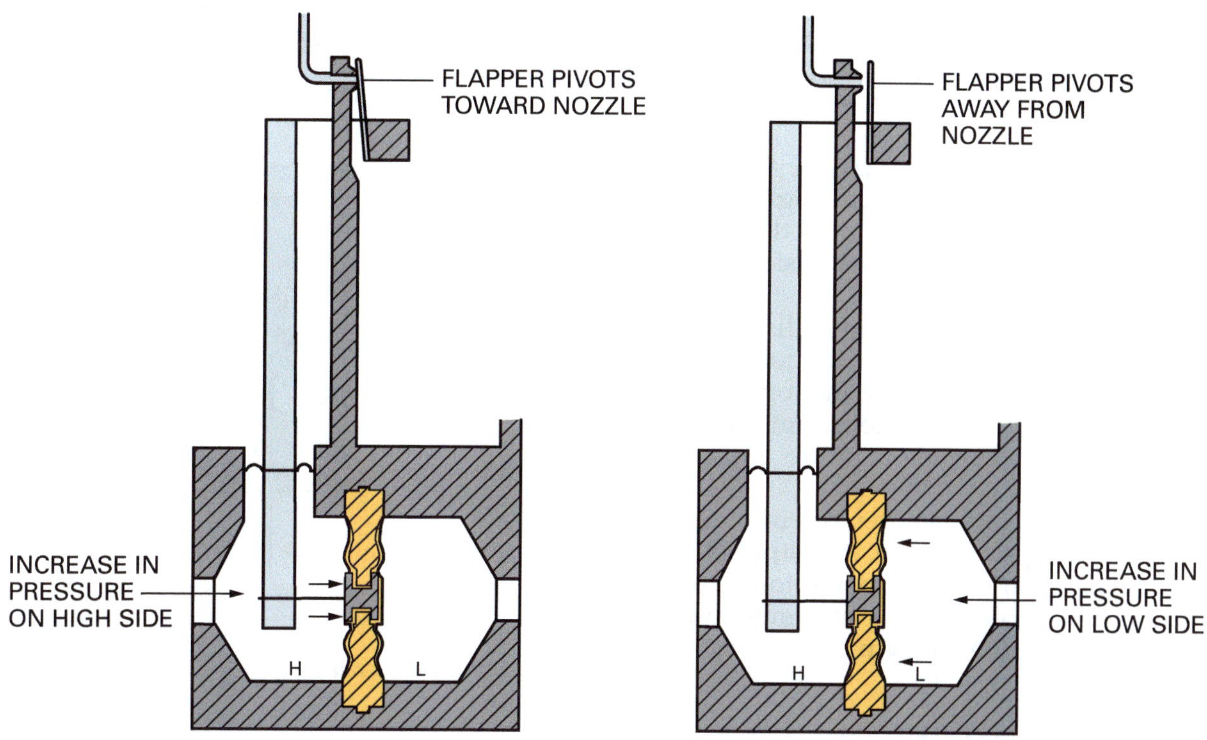

Figure 4 Diaphragm effect.

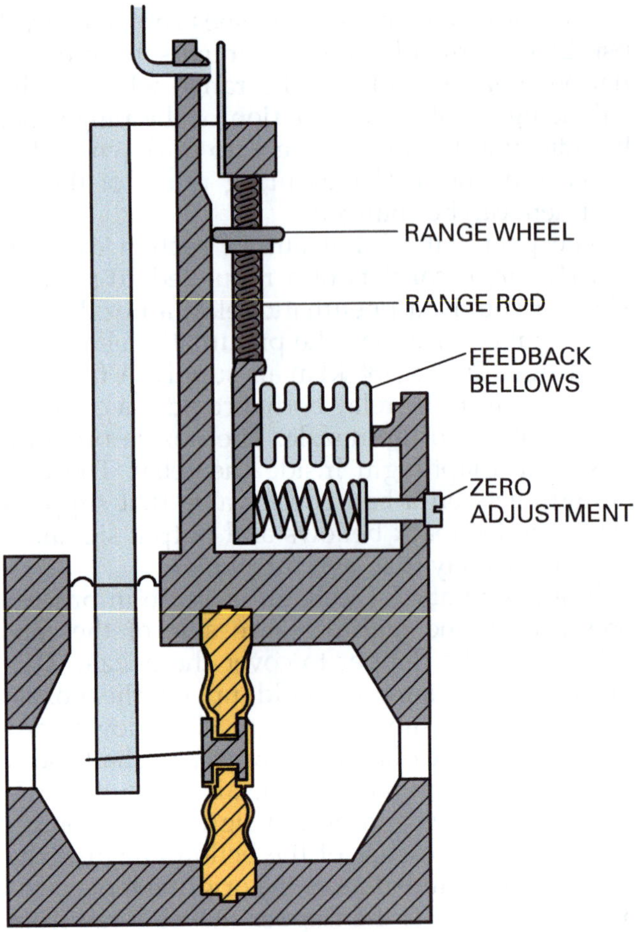

Figure 5 Balancing section.

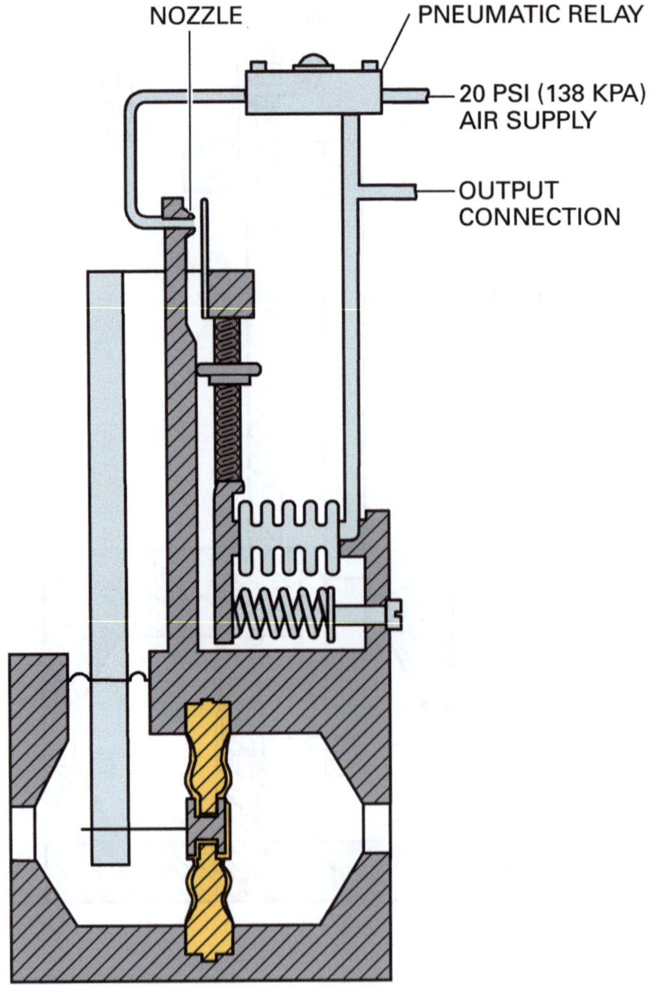

Figure 6 Pneumatic input/output section.

the nozzle, the diaphragm within the relay allows less air pressure to the feedback bellows and output signal. This action occurs until the unit is again in balance.

Motion balance transmitters (*Figure 8*) convert the input motion of basic measuring elements into a pneumatic output that is directly proportional to the measured variable. A basic measuring element such as a Bourdon tube, bellows, and diaphragm assembly can move distances of up to ½ inch (13 mm). Pneumatic motion balance transmitters convert this input motion into an output pressure. Motion balance transmitters are most commonly used in temperature and pressure measurement.

A motion balance transmitter receives an air supply input of 20 psi (138 kPa). Its output of 3 to 15 psi (21 to 103 kPa) is proportional to the measured value. The three sections of a motion balance transmitter are the following:

- Measuring section
- Link and flapper-nozzle section
- Bellows and relay section

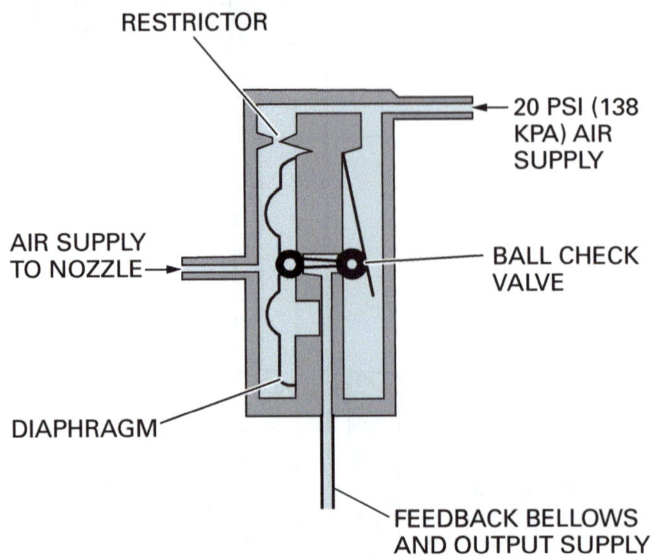

Figure 7 Pneumatic relay.

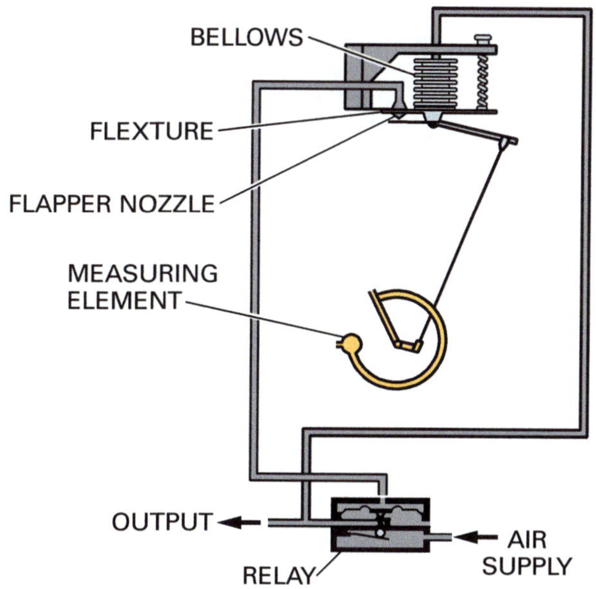

Figure 8 Motion balance transmitter.

The measuring section of a motion balance transmitter is the basic measuring element. In most motion balance transmitters, this basic element is either a Bourdon tube, a diaphragm assembly, or a bellows unit. Some common measuring elements used in motion balance transmitters are shown in *Figure 9*. These measuring elements rely on a link to transfer the measurement motion. The measuring element section provides the initial motion to the transmitters.

Changes in the measured variable move the connecting link on the measuring element. This motion is transferred to the link assembly. The link assembly moves up or down on its pivot point. This movement causes the flapper-nozzle relationship to change. The link and flapper-nozzle assembly is shown in *Figure 10*.

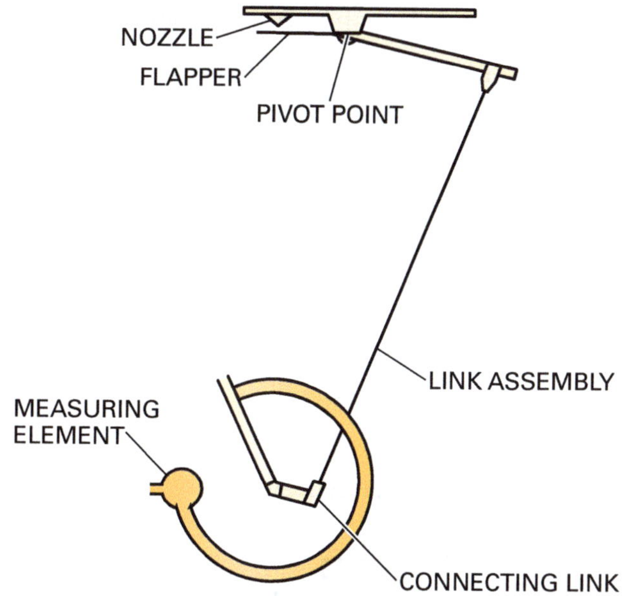

Figure 10 Link and flapper nozzle assembly.

Changes in the flapper-nozzle position affect the output of the pneumatic relay. At this point the action of the motion balance transmitter is the same as the action of a force balance transmitter. As the nozzle is covered or uncovered, the back pressure changes in the pneumatic relay assembly. These changes are fed back to the bellows. Any movement of the flapper toward or away from the nozzle causes a pressure change in the bellows. The pressure change continues until the unit is in balance.

As in a force balance transmitter, the feedback pressure is the output signal of the transmitter. A zero adjustment is provided for an initial 3 psi (21 kPa) output. The bellows and relay assembly is shown in *Figure 11*.

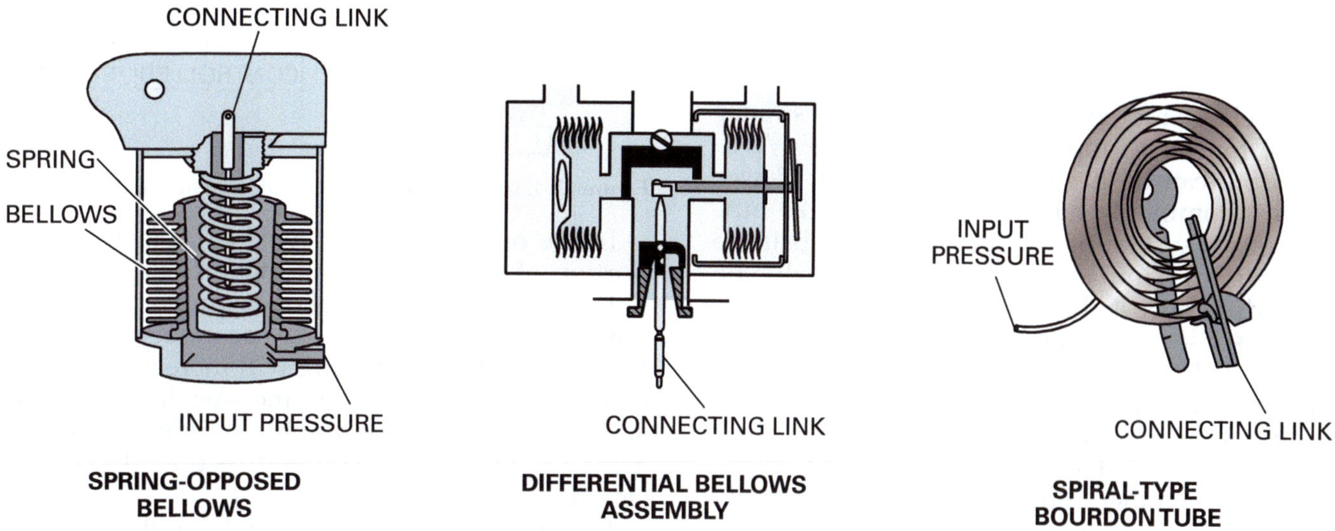

Figure 9 Common measuring elements.

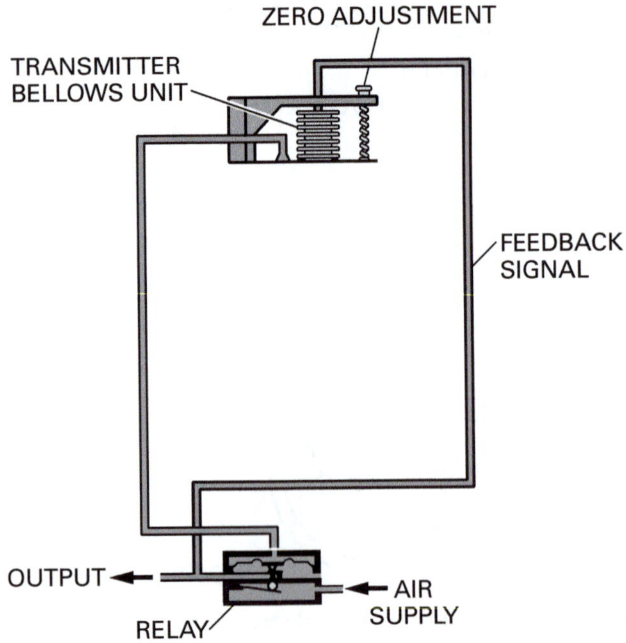

Figure 11 Bellows and relay assembly.

Another type of DP transmitter is a **high-static DP transmitter**. It is often used in very high pressure applications with harsh conditions, such as offshore oil and gas production. *Figure 12* shows an example of this type of transmitter. The transmitter may be configured to show tank liquid levels, usually with the high side connected to the process and the low side vented to atmosphere. The transmitter shown in *Figure 12* is normally calibrated in inches of water, spanned to correlate to tank height.

1.1.2 Electronic Transmitters

Electronic transmitters offer a higher degree of accuracy than pneumatic transmitters. They also have a faster response and a longer range transmission system. Their operation varies by the manufacturer. Two common electronic transmitter systems are the force balance system and the **variable capacitance cell**.

An electronic force balance DP transmitter (*Figure 13*) has a supply voltage of up to 95 VDC. Its output can be 10–50 mA or 4–20 mA. Newer systems commonly have an output of 4–20 mA. The sensor system of an electronic force balance transmitter is similar to a pneumatic DP cell. The major difference is the output device.

The sensor assembly of an electronic force balance DP transmitter is a sealed metallic diaphragm. Pressure on this diaphragm causes a small movement of the diaphragm link. This link movement is transferred to a force bar that transmits it to the range rod. The range rod assembly transmits the force to the armature, causing it to move. The sensor assembly is shown in *Figure 14*.

Figure 12 High-static DP transmitter.

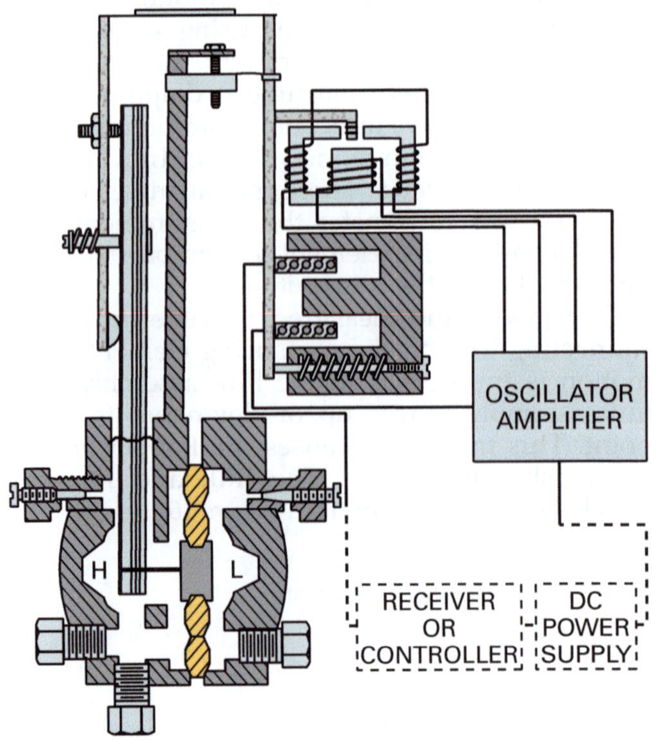

Figure 13 Electronic force balance DP transmitter.

In the output section (*Figure 15*), the armature portion of a range rod can move back and forth within the transformer. For example, if the armature is moved toward the secondary winding, the voltage at that winding will increase. This increase in voltage causes an increase in the output signal of an oscillator-amplifier. However, if the armature moves in the other direction, the output signal decreases.

In a variable capacitance cell, the output device is an electronic-capacitance bridge circuit. Small

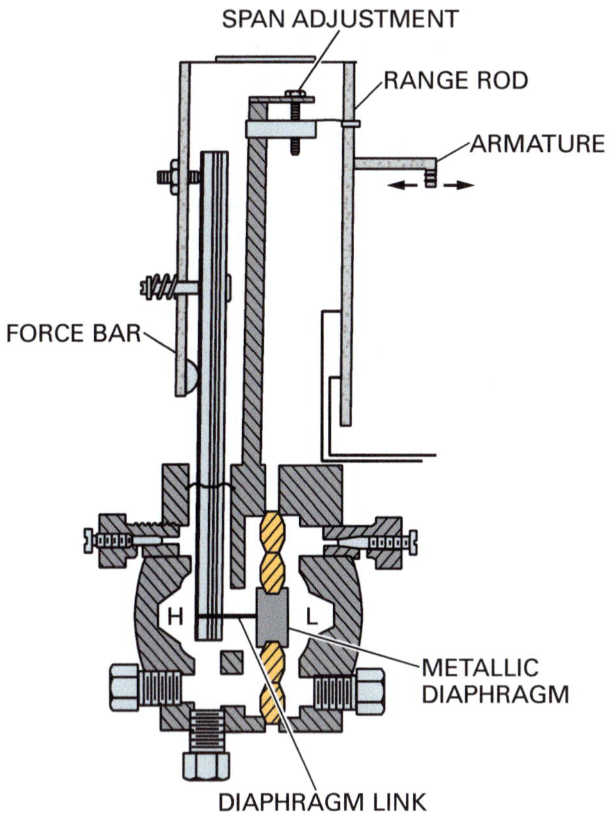

Figure 14 Sensor assembly.

lead wires from the circuit assembly are attached to the capacitor plates and sealed in the diaphragm. A constant electric current is passed through these wires to the capacitor plates. Changes in capacitance caused by pressure changes within the diaphragm cause the current to change. These changes are converted by the bridge circuit to a 4–20 mA signal that is a linear output directly proportional to the pressure changes.

1.1.3 Optical Transmitters

Fiber optics now play a significant role in transmitter technology due to their excellent low-noise properties, low signal loss, and clarity of signal. The two types of optical sources that have been widely used are the light emitting diode (LED) and the injection laser diode (ILD). These are solid-state devices that produce an increased light output with an increasing forward-bias current.

In these devices, electrons injected across a forward-biased PN junction combine with holes to produce light photons, which escape from the structure. Various semiconductor compounds of gallium, arsenic, indium, and phosphorous are used to produce devices that have a peak light emission at wavelengths from 600 nanometers (nm) in the visible range to 1,000 nm in the infrared range. Other gas and solid-state lasers used in optical communications are also included.

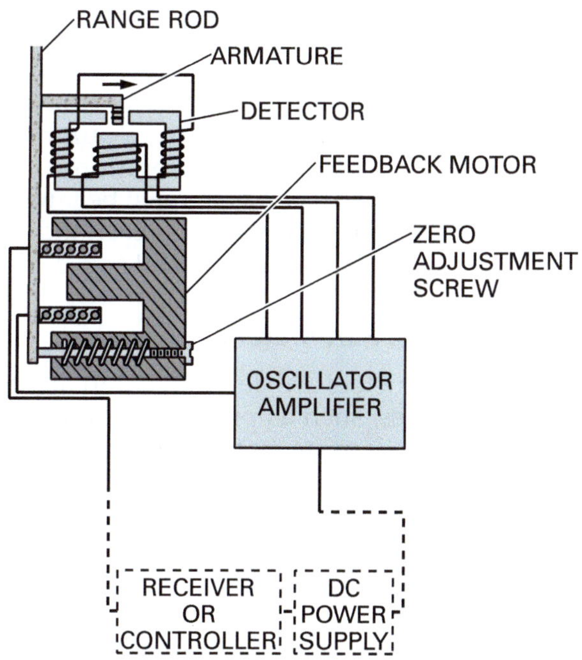

Figure 15 Output section.

Deciding between an LED or ILD optical source rests on understanding the differences between the devices. For example, the ILD has a shorter rise time than the LED, which makes the ILD more suitable for wide bandwidth and high data rate applications. In addition, more optical power can be coupled into a fiber with an ILD, which is important for long-distance transmission. However, the laser diode is strongly temperature dependent and has a significantly higher cost. Selecting between LED or ILD device types as well as what carrier wavelength to select depend on the transmission range, data rate, fiber type, detector, and total system cost.

The optical receiver must detect the received light signal, convert it to an electrical signal, and amplify the resulting electrical signal with minimum distortion. To accomplish this, the optical receiver consists of a photodetector and a pre-amplifier. Two general classes of photodetectors exist, the reverse-biased PN junction diode and the phototransistor. Common reverse-biased diodes are the PIN diode and the avalanche photodiode (APD). In a phototransistor, the light detection diode becomes the base-collector junction of the transistor.

1.2.0 Issues Associated with Pressure Transmitter Testing

Pressure transmitters serve a variety of functions in a pipeline system. Consequently, a number of issues surround testing and calibration activities. The pipeline E&I technician must be familiar with these issues and always assess each individual situation to see if any are relevant.

1.2.1 Protection Parameters

Protection parameters are developed and applied to control devices, such as process controllers or PLCs, to prevent damage from various conditions that could cause process upsets and/or dangerous operating conditions. Some examples of such conditions are: abnormal operating pressures, flow rate deviations, high or low temperature extremes, high or low vibration, and excessively high or low tank levels. Protection parameters enable a control device to initiate actions, such as an **alarm**, and to act as a **shutdown device** to protect process systems and their components, and to protect the personnel who operate the systems.

The following steps outline the general procedures used to verify and set protection parameters for a control device:

Step 1 Obtain verification of the target value(s) for the device setpoint or operating parameters that are to be implemented or changed. Follow a management of change (MOC) procedure, if company policy requires technicians to do so.

Smart Transmitters

Smart transmitters are increasingly being utilized in process systems. These transmitters operate differently from analog transmitters. A smart transmitter incorporates a microprocessor that stores data about the sensor's specific characteristics in response to pressure and temperature inputs and compensates for these sensor variations. Smart instrumentation transmitters commonly use the highway addressable remote transducer (HART®) digital communication protocol to communicate device configuration, status, measurement, and diagnostic information.

HART® devices require periodic calibration. However, unlike many conventional devices, the specific calibration requirements for any particular HART device depend on the application. Calibration of a HART® device involves the following three functions that are performed by a microprocessor within the device:

- Calibrating the process variable
- Scaling the process variable
- Producing the process signal

Smart instruments must always be calibrated by following the specific steps found in the user's manual for the instrument. These instruments are generally matched to loop parameters and cannot be calibrated using basic analog calibration procedures. When calibrating or testing smart instruments, always use the proper test and calibration equipment, such as the HART® communicator.

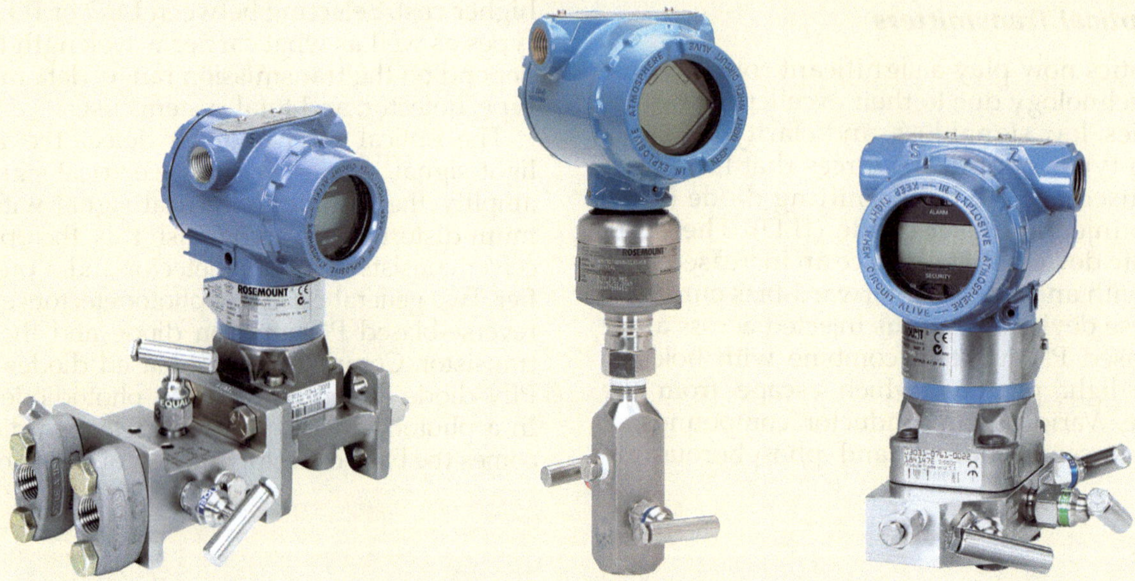

> **NOTE**
> Changes in control device parameter settings can be made for various reasons. For example, to protect refined product from tank contamination, the parameters for API gravity limits may be changed to correspond with changes in product gravity that resulted from refinery crude oil slate changes.

> **NOTE**
> Changes to parameter settings in timers, valve sequences, tank statuses, alarm sequences and statuses, or many other ladder logic rungs and elements may be temporary or they may be permanent.

Step 2 Ensure proper isolation of the affected section from the process system. Follow appropriate lockout/tagout procedures, as required.

Step 3 Verify data within the current program and document the data as found.

Step 4 Implement changes to the setpoints.

Step 5 Remove isolation from the affected section of the process system.

Step 6 Verify program acceptance through software or dynamic testing methods. Be sure to save program changes, and also to back up the changes on a different medium.

Step 7 Document system alterations as left.

1.2.2 Calibrating Pressure Transmitters in Explosive Atmospheres

When calibrating pressure transmitters in potentially explosive environments, use a pressure calibrator that is certified as **intrinsically safe**. Such devices are designed to be unable to release sufficient thermal or electrical energy to cause ignition of flammable material (e.g., gas, dust, particulates). An example of an intrinsically safe pressure calibrator is shown in *Figure 16*.

In addition to using an intrinsically safe pressure calibrator, strictly adhere to safe work practices for potentially explosive environments, including but not limited to the following:

- Follow the appropriate lockout/tagout procedures
- Secure the work area to prevent anyone from entering the area during the calibration
- Safely purge or vent the system to remove any remaining gases
- Use a gas detector to make sure the atmosphere is safe before starting the calibration

Figure 16 Intrinsically safe pressure calibrator.

Additional Resources

It is assumed that you will learn to perform these procedures under the direct supervision of your crew leader, foreman, or instructor. For independent study, refer to the recommended Additional Resources at the back of this module.

1.0.0 Section Review

1. The major difference between the sensor system of an electronic force balance transmitter and the pneumatic DP cell is the _____.
   a. sensor assembly
   b. supply voltage
   c. output device
   d. input device

2. A calibrator that can be safely used in an explosive atmosphere is known as being _____.
   a. intrinsically safe
   b. volatile sealed
   c. essentially flameproof
   d. pneumatically isolated

SECTION TWO

2.0.0 INSPECTING, TESTING, AND CALIBRATING PRESSURE TRANSMITTERS

Objective

Describe the procedures used to inspect, test, and calibrate pressure transmitters (CT25_2-17).

a. Identify the prerequisites, competencies, and abnormal operating conditions (AOCs) associated with inspecting, testing, and calibrating pressure transmitters.
b. Describe the procedures used to inspect, test, and calibrate pressure transmitters.

Performance Task

1. Inspect, test, and calibrate pressure transmitters (CT25_2-17).
   - Identify potential abnormal operating conditions (AOCs) that may occur during performance of this CT, and know the appropriate actions to take in response to them.
   - Utilize the appropriate personal protective equipment according to relevant company procedures.
   - Verify certification that test equipment has been calibrated prior to performing calibrations.
   - Verify range of transmitter prior to performing calibration.
   - Determine correct test medium and the range to be used for testing.
   - Notify control center and/or affected personnel before work begins.
   - Isolate the pressure transmitter from the process system and inspect all connections for leakage.
   - Apply test medium pressure and verify device setpoint or range "as found". Verify pressure transmitters at zero, mid, and full scale calibration points. Document results "as found."
   - Re-apply test medium for zero, mid-scale and full scale. Adjust if necessary according to device manufacturer's specifications to establish desired calibration.
   - Verify activation of digital control output and/or operation of the end device at the established setpoint. Document final setpoint settings "as left."
   - Notify control center and/or affected personnel after completion of work.
   - Complete appropriate documentation as required by operator's procedures.

Trade Terms

Abnormal operating conditions (AOCs): Conditions within a pipeline system, identified by the operator, that may indicate (1) a malfunction of a component or (2) a deviation from normal operations that may indicate a condition exceeding design limits or result in a hazard to persons, property, or the environment.

Test medium: The material used in the system being tested that allows reliable and repeatable test measurements.

Pressure transmitters require regular inspection, testing, and calibration to keep them functioning properly. Manufacturers usually specify inspection, testing, and calibration procedures for their instruments. To achieve accuracy and repeatability from the instruments, follow the procedures that are recommended by the manufacturer. The general procedures presented in this section should be applicable in most cases.

2.1.0 Prerequisites, Competencies, and Abnormal Operating Conditions

Personnel must possess a variety of skills in order to successfully inspect, test, and calibrate pressure transmitters. The various operations require a series of steps, that a pipeline E&I technician must be competent in performing. Personnel must also be aware of the particular abnormal operating conditions (AOCs) that may occur during the procedure and know the proper responses to them.

2.1.1 Prerequisites and Competencies

Personnel must possess the following basic knowledge and skills prior to inspecting, testing, and calibrating pressure transmitters:

- They must be skilled with test equipment and know how to operate the following types of equipment:
  - Pressure gauges of all types
  - Multifunction calibrators
  - Multimeters
  - Digital pressure modules

- They must be skilled with test media and the equipment used to manipulate it, including the following:
  - Hand pumps
  - Pressure regulators
  - Process connections
  - Air and nitrogen
  - Glycol
- They must be able to work safely with electrical systems.
- They must be familiar with the control system and its instrumentation, including alarm, control, indication, and recording functions.

They must also be familiar with the company safety, evacuation, and response procedures, as well as the proper use of personal protective equipment (PPE).

2.1.2 Abnormal Operating Conditions

Personnel inspecting, testing, and calibrating pressure transmitters must be able to identify abnormal operating conditions (AOCs) and know the proper response to each, should they occur. The following are some potential AOCs, and the appropriate action the E&I technician should take in response to them:

- *The device does not respond properly within its intended range or the device does not respond consistently* – In this situation, the pipeline E&I technician must alert appropriate personnel to take the requisite corrective action.
- *During testing, a safety or control device activates unintentionally, causing a shutdown or undesired valve movement* – Control loss or an overpressure condition results as a consequence. In this situation, the pipeline E&I technician must take appropriate action to restore normal conditions and then alert appropriate personnel as required by company procedure.
- *An isolation valve fails to isolate the process pressure due to incomplete closure or other malfunction* – In this situation, the pipeline E&I technician must halt the test and return the valve to normal in order to prevent a loss of control or other negative side effect. The technician must also alert appropriate personnel as required by company procedure.

2.2.0 Procedure

The following steps outline the general procedure for inspection, testing, and calibration of pressure transmitters:

Step 1 Determine the device's number and nameplate data in order to verify that it is the correct unit to calibrate/test and that any data coming from the unit will be properly identified.

> **NOTE**
> If the nameplate is missing or illegible, it must be replaced according to company policy and procedures.

Step 2 Before beginning the calibration process, check that the test equipment itself has been certified, calibrated, and verified. Some technicians will not use test equipment until it has been tested on a known source, even if its calibration is current and properly documented. Confirm that the equipment is appropriate for the transmitter in question.

Step 3 Check the appropriate company documentation to determine and verify input and output ranges for the transmitter in question.

Step 4 Identify and obtain the correct **test medium** (air, nitrogen, etc.) for the calibration procedure. Improper test media could cause inaccurate results and/or equipment damage.

Step 5 Before beginning the process, notify the control center, local personnel, and any other personnel who might be affected by the calibration/test procedure.

> **NOTE**
> The control center may have to provide feedback about the pressure transmitter's behavior, ID, or output data. It may also have to override automatic shutdowns or safety systems to protect the process from interruption. These events are often known as Critical Interlock and Trips (CIATs). They usually require higher authorization to override. Pipeline E&I technicians should be aware of these rules and know what they're allowed to do.

Step 6 Visually inspect the device and any associated equipment. Check for mechanical condition, corrosion, leakage, electrical connection problems, or problems with associated electronics.

WARNING!

Always exercise extreme caution when working with energized circuits to avoid contact and electrical shock. If shock occurs, call 911.

Step 7 Using appropriate measures, isolate the pressure transmitter from the process and depressurize it. Follow company procedure.

Step 8 Following the proper procedure, disconnect the transmitter from the process connections and attach the test equipment in its place. Confirm that there are no leaks as these would invalidate the test/calibration run.

Step 9 Apply pressure to the transmitter and cycle through its range to ascertain its current output response. Repeat this step as many times as necessary to determine the transmitter's repeatability. Based on the results, decide if the transmitter requires calibration. Follow company procedure.

Step 10 Document the results of Step 9 to provide historical data and the "as found" status of the device. Follow company procedure.

Step 11 Based on your conclusions from Step 9, calibrate the device if necessary, following the manufacturer's guidelines and company procedure. Repeat this step as often as required to establish a repeatable calibration.

Step 12 Document the results of Step 11 to provide historical data and the "as left" status of the device. Follow company procedure.

Step 13 Disconnect the test equipment and restore the unit's normal connections, confirming that they are leak free.

Step 14 Notify the control center, local personnel, and any affected personnel that the unit has been returned to normal operation.

Step 15 Complete all required documentation.

NOTE

Documenting data for compliance, historical review, and trending is an important element of valve maintenance and should not be neglected.

Additional Resources

It is assumed that you will learn to perform these procedures under the direct supervision of your crew leader, foreman, or instructor. For independent study, refer to the recommended Additional Resources at the back of this module.

2.0.0 Section Review

1. With which instrument should the pipeline E&I technician be skilled when working with a test medium?
   a. A thermocouple heat block
   b. A nitrogen strainer
   c. A hand pump
   d. A certified valve aligner

2. Before connecting the test equipment to the pressure switch, the technician must _____.
   a. calibrate the meter pressure module
   b. disable the alarm
   c. set SCADA to "Ignore" mode
   d. isolate the switch from the process

SUMMARY

Keeping pressure transmitters working reliably is key to consistent and safe pipeline operation. Pressure transmitters provide valuable data about the process both for control and safety purposes. Each function is important, so proper transmitter calibration is critical. This module summarized the various issues involved and provided a basic overview of the procedure for inspecting, testing, and calibrating these devices.

Review Questions

1. A device that was originally very specific in function and is used to prepare a signal to be sent from one location to another is a _____.
   a. switch
   b. transmitter
   c. communicator
   d. transponder

2. The most common type of pneumatic transmitter is the force balance differential-pressure transmitter, also called a _____.
   a. DP cell
   b. motion balance transmitter
   c. pneumatic relay bar
   d. FBDP

3. Settings programmed into a controller or PLC designed to prevent damage and process upsets are called _____.
   a. protection parameters
   b. shutdown thresholds
   c. lockout lines
   d. monitor alerts

4. A common test medium used for calibrating a pressure transmitter is _____.
   a. heliox
   b. argon
   c. hydrazine
   d. nitrogen

5. If a control device activates unintentionally during testing, what is the first thing that the pipeline E&I technician should do?
   a. Call the local 911.
   b. Alert appropriate personnel.
   c. Restore normal conditions.
   d. Follow emergency procedures.

6. What must the pipeline E&I technician do to determine if a pressure transmitter needs to be calibrated?
   a. Talk to the control room.
   b. Perform a fieldbus scan.
   c. Ping the transmitter with a handheld communicator.
   d. Cycle the test pressure through the transmitter's range.

Trade Terms Quiz

Fill in the blank with the correct term that you learned from your study of this module.

1. A(n) ________ is a mechanical pressure-measuring instrument that uses, for its sensing element, a curved or twisted metal tube that is flattened in cross section and closed.
2. A(n) ________ is a type of pneumatic transmitter that converts the input motion of basic measuring elements into a pneumatic output that is directly proportional to the measured variable.
3. A device that protects a pipeline in response to an unacceptable process condition is called a(n) ________.
4. A type of differential-pressure transmitter that is often used in very high pressure applications is called a(n) ________.
5. A(n) ________ is an electronic transmitter system that uses an electronic capacitance bridge circuit as its output device.
6. An audible and/or visual status event from the control system, notifying personnel that a process parameter has moved out of its normal range is called a(n) ________
7. A test instrument that is safe to use in an explosive atmosphere is considered ________ because it does not generate sufficient electrical or thermal energy to ignite in it.
8. A facility within which devices for measuring process variables such as temperature, pressure, level, and flow are installed is called a(n) ________.
9. A(n) ________ is a type of pneumatic transmitter that is operated through differential pressure. It is also called a DP cell.
10. Conditions within a pipeline system that may indicate (1) a malfunction of a component or (2) a condition exceeding design limits or resulting in a hazard to persons, property, or the environment are called ________.
11. The values that an instrument can measure, bounded by lower and upper limits are called the ________.
12. The material used in the system being tested that allows reliable and repeatable test measurements is the ________.

Trade Terms

Abnormal operating conditions (AOCs)
Alarm
Bourdon tube
Force balance transmitter
High-static DP transmitter
Intrinsically safe
Motion balance transmitter
Process system
Range
Shutdown device
Test medium
Variable capacitance cell

Trade Terms Introduced in This Module

Abnormal operating conditions (AOCs): Conditions within a pipeline system, identified by the operator, that may indicate (1) a malfunction of a component or (2) a deviation from normal operations that may indicate a condition exceeding design limits or result in a hazard to persons, property, or the environment.

Alarm: A status event from the control system, notifying personnel that a process parameter has moved out of its normal range; alarms can be audible, visual, or both.

Bourdon tube: A mechanical pressure-measuring instrument that uses for its sensing element a curved or twisted metal tube that is flattened in cross section and closed. It is also called a Bourdon pressure gauge.

Force balance transmitter: A type of pneumatic transmitter that is operated through differential pressure. It is also called a DP cell.

High-static DP transmitter: A type of differential-pressure transmitter that is often used in very high pressure applications.

Intrinsically safe: A test instrument that is safe to use in an explosive atmosphere because it does not generate sufficient electrical or thermal energy to ignite it.

Motion balance transmitter: A type of pneumatic transmitter that converts the input motion of basic measuring elements into a pneumatic output that is directly proportional to the measured variable.

Process system: A facility within which devices for measuring process variables such as temperature, pressure, level, and flow are installed.

Range: The values that an instrument can measure, bounded by lower and upper limits.

Shutdown device: A device that protects a pipeline in response to an unacceptable process condition.

Test medium: The material used in the system being tested that allows reliable and repeatable test measurements.

Variable capacitance cell: An electronic transmitter system that uses an electronic capacitance bridge circuit as its output device.

Additional Resources

The following recommended resources can provide additional helpful information related to the requirements for this covered task:

API Recommended Practice 1161, Third Edition, © 2014 American Petroleum Institute. API Publishing Services, 1220 L Street, NW, Washington, DC 20005. Distributed by Thompson Reuters (Scientific) LLC, **www.techstreet.com**.

NCCER Module AOCCC-17, *Abnormal Operating Conditions Control Center*

NCCER Module AOCFG-17, *Abnormal Operating Conditions Field and Gas*

NCCER Module 60102-02, *Pipeline E&I Safety*

NCCER Module 64202-02, *Use of Meters and Test Equipment*

NCCER Module 12402-16, *Instrument Calibration and Configuration*

Instrument Engineers Handbook, Volume 1: Process Measurement and Analysis. Béla G. Lipták (editor). Fourth Edition. 2003. Boca Raton, FL: CRC Press.

Instrument Engineers' Handbook, Volume 2: Process Control and Optimization. Béla G. Lípták (editor). Fourth Edition. 2005. Boca Raton, FL: CRC Press.

Instrumentation. Franklyn W. Kirk, Thomas A. Weedon, and Philip Kirk. Fifth Edition. 2010. Orland Park, IL: American Technical Publishers.

Instrumentation Reference Book. Walt Boyes (editor). Fourth Edition. 2009. Oxford, UK: Butterworth-Heinemann.

Measurement and Control Basics. Thomas A. Hughes. Fifth Edition. 2014. Research Triangle Park, NC: The International Society of Automation.

Purdy's Instrument Handbook. Ralph Dewey. 2015 Edition. Deer Park, TX: Glen Enterprises.

The Condensed Handbook of Measurement and Control. N.E. Battikha. Third Edition, 2006. Research Triangle Park, NC: The International Society of Automation.

The following websites offer resources for products and training:

Dwyer Instruments, Inc., **www.dwyer-inst.com**

Emerson Electric Co., **www.emerson.com**

The International Society of Automation, **www.isa.org**

Figure Credits

Bigelow-Liptak of Canada, Module Opener

Yokogawa Corporation of America, Figure 12

Fluke Corporation, Figure 16

Emerson Process Management, SA01

Section Review Answer Key

Answer	Section Reference	Objective
Section One		
1. c	1.1.2	1a
2. a	1.2.2	1b
Section Two		
1. c	2.1.1	2a
2. d	2.2.0	2b

NCCER CURRICULA — USER UPDATE

NCCER makes every effort to keep its textbooks up-to-date and free of technical errors. We appreciate your help in this process. If you find an error, a typographical mistake, or an inaccuracy in NCCER's curricula, please fill out this form (or a photocopy), or complete the online form at **www.nccer.org/olf**. Be sure to include the exact module ID number, page number, a detailed description, and your recommended correction. Your input will be brought to the attention of the Authoring Team. Thank you for your assistance.

Instructors – If you have an idea for improving this textbook, or have found that additional materials were necessary to teach this module effectively, please let us know so that we may present your suggestions to the Authoring Team.

NCCER Product Development and Revision

13614 Progress Blvd., Alachua, FL 32615

Email: curriculum@nccer.org

Online: www.nccer.org/olf

❑ Trainee Guide ❑ Lesson Plans ❑ Exam ❑ PowerPoints Other ______________________

Craft / Level: Copyright Date:

Module ID Number / Title:

Section Number(s):

Description:

Recommended Correction:

Your Name:

Address:

Email: Phone:

CT26_0-17

Verify or Set Protection Parameters for Programmable Controllers and/or Other Instrumentation Control Loops

Overview

Tuning (or adjusting) control loops is an essential part of maintaining process controllers. To perform this task, technicians need to have a good understanding of the basic components and functions of the various control loops that are widely used in the pipeline industry. This module reviews the different types of control loops, their functions, and the methods that are used to tune them. It also explains how to verify and set protection parameters, and covers how to perform continuity checks and troubleshoot loops, including troubleshooting an oscillating process.

Trainees with successful module completions may be eligible for credentialing through the NCCER Registry. To learn more, go to **www.nccer.org** or contact us at 1.888.622.3720. Our website has information on the latest product releases and training, as well as online versions of our *Cornerstone* magazine and Pearson's product catalog.

Your feedback is welcome. You may email your comments to **curriculum@nccer.org**, send general comments and inquiries to **info@nccer.org**, or fill in the User Update form at the back of this module.

This information is general in nature and intended for training purposes only. Actual performance of activities described in this manual requires compliance with all applicable operating, service, maintenance, and safety procedures under the direction of qualified personnel. References in this manual to patented or proprietary devices do not constitute a recommendation of their use.

CT26_0-17
Verify or Set Protection Parameters for Programmable Controllers and/or Other Instrumentation Control Loops

Objectives

When you have completed this module, you will be able to do the following:

1. Describe control and PID loops and explain how loops are tuned using various methods.
   a. Describe control loops.
   b. Describe PID loops.
   c. Describe the primary methods used to determine loop tuning values and identify factors that affect loop stability.
   d. Explain how to tune loops using closed-loop methods.
   e. Explain how to tune loops using open-loop methods.
2. Explain how to verify and set protection parameters (CT26_0-17).
   a. Identify prerequisites, competencies, and abnormal operating conditions (AOCs) associated with verifying and setting protection parameters.
   b. Explain the general procedure for verifying and setting protection parameters.
3. Explain how to troubleshoot loops.
   a. Explain how continuity checks are completed.
   b. Describe various troubleshooting procedures.
   c. Describe how to troubleshoot an oscillating process.

Performance Task

Under the supervision of the instructor, you should be able to do the following:

1. Verify and set protection parameters for programmable controllers and/or other instrumentation control loops (CT26_0-17).
   - Identify potential abnormal operating conditions (AOCs) that may occur during performance of this CT, and know the appropriate actions to take in response to them.
   - Utilize the appropriate personal protective equipment according to relevant company procedures.
   - Identify the controlling device.
   - Notify operations control center and/or affected personnel before work begins.
   - Verify that the device setpoint or operating parameters agree with documentation or drawings.
   - Ensure proper isolation of the controlling device from the process system. Verify and document data "as found".
   - Implement changes to parameters and verify acceptance of parameters.
   - Return controller parameters to desired settings for permanent operation.
   - Verify accuracy and scaling for the analog input device.
   - If a loop controller or PLC is used for overpressure protection, verify activation of digital control output and/or operation of the end device at the established setpoint.
   - If pipeline pressure is maintained using a control valve or speed control, verify operation of analog output and control devices, such as control valve or throttle actuator.
   - Document system parameters "as left."
   - Notify operations control center and/or affected personnel.
   - Complete appropriate documentation as required by operator's procedures.

Trade Terms

Abnormal operating conditions (AOCs)
Critical frequency
Dampened oscillation method
Deadtime
Notch method
Offset
PID
Process variables
Proportional band (gain)
Rate (derivative)
Reaction rate method
Reset (integral)
Steady-state
Supervisory Control and Data Acquisition (SCADA)
Time constant method
Tolerance
Ultimate gain
Ultimate period
Ultimate period method
Ziegler-Nichols equations

Industry Recognized Credentials

If you are training through an NCCER-accredited sponsor, you may be eligible for credentials from NCCER's Registry. The ID number for this module is CT26_0-17. Note that this module may have been used in other NCCER curricula and may apply to other level completions. Contact NCCER's Registry at 888.622.3720 or go to **www.nccer.org** for more information.

Contents

1.0.0 Controllers 1
1.1.0 Control Loops 2
1.1.1 Sensing Component 2
1.1.2 Transmitting Component 2
1.1.3 Controlling Component 2
1.1.4 Final Control Component 3
1.1.5 Control Loop Operation 3
1.2.0 PID Loops 4
1.2.1 Proportional Band 4
1.2.2 Reset (Integral) 4
1.2.3 Rate (Derivative) 4
1.3.0 Tuning Loops 4
1.3.1 Factors Affecting Stability 5
1.4.0 Closed-Loop Methods 5
1.4.1 Notch Method 5
1.4.2 Ultimate Period Method 11
1.4.3 Dampened Oscillation Method 14
1.5.0 Open-Loop Methods 16
1.5.1 Time Constant Method 17
1.5.2 Reaction Rate Method 20
2.0.0 Verifying And Setting Protection Parameters 25
2.1.0 Prerequisites, Competencies, and Abnormal Operating Conditions 25
2.1.1 Prerequisites and Competencies 25
2.1.2 Abnormal Operating Conditions 26
2.2.0 Procedure 26
3.0.0 Troubleshooting Loops 28
3.1.0 Continuity Checks 28
3.1.1 Continuity Check of Pneumatic Loops 28
3.2.0 Seven-Step Troubleshooting Procedure 29
3.2.1 Symptom Recognition 29
3.2.2 Symptom Definition 31
3.2.3 Listing of Probable Faulty Functions 31
3.2.4 Localization of Faulty Functions 31
3.2.5 Localization of the Trouble to the Circuit 33
3.2.6 Failure Analysis 33
3.2.7 Retesting 34
3.3.0 Troubleshooting Oscillating Processes 34
3.3.1 Verifying That a Problem Exists 34
3.3.2 Gathering Information 34
3.3.3 Identifying Possible Causes 34
3.3.4 Checking Instruments in the Field 36
3.3.5 Checking the Transmitter 36
3.3.6 Locating the Problem 37
3.3.7 Using a Troubleshooting Flowchart 38

Figures

Figure 1 Frequency response curve for a three-mode controller 6
Figure 2 Adjustment of controller gain for a three-mode controller 7
Figure 3 Effects of reset and rate adjustments 7
Figure 4 Block diagram of a process being notch tuned 8
Figure 5 Controller with a wide and deep notch 9
Figure 6 Controller gain at the gain limit due to proportional adjustment 9
Figure 7 Controller gain after final adjustment of proportional band 10
Figure 8 Controller gain at the gain limit due to the reset adjustment 10
Figure 9 Controller gain after final setting of the reset adjustment 11
Figure 10 Controller gain at the gain limit due to the rate adjustment 11
Figure 11 Controller gain after the final setting of the rate adjustment 12
Figure 12 Control system at the ultimate gain or proportional band 13
Figure 13 Responses of the measured variable to a supply disturbance 15
Figure 14 Test data for ultimate tuning method 15
Figure 15 Test data for dampened oscillation method 16
Figure 16 Block diagram of a process being open-loop tuned 17
Figure 17 Process reaction curve for an open-loop-tested process 19
Figure 18 Time constant method for a single capacity with deadtime approximation 20
Figure 19 Test data for open-loop time constant method 21
Figure 20 Reaction rate method for a single capacity with deadtime approximation 22
Figure 21 Test data for open-loop reaction rate method 23
Figure 22 Multimeter 28
Figure 23 Common signal paths 30
Figure 24 Bracketing technique 32
Figure 25 Panel graphic 35
Figure 26 Piping and instrumentation diagram 35
Figure 27 Process diagram 37
Figure 28 Flowchart 38

Section One

1.0.0 Controllers

Objective

Describe control and PID loops and explain how loops are tuned using various methods.

a. Describe control loops.
b. Describe PID loops.
c. Describe the primary methods used to determine loop tuning values and identify factors that affect loop stability.
d. Explain how to tune loops using closed-loop methods.
e. Explain how to tune loops using open-loop methods.

Trade Terms

Critical frequency: The frequency at which a 360-degree phase shift around the control loop exists.

Dampened oscillation method: A closed-loop tuning method that is a modification of the ultimate period method; it is used for processes that cannot tolerate continuous oscillations.

Deadtime: The delay interval between the point when the controlled variable changes and the point when the controller begins to respond to the change.

Notch method: A closed-loop tuning method that uses the frequency response of a controller to determine the point at which the loop gain is less than one at the critical frequency.

Offset: The difference between the point of stabilization and the setpoint.

PID: A method of control made up of proportional band, integral, and derivative control. (It is an acronym for *proportional + integral + derivative* control.)

Process variables: Any physical conditions of a process that need to be controlled, such as pressure, speed, level, temperature, or another measurable quantity.

Proportional band (gain): A method of control based on the range of values of a process variable that requires proportional (gain) control; usually shown as a percentage of the full-scale range of the controller.

Rate (derivative): A method of control based on derivative, or rate of change, action.

Reaction rate method: An open-loop tuning method based on the slope of the maximum rate of rise line (the reaction rate) from the process reaction curve and the deadtime.

Reset (integral): A method of integral control based on reset rate in repeats per minute required in proportion to the size and duration of an error, or offset.

Steady-state: The condition in which a process variable is stabilized at an acceptable value.

Supervisory Control and Data Acquisition (SCADA): A computer-based system used remotely by a controller in a control room that displays information about a pipeline facility, and which may have the ability to send commands back to the facility.

Time-constant method: An open-loop tuning method that is based on the time constant, process gain, and deadtime from the process reaction curve.

Tolerance: The acceptable amount by which a process variable can deviate from its setpoint, such as 68 degrees ±3.

Ultimate gain: The maximum allowable value of gain or proportional band at which a system is stable.

Ultimate period: The period of response during which either the gain or proportional band is set at the ultimate value.

Ultimate period method: A closed-loop tuning method that requires the determination of the ultimate gain or proportional band and the ultimate period.

Ziegler-Nichols equations: Equations developed in the early 1940s by J.G. Ziegler and N.B. Nichols that are still used today as the basis of some of the most common methods of loop tuning.

The **Supervisory Control and Data Acquisition (SCADA)** system for a process is connected to the equipment or machines that are being controlled and monitored through devices called *field instruments*. Field instruments include the actuators for controlling certain aspects of the process, as well as the sensors that monitor **process variables** and their defined parameters.

These instruments convert physical parameters (such as fluid flow, velocity, and level) into electrical signals (voltage or current). These electrical signals are then transmitted to and read by the remote station equipment.

The outputs of the field devices can be either analog or digital. Voltage outputs are used when the sensors are installed near the controllers, either remote terminal units (RTUs) or programmable logic controllers (PLCs). Current outputs are used when the sensors are located far from the controllers. Some of the industry standard analog outputs of these sensors are 0–5 V, 0–10 V, 4–20 mA, and 0–20 mA.

Most process instruments, sensors, and actuators offer a 4–20 mA direct current (DC) input or output as an option. A transmitter accepts a pneumatic, voltage, current, frequency, or pulse input from a field sensor and converts it to a 4–20 mA analog signal. In process applications, transmitters are usually mounted at the sensor, such as when a differential-pressure flow transmitter is mounted directly on a process line or on a pipe stand next to a pipeline. An enclosure rated by the National Electrical Manufacturer's Association (NEMA) is usually required for housing the transmitters and sensor electronics, particularly if the equipment is inside a classified or hazardous area.

Signal converters translate process signals from one form into another and often perform signal conditioning at the same time. Typical converters include the following:

- *P/I* – Pneumatic to electric current; converts 3 to 15 psi (21 to 103 kPa) to 4–20 mA DC
- *I/P* – Electric current to pneumatic; converts 4–20 mA DC to 3 to 15 psi (21 to 103 kPa)
- *A/D* – Analog to digital; converts analog current or voltage input signal to digital for computers
- *D/A* – Digital to analog; converts digital output from computers to analog current or voltage

All of these instruments must be calibrated and tuned to perform properly and safely. In turn, pipeline E&I technicians must be able to perform instrument calibrations and tuning correctly and follow the manufacturer's recommended procedures to maintain a pipeline safely and efficiently.

Pipeline E&I technicians should always be aware of hazardous, or classified, areas and follow all guidelines and procedures for testing equipment within these areas.

> **WARNING!**
>
> Never test an isolation device with an ohmmeter or other instrument while it is connected in circuit. This practice bypasses isolation and could allow higher voltages to enter the hazardous area, potentially causing an ignition or explosion.

1.1.0 Control Loops

Control loops consist of several components that work together to maintain a process variable within desired limits. The components measure the actual quantity, compare it to a desired setpoint or range, and adjust the actual quantity to agree with the desired quantity. As the process variable changes, the control loop continues to make adjustments to maintain the variable within the desired range.

Process control loops have four basic components:

- A sensing component
- A transmitting component
- A controlling component
- A final control component

In some systems, one instrument/device serves as both the sensor and the transmitter.

1.1.1 Sensing Component

The sensing component can be almost any device that can accurately determine the condition of the process variable. Sensing components include those that measure speed, pressure, level, temperature, number, consistency, and motion. The sensor is usually, but not always, in direct contact with the process. Examples of sensors are devices such as probes, floats, thermocouples, and thermistors.

1.1.2 Transmitting Component

The transmitting component, or the transducer, measures the amount of change determined by the sensing component and transmits a proportional signal that indicates the value of the process variable. Some transducers can also convert analog signals into digital signals for use by a computer or a PLC.

1.1.3 Controlling Component

A controlling component (controller) performs a series of basic functions, essential to the operation of a control loop:

1. Measures the signal from the transmitter
2. Compares the signal to the desired setpoint or range
3. Computes the difference between the actual (measured) value and the desired value
4. Sends a signal to the final control component to correct the measured value and achieve the desired setpoint.

GOING GREEN

Electro-Pneumatic Transducers and EPA Emission Limits

Pneumatic controllers installed at oil and gas well sites and similar locations are typically powered by high-pressure natural gas. The controllers release supply gas, which includes volatile organic compounds (VOCs) and methane, intermittently with every valve movement or, in many cases, continuously as part of normal controller operation. The allowable bleed limit that the US Environmental Protection Agency (EPA) has set for such emissions is 6 standard cubic feet of gas per hour (SCFH). Oil and gas production facilities increasingly use electro-pneumatic transducers to help them meet EPA emission control and reporting requirements. The transducer, which is instrumental in feeding the operating natural gas medium to the controller, is specifically designed to provide a maximum, steady state pneumatic output of less than the 6 SCFH allowable bleed limit.

Figure Credit: © rCarner/Shutterstock.com

If many inputs from one area are used, the controlling component may be a PLC, in which case a software program is used to compile the inputs and determine the amount of correction. Sometimes several programmable controllers are connected so that they do the same work as a large computer.

Information is sometimes received from many inputs, covering several process variables over a large area or even an entire process, and each input has an effect on the others. The calculations are complex and must be made very quickly, so a wide-ranging computer system, called a distributed control system (DCS), is used to compute the inputs in relation to each other. Then a computer sends signals to the final components to make the necessary corrections to the process.

1.1.4 Final Control Component

The final control element receives the signal from the controller and translates it into a corrective action, such as opening a valve or increasing speed. The final control component can be either an on/off device or a proportional device, such as a flow control valve.

1.1.5 Control Loop Operation

More than one sensing component can provide input information to the controller in a control loop. For example, the temperature in a scrubber column may vary considerably from top to bottom, so there might be three or more sensors at strategic points in the column. The controller processes all the inputs and then acts on a single output.

In closed loops, information about the process variable (the actual value[s]) is sent through the loop components that monitor and control the variable so that changes or disturbances can be acted upon to maintain the variable within acceptable limits (desired value[s]).

In an open loop, no comparison between the actual value and the desired value takes place. The system is designed for certain events to take place during the process, and no feedback is necessary. This type of system is effective in a noncritical process. Often, the events are initiated by timers, such as in a production line.

Controlling temperature through a heater or furnace is an example of a closed control loop. Assume that the temperature on the thermostat is set at 70°F (21°C) and the system has been set up for a tolerance of plus or minus 5°F (3°C). A sensing device, such as a thermometer, placed in a crucial location determines the actual temperature, 64°F (18°C). The measuring device, or transducer, takes the information from the thermometer, converts it into an electric current or voltage, and sends it to the controller. The controller receives the voltage, interprets it, and compares the actual temperature to the setpoint. The temperature is outside the acceptable range, so the controller turns on a relay, the final control component, to start the furnace. When the signal to the controller indicates that the temperature has reached the setpoint of 70°F (21°C), the controller turns off the relay. The temperature may continue to rise for the length of time it takes for the change in temperature to reach the sensing device and register.

1.2.0 PID Loops

The acronym **PID** is taken from **proportional band (gain)**, **reset (integral)**, and **rate (derivative)**, which make up the three modes of control. PID control is sometimes defined as *proportional-plus-integral-plus-derivative control*. PID loops are more complex than standard control loops; they are used in process applications that depend on response time or that need to allow for lag or lead time.

1.2.1 Proportional Band

The first mode of control, proportional band, is the range of values of the process variable that requires proportional control and is usually shown as a percentage of the full-scale range of the controller. For example, if the full range of a pressure controller is from 0 to 100 psi (0 to 690 kPa) and the proportional control is over a 25 psi (172 kPa) range, then the proportional band is 25 percent. If the proportional control action were over the entire 100 psi (690 kPa) range, the proportional band would then be 100 percent. The width of the proportional band determines the amount of valve motion for any given change in a controlled variable.

1.2.2 Reset (Integral)

The second mode of control is reset. With a simple proportional controller, the process can stabilize at a point other than the setpoint. The difference between the point of stabilization and the setpoint is referred to as **offset**. Small offsets may be tolerated, but larger ones may have an undesirable effect on the process. In this case, two-mode control (that is, proportional plus reset or PI) is required to adjust the output signal and bring the offset back into an acceptable range. The reset rate is usually indicated in repeats per minute (rpm), as tuning is a form of timing. The amount of reset required is in proportion to the size and duration of the error, or offset.

1.2.3 Rate (Derivative)

The third mode of control is the derivative, or rate action. Delayed response and lags can make two-mode control unsuitable, so the third mode of control must be added. Derivative action provides a continuous relationship between the rate of change of the process variable and the output signal. Sometimes a large initial overcorrection results when correcting a large process lag. Derivative time is the interval by which derivative action advances the time of proportional action, and as with reset adjustment, the derivative mode of adjustment (tuning) is a form of timing.

1.3.0 Tuning Loops

The most difficult problem in dealing with process control is the setting of the adjustments that are located on the controller. Information concerning the controller, the measuring element, and the final control element is usually available from the manufacturer. If not, it is easily obtained through simple calculations and measurements.

Items such as the gain, **deadtime**, and time constant are not difficult to determine for the control system components. However, the same characteristics for the process are extremely difficult, if not impossible, to determine. In general, there are two means of obtaining this information: the mathematical method, and the testing method.

The mathematical method is used to calculate the values of the characteristics. This method is accomplished by applying a mathematical analysis to the process and the control system. This requires writing mass and energy balance equations and then solving the equations to determine the values for the characteristics that are of interest. This method requires a high degree of skill in mathematics and is extremely time consuming.

The testing method is, presented in detail in this module, involves performing simple tests on the control system. From these tests, technicians obtain information that gives values for process characteristics. The technician then uses this data in equations to determine the settings for the controller. The settings that are covered in this module are for a proportional plus reset plus rate controller (PID), which is the most common controller in use.

1.3.1 Factors Affecting Stability

One of the primary points of interest in the overall performance of a control system is the response of the control loop and the process acting together. The term *response* is usually used to describe the dynamic behavior of the entire system after a disturbance that may be either supply or demand.

If the supply is uneven or subject to excessive fluctuations in flow or pressure, the control system may try to overcompensate, producing surges (also known as *swinging* or *oscillation*). Once the system is surging and out of balance, it may not be able to recover. In this case, the system must be shut down before the equipment is damaged. Once the uneven supply problem is corrected, the system can be restarted. The same is true for uneven demand on the system. If the demand fluctuates rapidly enough, it can upset the balance and cause the system to begin surging as it tries to meet the constantly changing demand.

The overall quality of the response is judged by the speed with which the controlled variable returns to the setpoint, by the amount of overshoot that occurs, and by the stability of the system during the transient condition.

The performances that describe the quality of response for a control system tend to counteract each other. Generally, an increase in the speed of the response decreases the stability of the system. An attempt at achieving too fast a speed of response may actually create a totally unstable system. A control system that is stable is one that is undergoing oscillations that are constant or decreasing in amplitude.

An unstable control system is one in which the oscillations are increasing in amplitude. These oscillations normally occur at a certain frequency called the critical frequency. This critical frequency is the one at which a 360-degree phase shift around the control loop exists. The value of the critical frequency depends upon the physical properties of the individual components within the control system. The component whose physical properties are the most difficult to determine or measure is the process itself. The 360-degree phase shift is critical because, under this condition, energy or material enters the process in such a way that the oscillations are maintained.

Another factor must exist besides the 360-degree phase shift to sustain the oscillations. This factor is the gain of the control loop. The gain of the control loop must be equal to 1 in order to sustain oscillations at the critical frequency. A control loop gain equal to one means that enough energy or material is entering the process to overcome losses. To sustain oscillations, not only must the energy or material enter the process at the right time, but enough must also enter so that the oscillations are constant in amplitude. The control loop gain is adjustable and the adjustment is normally made at the controller.

Tuning a controller is the process of adjusting it to match the process. The goal is to achieve a condition in which the control loop gain is less than 1 at the critical frequency. This produces a stable response of the control system (that is, oscillations die out) during a supply or demand disturbance. Generally, the response of a control system that is considered satisfactory is the minimum area criteria. The procedures for controller tuning that are presented in this module are designed to produce a one-quarter dampened wave following a supply or demand disturbance.

There are two general categories of loop-tuning methods: closed-loop methods, and open-loop methods. Each of these methods has several variations.

1.4.0 Closed-Loop Methods

Closed-loop methods require that the controller be set to automatic. The characteristics of the loop are determined from the closed-loop response of the system. The following are some common variations on the closed-loop method:

- Notch method
- Ultimate period method
- Dampened oscillation method

1.4.1 Notch Method

One of the simplest closed-loop methods of tuning a controller is the notch method. This method uses the frequency response of a controller to determine the point at which the loop gain is less than 1 at the critical frequency.

Some rather interesting and useful information can be obtained by plotting the gain of a three-mode controller against the frequency of the signals to which it must respond. A plot of this type is known as a frequency response curve. *Figure 1* shows the frequency response curve for a three-mode controller. Notice that the x-axis (the horizontal line) shows frequency, rather than time. The y-axis (the vertical line) represents the controller's gain. (The term *gain* describes the relationship between input and output in terms of energy lost or gained.)

The curve can be divided into three major sections: proportional (gain), reset (integral), and rate (derivative). These three sections make up the *notch.*

The gain in the middle portion of the curve is controlled by the proportional adjustment. Normally, this gain is the lowest of the three sections. The gain at the lower frequencies is controlled by the reset adjustment. This gain starts at the proportional value and increases with decreasing frequency, in a linear (direct and straight) manner, until a maximum point is reached. Reset gain then levels off at this point.

The maximum gain value is determined by the internal components of the controller. Usually, the values of the components that comprise the reset portion of the controller are the limiting items. High-frequency gain is controlled by the rate adjustment. This gain value also starts at the proportional level and rises linearly with increasing frequency until it is at a maximum point, and then levels off. The values of the components within the controller determine this maximum gain. Normally, these components are the ones that comprise the rate portion of the controller.

Adjusting the controller's gain (or proportional band) causes the frequency response curve to change, as shown in *Figure 2*. The curve shown by the upper dotted line results from changing the proportional adjustment. The adjustment that creates this curve is referred to as making the notch shallower. The notch is the area between the reset and rate gain lines. Making the notch shallower is the same as increasing the gain or narrowing the proportional band. Increasing the proportional band or decreasing the gain has the effect of deepening the notch; this is shown as the lower dashed line in *Figure 2*. Note that the width of the proportional portion of the curve is essentially unchanged; only the position of the curve is changed.

The width of the notch is set by the controller reset and rate adjustments. The effects of these adjustments are shown in *Figure 3*. The left side of the notch is controlled by the reset adjustment. Increasing the reset adjustment, which is increasing the number of repeats per minute (or decreasing the integral time), causes the reset portion of the curve to shift to the right. The reset gain therefore increases. A left shift of the curve is a decrease in the reset gain. Both of these adjustments are shown as dotted lines on the left side of *Figure 3*. Adjustment of the rate has similar effects. An increase in rate time produces an increase in rate gain. The increase in rate gain is shown as a movement of the rate portion of the curve to the left; movement to the right is a decrease in rate

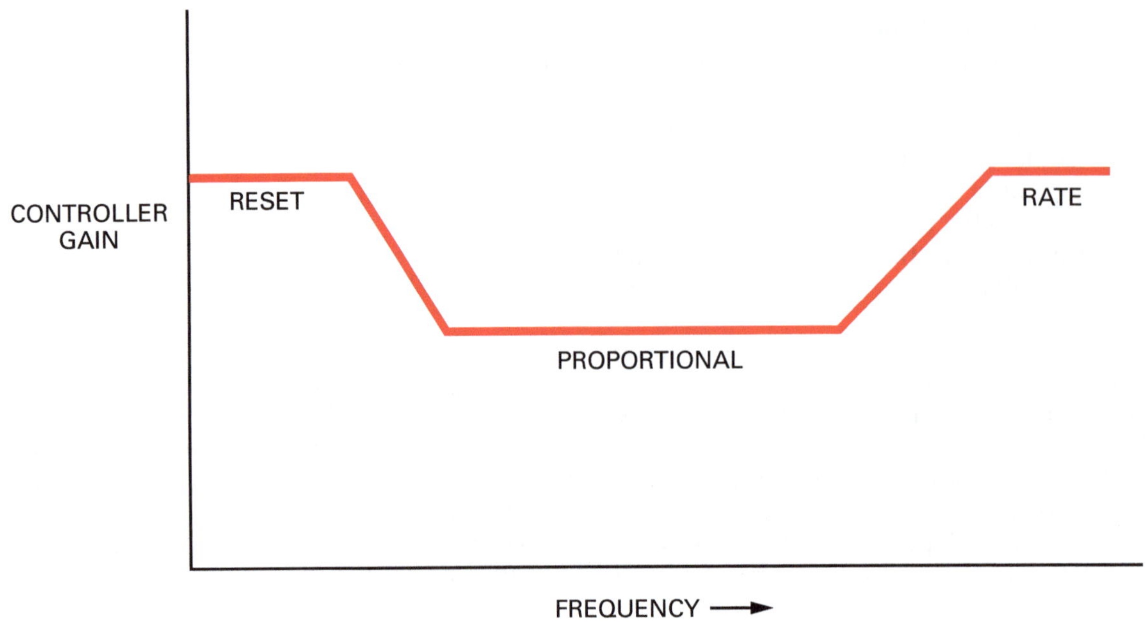

Figure 1 Frequency response curve for a three-mode controller.

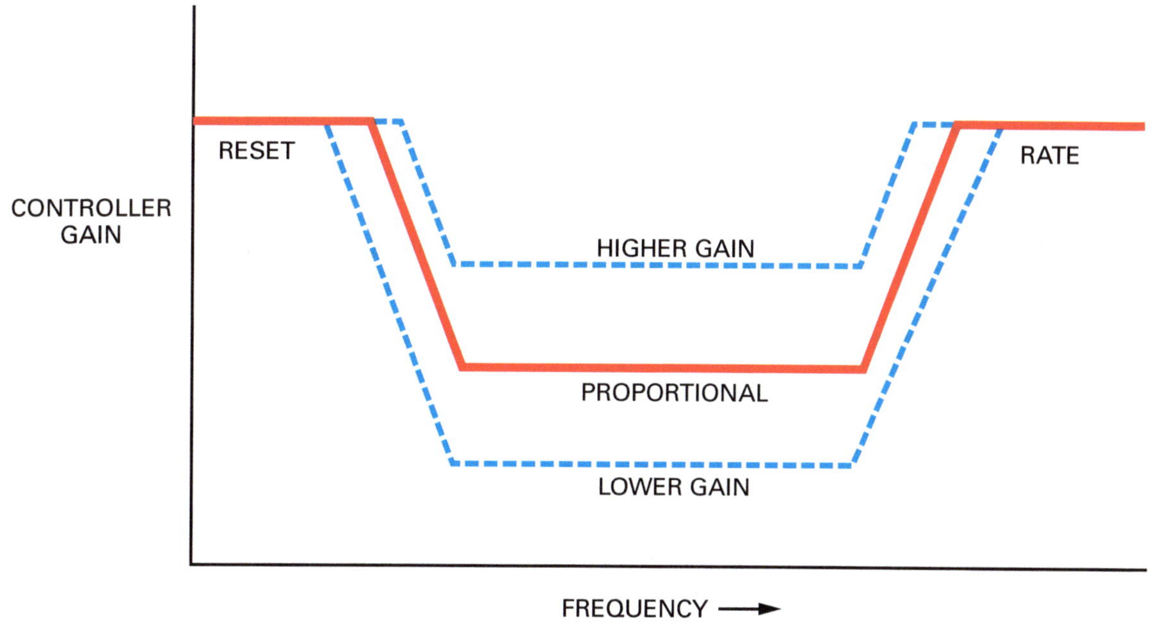

Figure 2 Adjustment of controller gain for a three-mode controller.

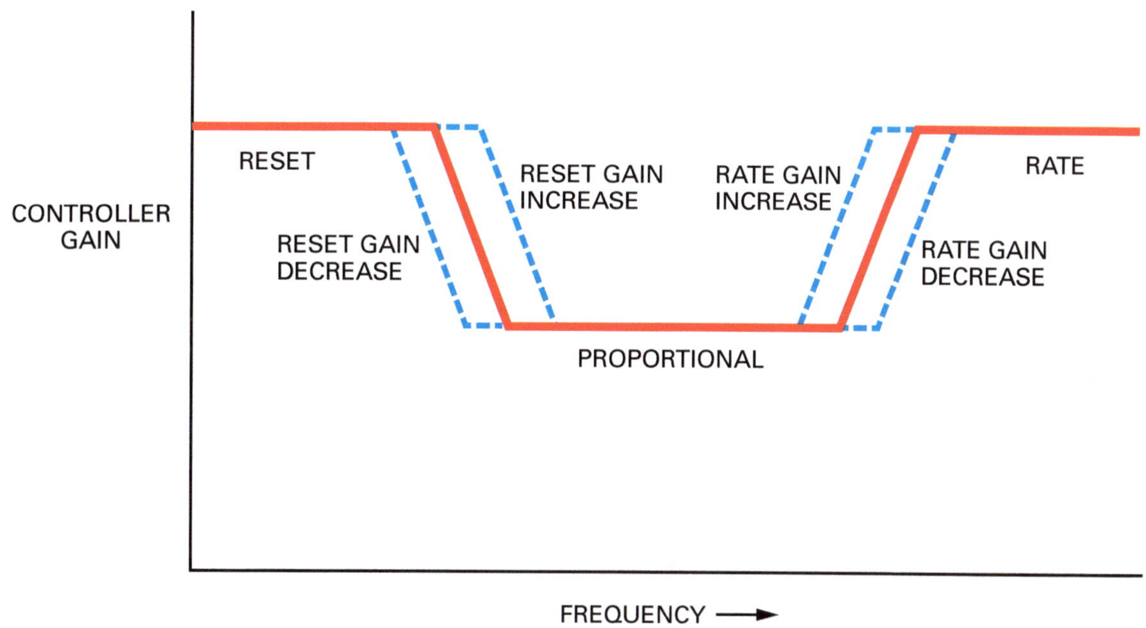

Figure 3 Effects of reset and rate adjustments.

gain. *Figure 3* shows these adjustments as dotted lines on the right side of the figure.

The examples of response curves shown in *Figure 1*, *Figure 2*, and *Figure 3* illustrate that the size and shape of the notch can be changed by the adjustments of the controller. The depth of the notch is controlled by the gain (proportional) adjustment, and the width of the notch is manipulated on the left by the reset adjustment and on the right by the rate adjustment.

Matching the controller to the process is now a matter of having the loop gain less than 1 at the critical frequency. If the gain of the controller is sufficiently less than 1, then the loop gain is less than 1. The notch method of tuning strives to place the critical frequency within the notch. With the critical frequency within the notch, the gain of the controller is reduced to ensure stability.

The notch tuning procedure is normally performed with the controller in service. *Figure 4* shows a block diagram of a control loop upon which the notch tuning method is being employed. This block diagram has been modified by the addition of a recorder. The recorder indicates the value of the setpoint and the measured

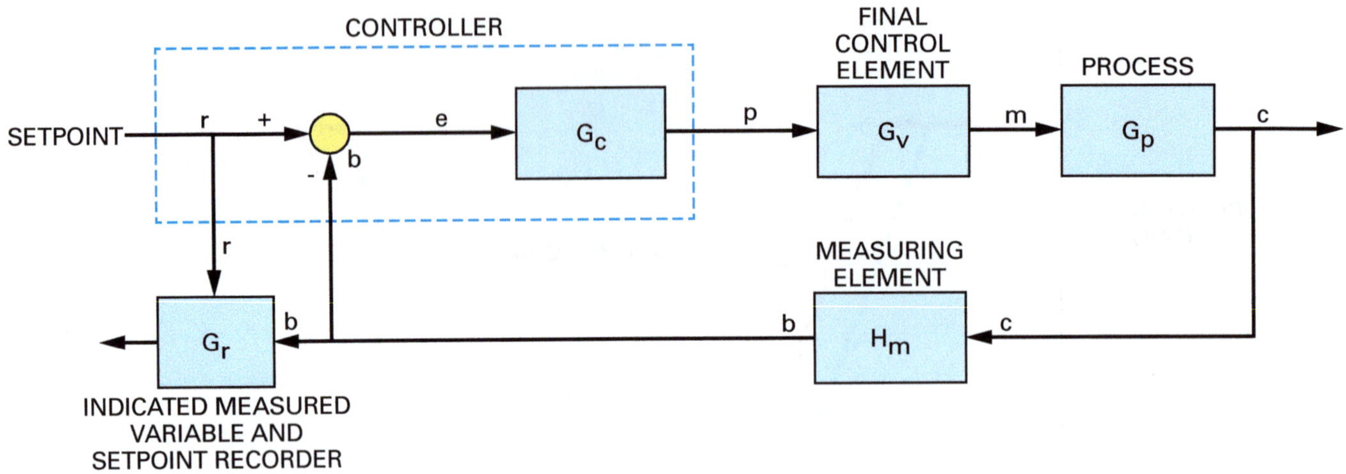

Figure 4 Block diagram of a process being notch tuned.

variable. The required information needed for the tuning process is obtained from the recorder.

The initial step is to broaden the notch. This is accomplished by decreasing the reset and rate gains with the controller in manual. Set the reset adjustment to either minimum repeats per minute or maximum integral time. Set the rate adjustment to minimum. Placing the reset and rate controls to their minimum values reduces their respective gains. This widens the notch and ensures that the critical frequency is somewhere within the notch.

The next step is to deepen the notch. This is done by decreasing the gain or widening the proportional band. With the notch widened, the gain of the controller is sufficiently low to prevent the process from oscillating. The controller gain that causes oscillation is referred to as *gain limit. Figure 5* shows the effect of widening and deepening the notch. The critical frequency is drawn in the center of the notch. Knowing the exact location of the critical frequency within the notch is not as important as ensuring that the critical frequency is not outside of the notch.

After establishing the initial conditions, place the controller in automatic. The next step is to narrow the proportional band, or increase the gain. Then, introduce a small disturbance into the control loop. This disturbance is normally a supply disturbance and is introduced by changing the setpoint. After introducing the disturbance, monitor the response of the measured variable. Initially, the response of the system is sluggish, but as the proportional band is made narrower, the system oscillates more. Repeat this narrowing of the proportional band and disturbing of the system until the system continues to cycle after a disturbance; the bottom of the notch will have been raised to the limiting value of the gain. *Figure 6* shows a process in this condition. The gain of the controller has been raised high enough so that the loop gain of the control system is equal to 1. This is the gain necessary to sustain oscillations.

Once this condition is reached, double the proportional band (reduce the gain by half). This change results in the system once again becoming stable. The outcome of this adjustment is shown in *Figure 7*. The gain of the controller is now sufficiently low that the loop gain is less than 1.

Once the proportional band has been adjusted, the next step is to adjust the gain of the reset portion of the curve. Again, increase the number of repeats per minute or decrease the integral time. This adjustment increases the reset gain portion of the response curve. After making the adjustment, introduce a small disturbance into the process. In this example, a setpoint change is used. Monitor the value of the measured variable for its response. As before, the process will respond sluggishly. As the number of repeats per minute increase, the process begins to oscillate. Continue increasing the repeats per minute followed by a disturbance until continuous oscillation of the measured variable is again achieved. During this procedure, the notch is being narrowed.

Figure 8 shows the result of the notch being narrowed until the reset gain is at the limiting value for the gain. The controller gain is again large enough to provide a loop gain of 1. Once this condition is reached, reduce the number of repeats per minute to a value that is one third of the one that placed the reset gain portion of the curve at the gain limit. This adjustment makes the system stable. The reset portion of the response curve is now sufficiently to the left of the gain limit so that the gain of the controller is low enough to ensure that the loop gain is less than one at the critical frequency. The control system is now in a stable condition. The result of this adjustment is shown

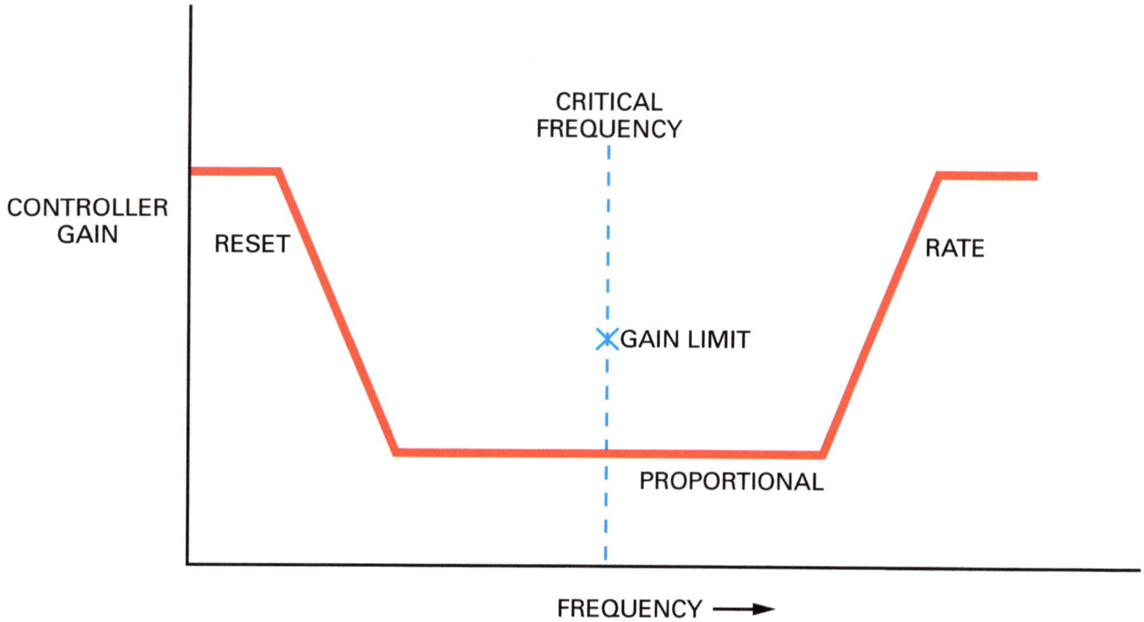

Figure 5 Controller with a wide and deep notch.

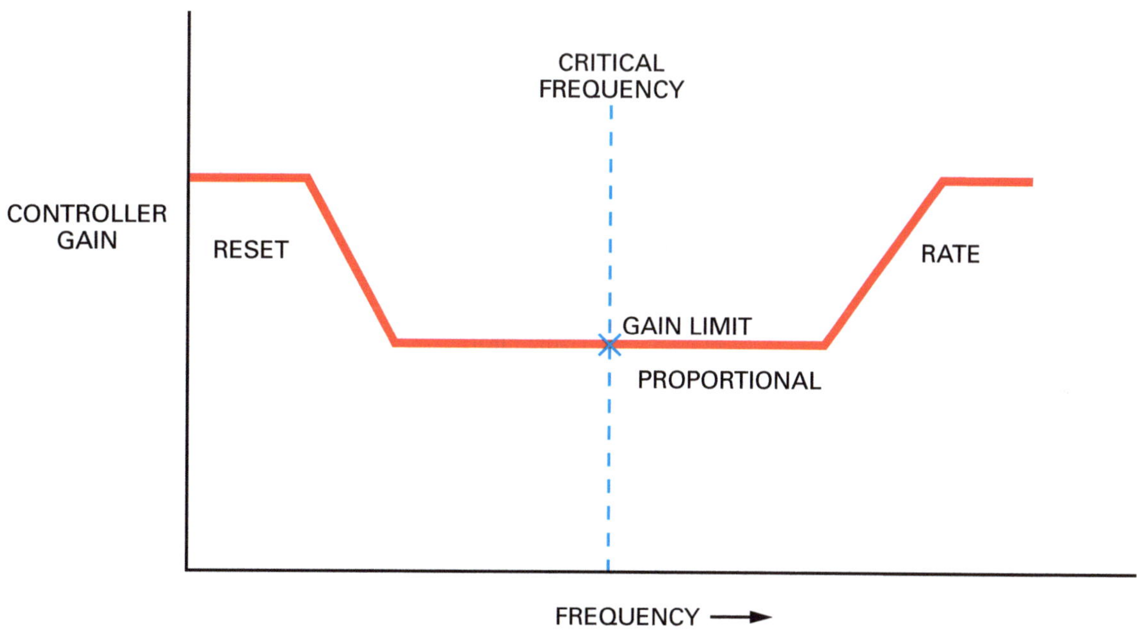

Figure 6 Controller gain at the gain limit due to proportional adjustment.

in *Figure 9*. The reset gain portion of the response curve is now closer to the critical frequency than the rate portion of the response curve is.

After completing the adjustment of the proportional band and the repeats per minute, the next step is to adjust the rate. This time, increase the rate time. This action increases the rate gain portion of the response curve. Introduce a small disturbance, such as a setpoint change, into the process. The recorder monitors the effect of the disturbance on the measured variable. As before, the process responds sluggishly. However, as the rate time increases, the response of the process becomes more oscillatory. Continue to increase the rate time followed by the introduction of a disturbance until the measured variable breaks into continuous oscillation. As this part of the procedure is being performed, the notch is being narrowed. *Figure 10* shows the effect of the notch being narrowed until the rate gain is at the limiting value of the gain. The gain of the controller is now large enough to establish a loop gain of 1.

After this condition is reached, the rate time is reduced to a value that is one third of the value that placed the rate gain portion of the curve at the gain limit. This adjustment makes

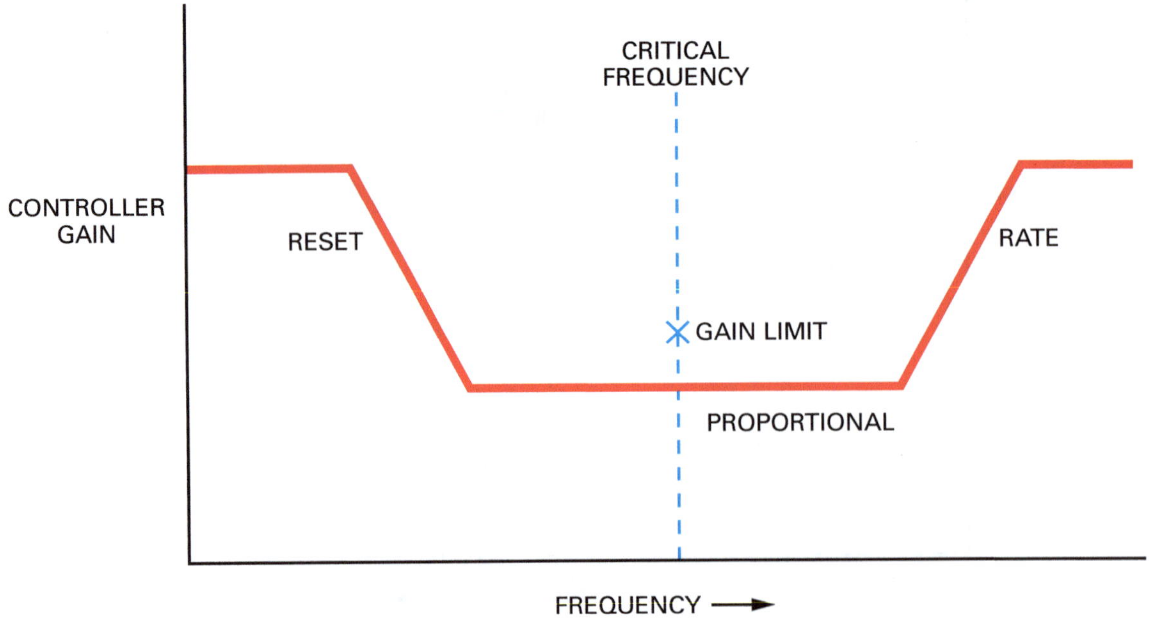

Figure 7 Controller gain after final adjustment of proportional band.

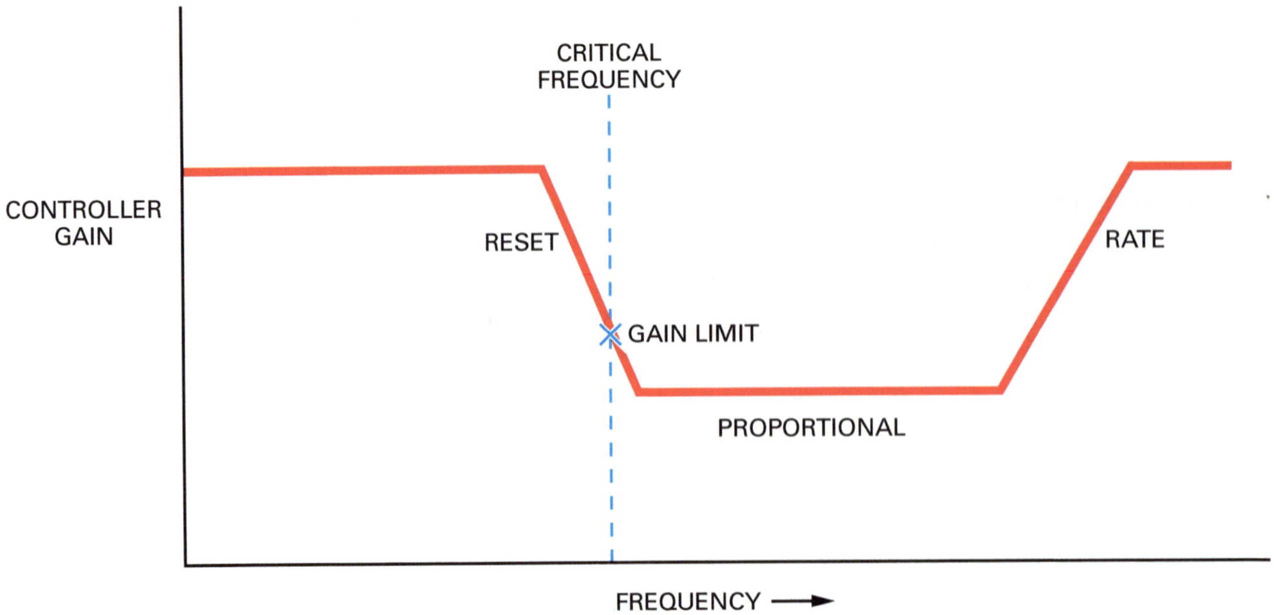

Figure 8 Controller gain at the gain limit due to the reset adjustment.

the system stable again. The rate portion of the response curve has now been moved far enough to the right of the gain limit that the gain of the controller is low enough at the critical frequency to ensure that the loop gain is less than one. *Figure 11* shows the position of the rate portion of the response curve after the rate adjustment is completed.

The final result is that the critical frequency is directly in the middle of the frequency response curve. Due to the proportional part of the response curve, the gain of the controller is low enough to ensure that the loop gain is less than 1 at this point. In order to improve performance, the loop gain should be higher both below and above the critical frequency. This is accomplished by narrowing the notch as far as possible while still maintaining stability. The narrowing of the notch is achieved by the reset and rate portions of the response curve. Upon completion of the notch tuning method, the response of the process to a disturbance appears as a one-quarter dampened wave in the value of the measured variable.

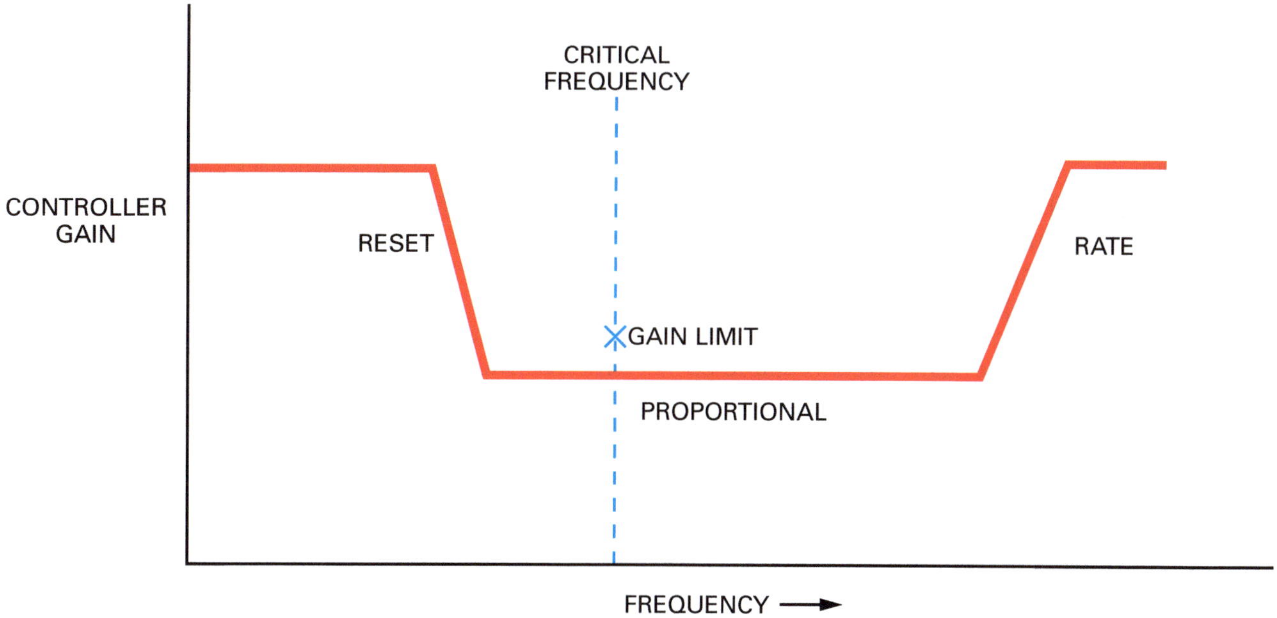

Figure 9 Controller gain after final setting of the reset adjustment.

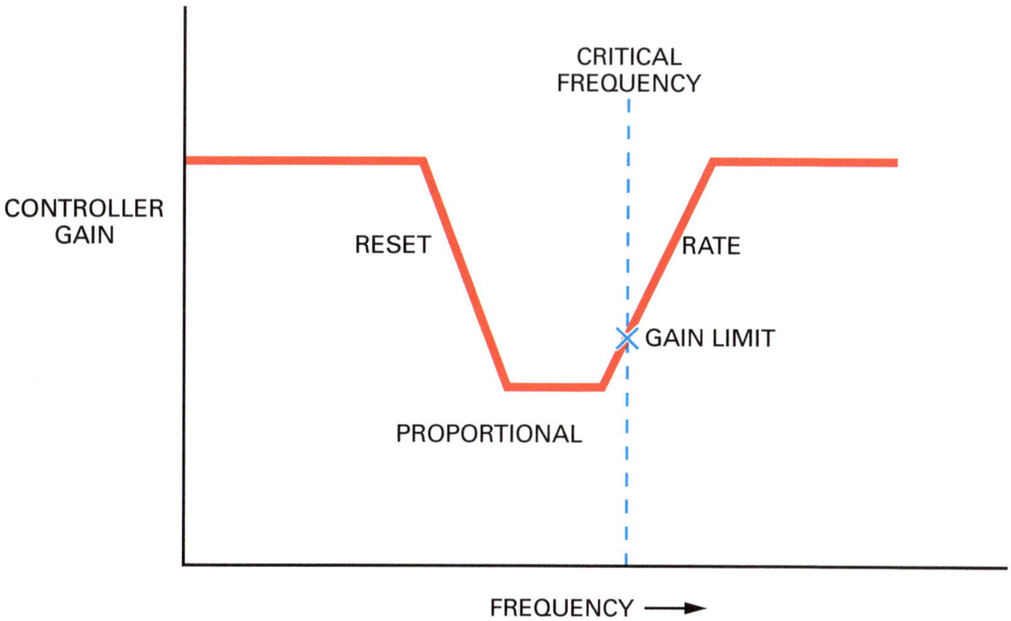

Figure 10 Controller gain at the gain limit due to the rate adjustment.

One of the main disadvantages of the notch tuning method is that the process must be placed into continuous oscillation three times. Some processes that cannot tolerate being placed into continuous oscillation this many times. Another closed-loop tuning method is available that only requires placing the process into continuous oscillation once. This is the ultimate period method (described in the following sections).

1.4.2 Ultimate Period Method

The most common methods of loop tuning in use today were developed in the early 1940s by J.G. Ziegler and N.B. Nichols. The Z-N method, as it is sometimes called, is also the basis of many variations, including the ultimate period method. The term *ultimate* is used with this method because it requires the determination of the **ultimate gain** (or ultimate proportional band) and the **ultimate period**.

The ultimate gain, K_u (or ultimate proportional band, PB_u) is the maximum allowable value of gain or proportional band at which the system is stable. The ultimate period, P_u, is the period of response during which either the gain or proportional band is set at the ultimate value. During the determination of the ultimate gain or proportional band and the ultimate period, the controller is operated as a purely proportional controller. *Figure 12* shows the response of a control system that is at the ultimate gain or proportional band. As shown in *Figure 12*, the controlled variable oscillates at a constant frequency; this is the critical frequency.

A 360-degree phase shift exists. The amplitude of the oscillations is constant. The gain of the controller is sufficient to produce a loop gain of 1. The controller gain is the ultimate gain or proportional band. *Figure 12* also shows the ultimate period. The ultimate period in this case is the time period that runs from when the controlled variable is at its maximum value to when it is at its maximum value again.

The use of these two pieces of information (the ultimate gain and the ultimate period) is the basis for the ultimate period method of controller tuning. The values of the ultimate gain or proportional band and ultimate period are used in mathematical formulas to determine the controller settings. However, a method is first needed to obtain the necessary information from the control system.

As with the notch tuning method, technicians perform the ultimate period method with the controller in service. An arrangement such as the block diagram shown earlier in *Figure 4* is still used. Technicians obtain the required information for the tuning of the controller by observing the waveform on the recorder.

The first step is to set the controller for proportional control action only. This is needed so that only one of the three possible control actions will cause the system to break into continuous oscillations. The formulas are based only on proportional action causing the oscillations. To establish proportional-only control set the reset adjustment either to minimum repeats per minute or to maximum integral time. Set the rate time to minimum, and set the gain of the controller at a low value. (This is the same as a wide proportional band.) Make these adjustments with the controller in manual.

Then, after placing the controller in automatic, introduce a disturbance. The easiest way is to introduce a supply disturbance. This is normally done by causing a change in the setpoint. Observe the effect of the supply disturbance upon the measured variable as shown on the recorder.

Generally, the response to the disturbance is similar to the one shown in *Figure 13.* As shown in the figure, the oscillations tend to decrease as time continues. This is demonstrated by Curve A. However, in some cases, the measured variable responds as indicated by curve B. This curve represents an unstable response. If this response is obtained, the gain has to be decreased or the proportional band widened.

A response such as the one shown in Curve A is required as the response after the initial disturbance. Continue increasing the gain or narrowing

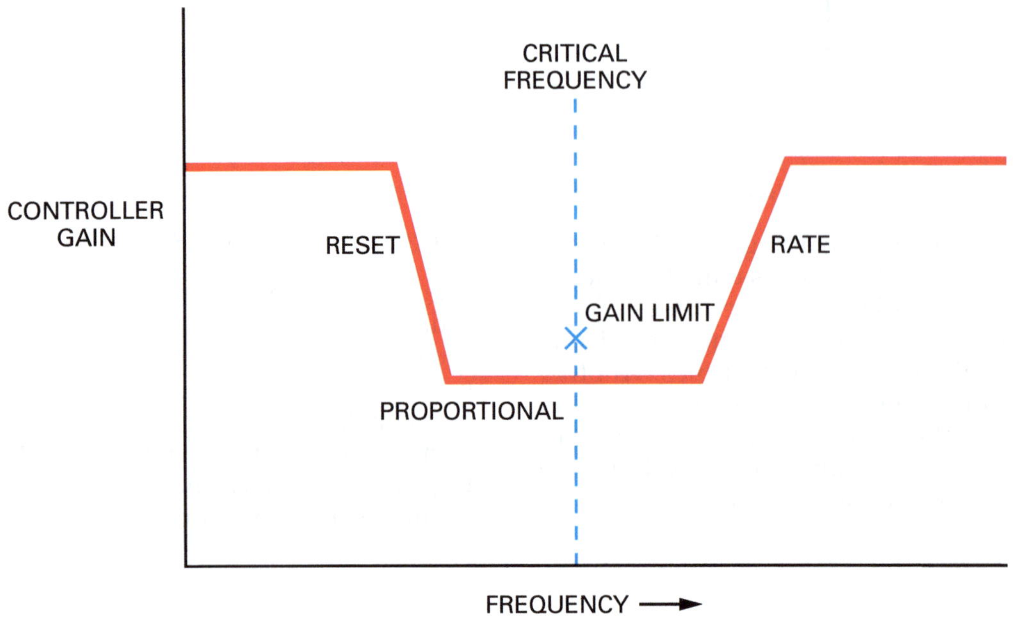

Figure 11 Controller gain after the final setting of the rate adjustment.

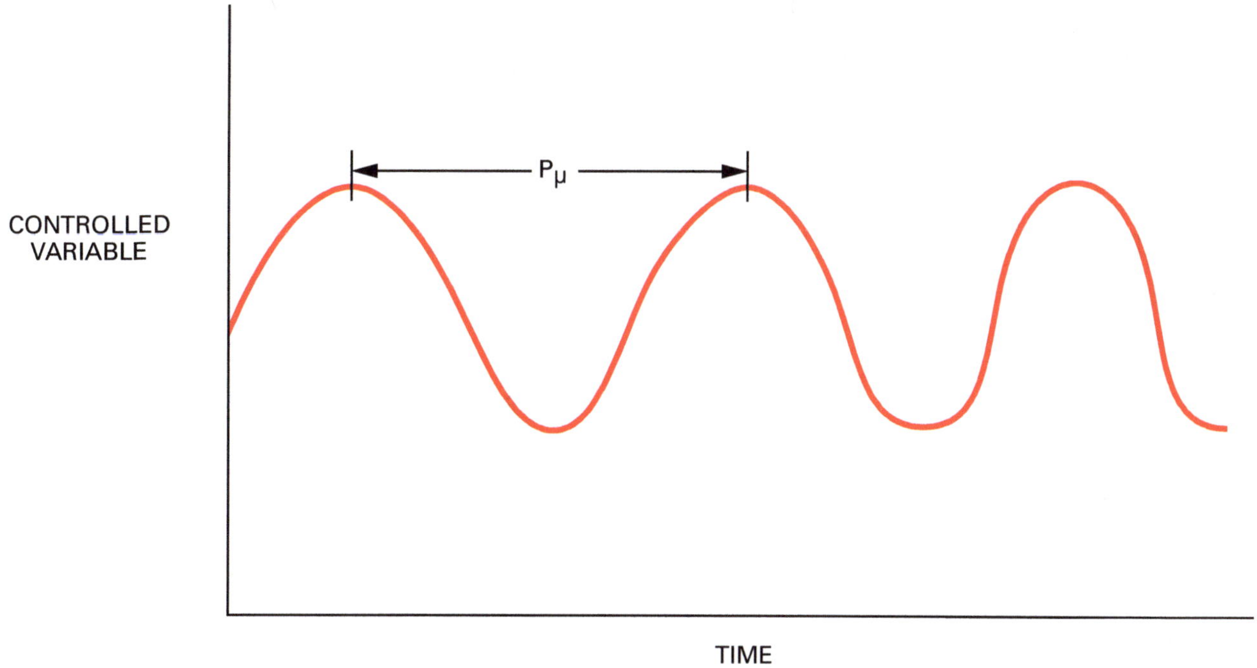

Figure 12 Control system at the ultimate gain or proportional band.

the proportional band and inserting a supply disturbance until continuous constant amplitude oscillations are achieved, as shown in *Figure 12*. Between each adjustment and disturbance, it may be necessary to stabilize the process. The setting of the gain or proportional band adjustment is recorded. This is the ultimate gain or proportional band. The ultimate period is measured from the information contained on the recorder.

With the ultimate gain or proportional band and the ultimate period, use the following equations to determine the settings that provide a one-quarter dampened wave of the measured variable:

Proportional (P):

$K_p = 0.5 \times K_u$ $\quad$ $PB = 2 \times PB_u$

Where:

K_p = gain
K_u= ultimate gain
PB = proportional band
PB_u = ultimate proportional band

Proportional and reset (PI):

$K_p = 0.45 \times K_u$ $\quad$ $PB = 2.2 \times PB_u$

$T_i = P_u \div 1.2$ $\quad$ $rpm = 1.2 \div P_u$

Where:

T_i = reset
rpm = repeats per minute
P_u= ultimate period

Proportional and rate (PD):

$K_p = 0.6 \times K_u$ $\quad$ $PB = 1.66 \times PB_u$

$T_d = P_u \div 8$

Where:

T_d = rate

Proportional, reset, and rate (PID):

$K_p = 0.6 \times K_u$ $\quad$ $PB = 1.66 \times PB_u$

$T_i = P_u \div 2$ $\quad$ $rpm = 2 \div P_u$

$T_d = P_u \div 8$

Bear in mind that these equations are empirical. That is, they are based on data that were obtained by observation only. Consequently, using these equations alone to tune a control system gives only a very close approximation of a one-quarter dampened wave for the measured variable. Technicians must do a certain degree of trimming after the controller has been set by these equations to obtain the one-quarter dampened wave exactly.

As an example, consider the test results shown in *Figure 14*. As shown in the figure, the sustained continuous oscillation occurs for this process when the gain of the controller is 4. This is the same as a proportional band of 25 percent ($\frac{1}{4}$).

A measurement of the time between maximum values of the controlled variable gives an ultimate period of 16 minutes.

Using the appropriate equations produces the following settings (equations to be used are presented in **bold**):

Proportional (P):

$\mathbf{K_p = 0.5 \times K_u}$
$K_p = 0.5 \times 4$
$K_p = 2$

$\mathbf{PB = 2 \times PB_u}$
$PB = 2 \times (25\%)$
$PB = 50\%$

Proportional and reset (PI):

$\mathbf{K_p = 0.45 \times K_u}$
$K_p = 0.45 \times 4$
$K_p = 1.8$

$\mathbf{PB = 2.2 \times PB_u}$
$PB = 2.2 \times (25\%)$
$PB = 55\%$

$\mathbf{T_i = P_u \div 1.2}$
$T_i = 16 \text{ min} \div 1.2$
$T_i = 13.3 \text{ min}$

$\mathbf{rpm = 1.2 \div P_u}$
$rpm = 1.2 \div 16 \text{ min}$
$rpm = 0.075 \text{ repeats/min}$

Proportional and rate (PD):

$\mathbf{K_p = 0.6 \times K_u}$
$K_p = 0.6 \times 4$
$K_p = 2.4$

$\mathbf{PB = 1.66 \times PB_u}$
$PB = 1.66 \times (25\%)$
$PB = 41.5\%$

$\mathbf{T_d = P_u \div 8}$
$T_d = 16 \text{ min} \div 8$
$T_d = 2 \text{ min}$

Proportional, reset, and rate (PID):

$\mathbf{K_p = 0.6 \times K_u}$
$K_p = 0.6 \times 4$
$K_p = 2.4$

$\mathbf{PB = 1.66 \times PB_u}$
$PB = 1.66 \times (25\%)$
$PB = 41.5\%$

$\mathbf{T_i = P_u \div 2}$
$T_i = 16 \text{ min} \div 2$
$T_i = 8 \text{ min}$

$\mathbf{rpm = 2 \div P_u}$
$rpm = 2 \div 16 \text{ min}$
$rpm = 0.125 \text{ repeats/min}$

$\mathbf{T_d = P_u \div 8}$
$T_d = 16 \text{ min} \div 8$
$T_d = 2 \text{ min}$

As noted earlier, these settings give a close approximation of a one-quarter dampened wave. Further training may be necessary for technicians to achieve an exact one-quarter dampened wave.

1.4.3 Dampened Oscillation Method

A modification of the ultimate period method is the dampened oscillation method. This method was developed for those processes that cannot tolerate continuous oscillations. In this method, technicians use steps that are similar to those used in the ultimate period method to adjust the gain or proportional band until they obtain a response curve with a one-quarter decay ratio. This response is shown in *Figure 15.*

Technicians obtain this response by using only the proportional mode of the controller. They adjust the reset and rate portions to minimum values.

The data obtained by this method is only the period of the oscillation. Knowing this, use the following equations to set the reset and rate adjustments:

Reset (I):

$T_i = P \div 1.5$ $\qquad$ $rpm = 1.5 \div P$

Rate (D):

$T_d = P \div 8$

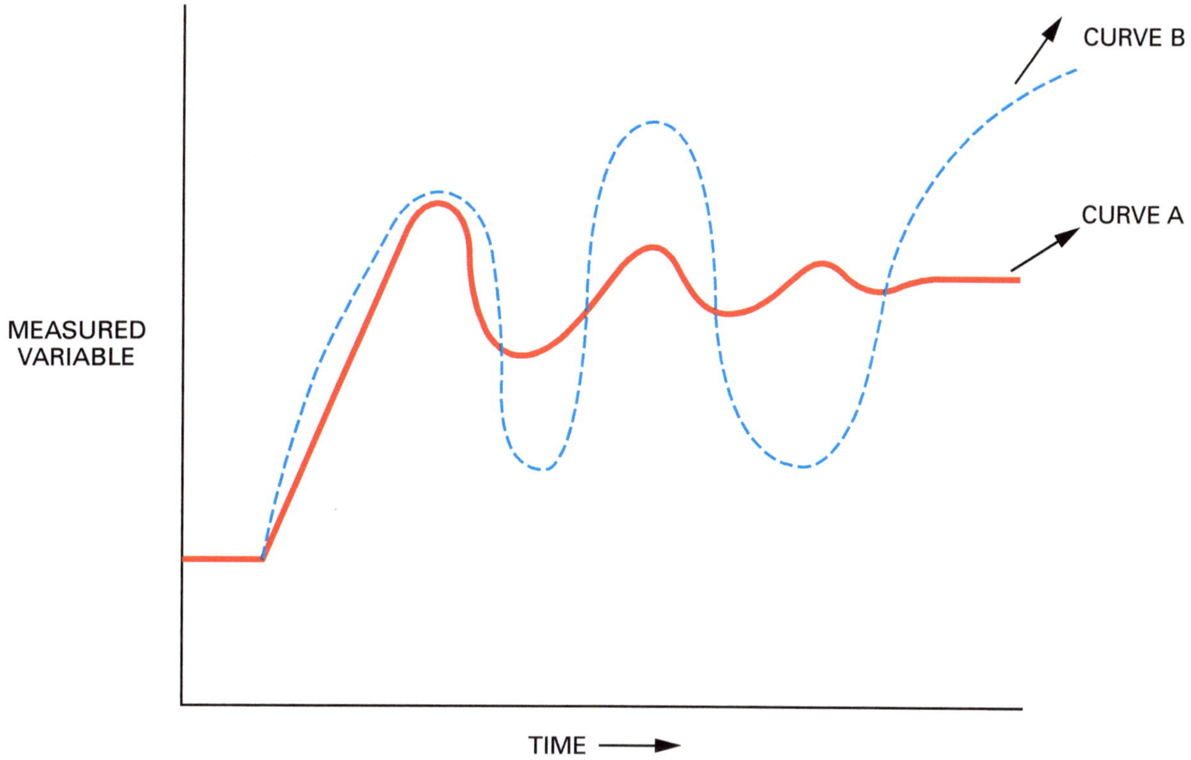

Figure 13 Responses of the measured variable to a supply disturbance.

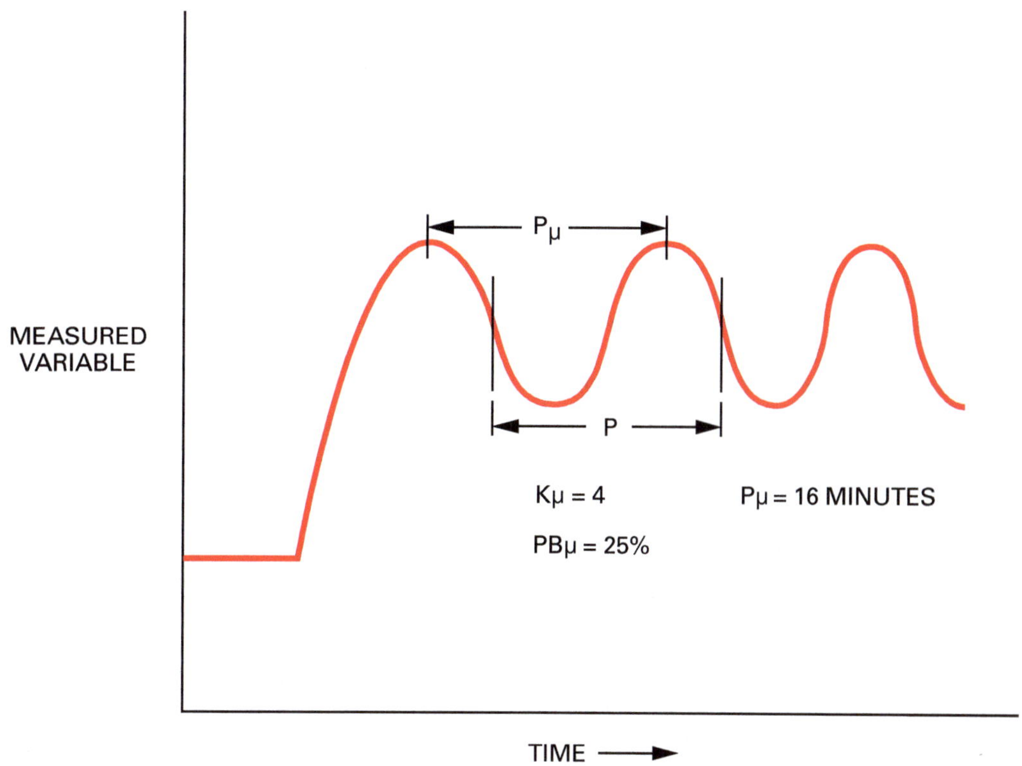

Figure 14 Test data for ultimate tuning method.

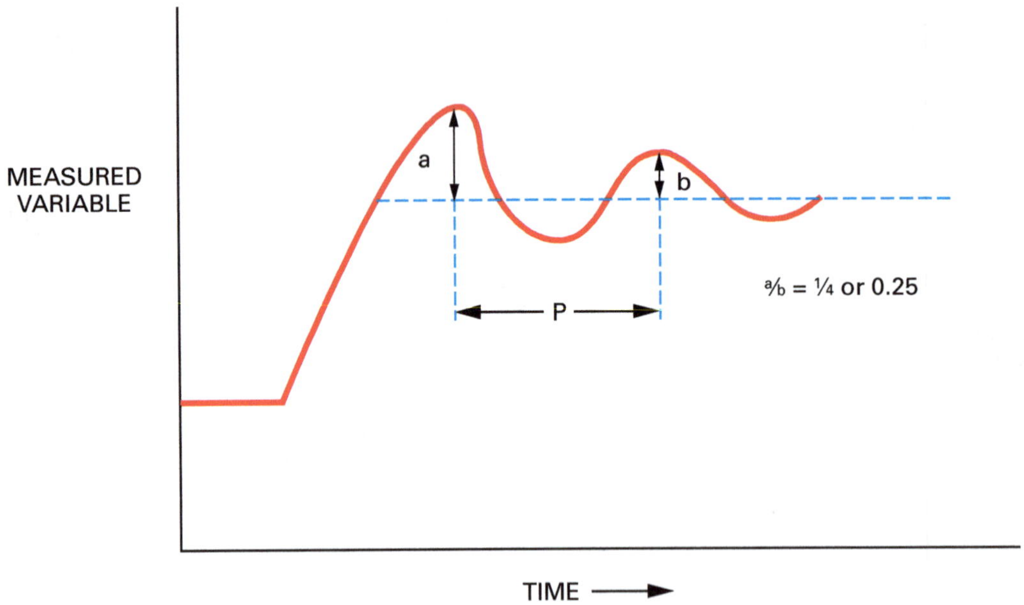

Figure 15 Test data for dampened oscillation method.

Once the reset and rate adjustments are made, readjust the gain or proportional band to provide a one-quarter dampened wave response of the measured variable.

1.5.0 Open-Loop Methods

In contrast to closed-loop methods, open-loop techniques require that only one disturbance be imposed upon the process. In fact, the controller is not in the loop at all during the testing of the process. Open-loop methods seek to determine the characteristics of the process.

Technicians obtain the proper values for the controller settings from the process characteristics. In general, most processes are multiple capacity. So, obtaining the controller settings requires the use of an approximation. The most common approximation is called the *single capacity with deadtime approximation*.

Technicians obtain the process information needed for open-loop tuning methods from the process reaction curve. The generation of this curve requires introducing a disturbance into the control system (typically, a supply disturbance). The block diagram for a process that is being tuned by the open-loop method is shown in*Figure 16*.

The block diagram in *Figure 16* does not show much difference between this arrangement and the one for closed-loop tuning; the major difference is that the controller is in manual mode (and not automatic). With this condition, the control loop is essentially broken. This is because the automatic signal does not reach the final control element. The system cannot respond to changes in the controlled variable when in this configuration. The other difference is that, other than indicating the measured variable, the recorder indicates controller output rather than setpoint. Technicians can use an arrangement of this type to obtain the process reaction curve for a process.

To determine the process reaction curve, bring the process to a steady-state level. Ideally, the steady-state level should be one where the controlled variable is at the setpoint. Then place the controller in manual. Introduce a supply disturbance by changing the manual output of the controller. The response of the process to the supply disturbance is recorded by the recorder that receives the measured variable signal. The output of the controller is also recorded to provide a reference for when the disturbance was initiated. After the process reaction curve has been recorded, manually return the controlled variable to the setpoint. It is easier to obtain a process reaction curve than it is to determine the ultimate gain.

Figure 17 shows a typical process reaction curve for a process that has been open-loop tested. The curve is one for a multiple-capacity process. This curve is approximated, as shown in *Figure 17.*

Two methods are employed in extracting the information provided by the approximated curve. The methods only differ in the way that the time constant is determined.

In open-loop methods, the process reactive curve is used to determine the loop characteristics. The controller is not required for open-loop tuning, because only one disturbance is imposed

Wireless Process Control

Wireless transmitters and actuators are increasingly being used in the process industry not only to monitor processes but also for control. The challenge has been devising a way to incorporate wireless measurements into closed-loop control.

Because most wireless transmitters are battery-powered, they are designed to minimize power consumption by transmitting a new measurement infrequently and/or only if the measurement has changed significantly. Consequently, wireless control relies on slower, non-periodic measurement updates and communication interruptions that would not be acceptable in control loops using traditional, wired transmitters. To effectively use a wireless measurement for closed-loop control, the feedback control must be designed to work with slow periodic and/or non-periodic measurement updates as well as communication interruptions. To accomplish this, wireless devices are available that employ standard communication protocols such as WirelessHART® and a modified form of PID control.

The modified PID control can employ either of two communication techniques for implementing control using a wireless transmitter: the continuous (periodic) update technique and the window technique. With continuous (periodic) update, the transmitter wakes up at a configured update period, senses the measurement, and then communicates the value. For the window technique, the transmitter wakes up at a configured update period, senses the measurement, and then communicates the measurement if the change in value since the last communication exceeds a specified amount. Window communication is generally the preferred technique because it requires less power for the same update period, in comparison to the continuous (periodic) update technique.

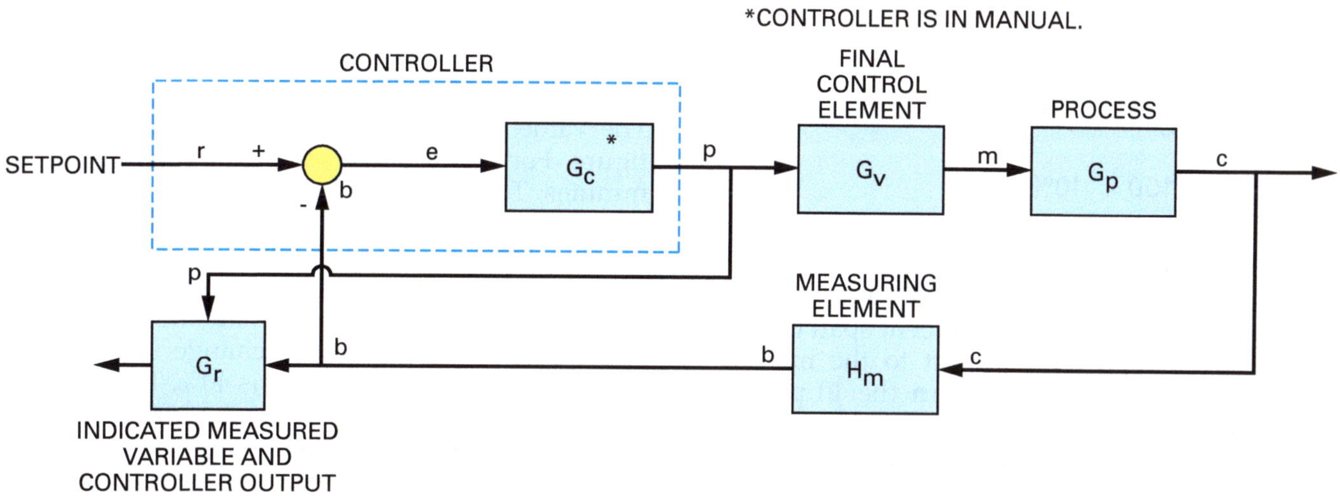

Figure 16 Block diagram of a process being open-loop tuned.

on the process. The more common variations on the open-loop method, as explained in the following sections, are:

- Time constant method
- Reaction rate method

1.5.1 Time Constant Method

The time constant method is shown in *Figure 18.* The information required from the process reaction curve is the time constant, process gain, and deadtime. The time constant is the time that it takes the process to reach 63.2 percent of its total change. The process gain is the change in the process output divided by the change in the process input. The change in output is the change that results in the measured variable.

This change must be converted to percent of span. The change as a percentage of the span is more accurate than the units of the measured variable because. For example, a change of 20°C for a 100°C span is greater (more drastic) than a change of 20°C for a 200°C span. Therefore, a

20°C change in the measured variable does not indicate the magnitude of the change as well as the percentage of the span does.

The exact percent of span can be calculated using the following equation:

$$\frac{\text{changed value} - \text{initial value}}{\text{span}} \times 100$$

Using the equation, calculate a temperature increase from 30°C to 50°C for a span of 100°C as a percentage:

$$\frac{\text{changed value} - \text{initial value}}{\text{span}} \times 100$$

$$\frac{50°C - 30°C}{100°C} \times 100 =$$

$$\frac{20}{100} \times 100 = 20\%$$

Now, determine the same change (from 30°C to 50°C) for a span of 200°C:

$$\frac{\text{changed value} - \text{initial value}}{\text{span}} \times 100$$

$$\frac{50°C - 30°C}{200°C} \times 100 =$$

$$\frac{20}{200} \times 100 = 10\%$$

A 20°C change for a 100°C span is 20 percent of span, whereas a 20°C change for a 200°C span is 10 percent of span. The 20 percent span change is a larger change with respect to the measuring span of the instrument than the 10 percent change, even though both changes are 20°C. The process input change is equal to the percent of the span change in the controller's output. (Change in controller output is often abbreviated ΔCO.) Using the percent of span requirement for the changes, the process gain may be calculated as follows:

$$K = \frac{\text{change in measured variable (\%)}}{\Delta CO\ (\%)}$$

Where:

K = process gain

The deadtime is determined from the curve. With this information, use the following equations to tune the controller for a one-quarter dampened wave response of the measured variable:

Proportional (P):

$K_p = t \div (t_d \times K)$ PB = 100% × K_p

Proportional and reset (PI):

$K_p = 0.9t \div (t_d \times K)$ PB = 100% × K_p

$T_i = 3.33 \times t_d$ rpm = $1 \div 3.33t_d$

Proportional, reset, and rate (PID):

$K_p = 1.2t \div (t_d \times K)$ PB = 100% ÷ K_p

$T_i = 2.0 \times t_d$ rpm = $1 \div 2.0t_d$

$T_d = 0.5t_d$

As an example, consider the results that are shown in *Figure 19*. As indicated in the figure, the total change in the measured variable is 40°C. The controller output (Δ INPUT on the graph) has a range of 20 to 100 kPa (an 80 kPa span). The change in the controller output is 7 kPa (62 – 55 = 7). To determine the time constant, the value of the measured variable at 63.2 percent of the total change needs to be known. Since the total change is 40°C, 63.2 percent of this change is 25.3°C. The 63.2 percent value of the measured variable occurs when the temperature reaches 65.3°C. This value is plotted on the vertical axis in *Figure 19*. The value of the time constant is read from the figure. For this example, the time constant is 1.6 minutes. The deadtime can also be read from the figure.

From the figure, the deadtime is 0.2 minutes. The only piece of information remaining is the process gain. Since, as noted on the graph, the span is 100°C and the total change is 40°C, the change in the percent of span is 40%. The controller output has changed 7 kPa of its 80 kPa span, or 8.75 percent of the span. Using the following equation, the process gain can be determined:

$$K = \frac{\text{change in measured variable (\%)}}{\Delta CO\ (\%)}$$

$$K = \frac{40\% \text{ of span}}{8.75\% \text{ of span}}$$

$$K = 4.57$$

Using Ziegler-Nichols equations along with the information from the process reaction curve, the following settings can be determined (the equations are presented in **bold**):

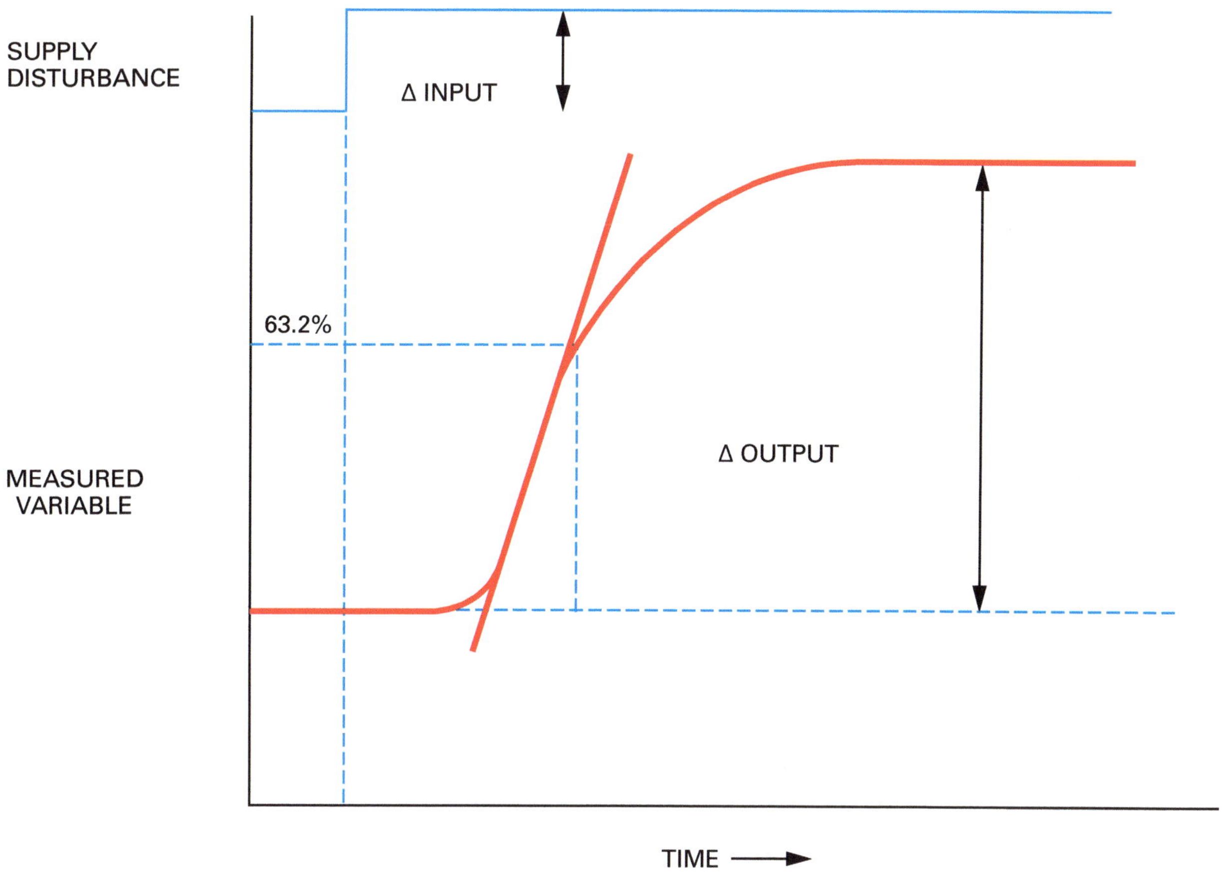

Figure 17 Process reaction curve for an open-loop-tested process.

Proportional (P):

$\mathbf{K_p = t \div (t_d \times K)}$
$K_p = 1.6 \text{ min} \div (0.2 \text{ min} \times 4.57)$
$K_p = 1.6 \text{ min} \div 0.914$
$K_p = 1.75$

$\mathbf{PB = 100\% \div K_p}$
$PB = 100\% \div 1.75$
$PB = 57\%$

Where:

t_d = deadtime

Proportional and reset (PI):

$\mathbf{K_p = (0.9 \times t) \div (t_d \times K)}$
$K_p = (0.9 \times 1.6 \text{ min}) \div (0.2 \text{ min} \times 4.57)$
$K_p = 1.44 \div 0.91$
$K_p = 1.58$

$\mathbf{PB = 100\% \div K_p}$
$PB = 100\% \div 1.58$
$PB = 63.5\%$

$\mathbf{T_i = 3.33 \times t_d}$
$T_i = 3.33 \times 0.2 \text{ min}$
$T_i = 0.666 \text{ min}$

$\mathbf{rpm = 1 \div (3.33 \times t_d)}$
$rpm = 1 \div (3.33 \times 0.2)$
$rpm = 1 \div 0.666$
$rpm = 1.5$ repeats/min

Proportional, reset, and rate (PID):

$\mathbf{K_p = (1.2 \times t) \div (t_d \times K)}$
$K_p = (1.2 \times 1.6 \text{ min}) \div (0.2 \text{ min} \times 4.57)$
$K_p = 1.92 \div 0.914$
$K_p = 2.10$

$\mathbf{PB = 100\% \div K_p}$
$PB = 100\% \div 2.10$
$PB = 47.6\%$

$\mathbf{T_i = 2.0 \times t_d}$
$T_i = 2.0 \times 0.2 \text{ min}$
$T_i = 0.4 \text{ min}$

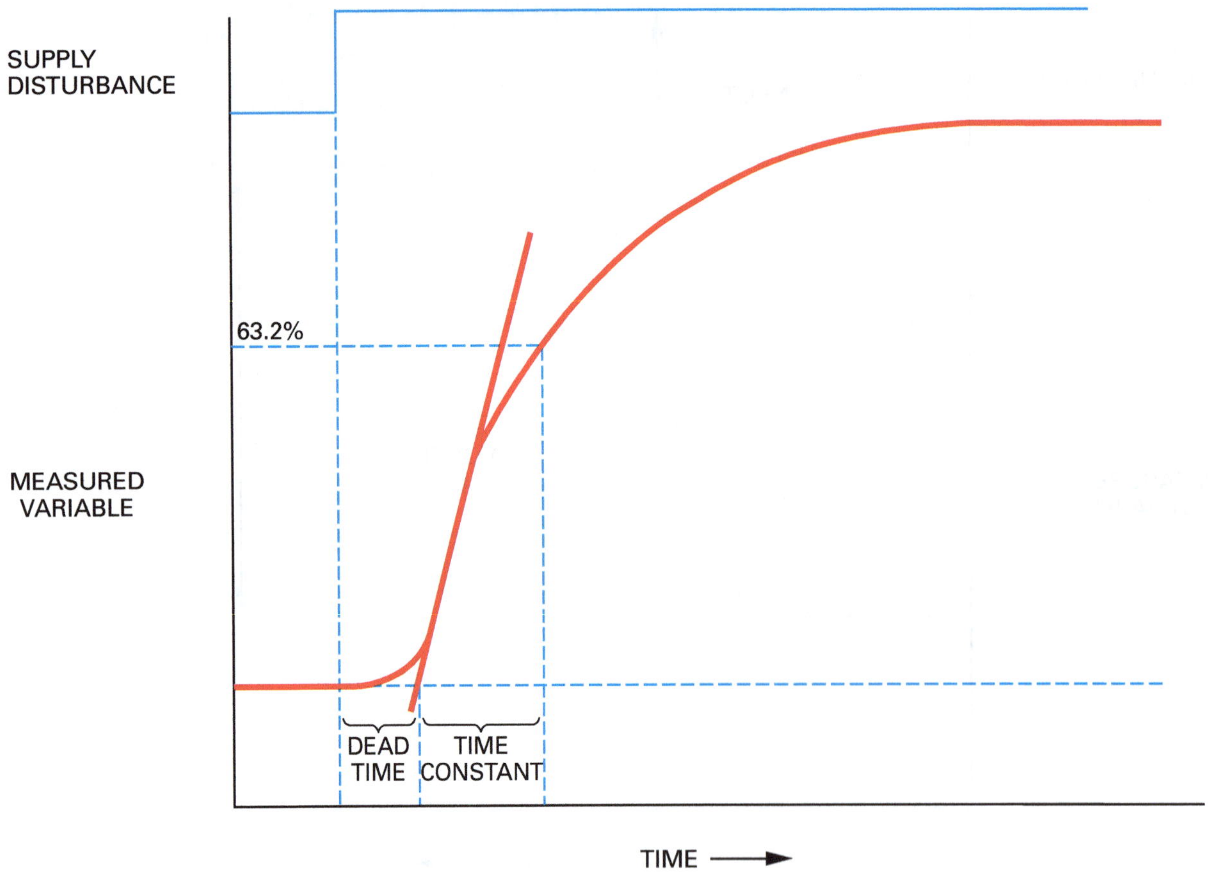

Figure 18 Time constant method for a single capacity with deadtime approximation.

rpm = 1 ÷ (2.0 × t_d)
rpm = 1 ÷ (2.0 × 0.2 min)
rpm = 1 ÷ 0.4
rpm = 2.5 repeats/min

T_d = 0.5 × t_d
T_d = 0.5 × 0.2 min
T_d = 0.1 min

1.5.2 Reaction Rate Method

The second method of open-loop tuning is known as the *reaction rate method* and is shown in *Figure 20*. In this method, the information required from the process reaction curve is the slope of the maximum rate-of-rise line, which is termed the reaction rate. The deadtime is also required. The reaction rate (R_r), as shown in *Figure 20*, is a change in the measured variable in percent of span divided by a change in time. However, the reaction rate not only depends upon the process, but is also affected by the magnitude of the change in the disturbance. A large disturbance produces a steeper reaction rate.

To make the reaction rate independent of the magnitude of the disturbance, the magnitude of the change in the measured variable is divided by the disturbance. Normally the disturbance is a change in the controller output (ΔCO) in percent of span. The resulting modified reaction rate equation is therefore:

$$R_r = \frac{\text{changed in measured value}}{\text{change in time} \times \Delta CO}$$

Where:

R_r = reaction rate

As before, the deadtime is read directly from *Figure 20*. With the reaction rate and the deadtime, use the following equations to determine the controller settings that produce a one-quarter dampened wave in the measured variable:

Proportional (P):

$K_p = 1 \div (R_r \times t_d)$ PB = 100% × K_p

Proportional and reset (PI):

$K_p = 0.9 \div (R_r \times t_d)$ PB = 100% × K_p

$T_i = 3.33 \times t_d$ rpm = 1 ÷ 3.33t_d

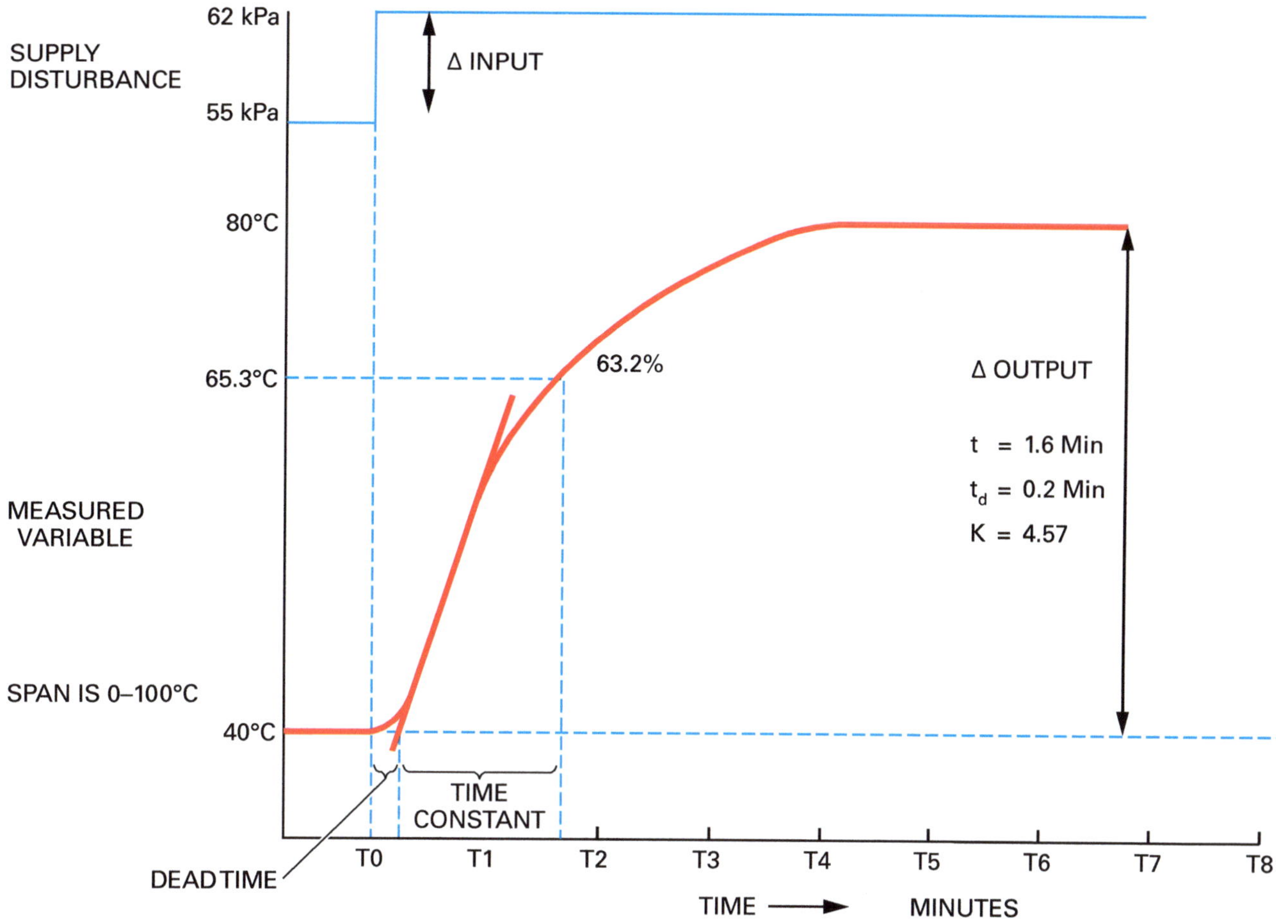

Figure 19 Test data for open-loop time constant method.

Proportional, reset, and rate (PID):

$K_p = 1.2 \div (R_r \times t_d)$ PB $= 100\% \div K_p$

$T_i = 2.0 \times t_d$ rpm $= 1 \div (2.0 \times t_d)$

$T_d = 0.5t_d$

An example of this tuning method is given in *Figure 21*. As shown in the figure, the total change in the measured variable is 40°F.

From the diagram, you can see that there is a 10°F change in the measured variable. The span is 100°F, and the controller operates over 3 to 15 psi range (a 12 psi span).The controller changes by 1 psi (9 – 8 = 1). The supply disturbance (process input) is therefore a change of 8.33 percent of span ([1 / 12] × 100% = 8.33%). The time period for the 10°F change is 0.5 minutes. This time is determined by using the maximum rate of rise line. The deadtime is read directly from the figure and is 0.2 minutes. Determine the reaction rate with the following equation:

$$R_r = \frac{\text{changed in measured value}}{\text{change in time} \times \Delta CO}$$

$$R_r = \frac{20\%}{5 \text{ min} \times 8.33\%}$$

$$R_r = \frac{10\%}{4.165}$$

$$R_r = 2.4\%/\text{min}$$

Use the information just obtained from the process reaction curve to determine the following settings. The equations used here are also Ziegler-Nichols equations (presented in **bold**):

Proportional (P):

$\mathbf{K_p = 1 \div (R_r \times t_d)}$
$K_p = 1 \div (2.4\%/\text{min} \times 0.2 \text{ min})$
$K_p = 2.1$

$\mathbf{PB = 100\% \div K_p}$
$PB = 100\% \div 2.1$
$PB = 47.6\%$

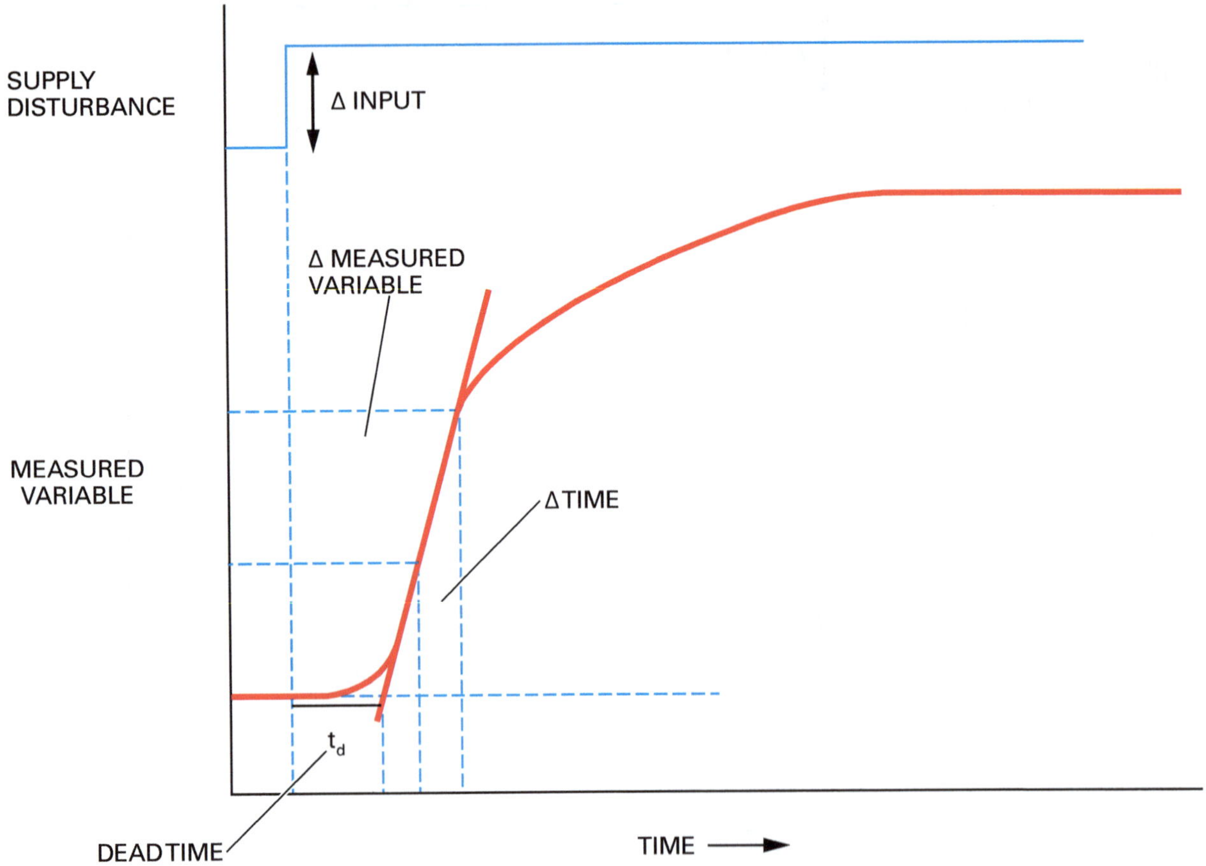

Figure 20 Reaction rate method for a single capacity with deadtime approximation.

Proportional and reset (PI):

$\mathbf{K_p = 0.9 \div (R_r \times t_d)}$
$K_p = 0.9 \div (2.4\%/\text{min} \times 0.2\ \text{min})$
$K_p = 0.9 \div 0.48$
$K_p = 1.875$

$\mathbf{PB = 100\% \div K_p}$
$PB = 100\% \div 1.875$
$PB = 53.3\%$

$\mathbf{T_i = 3.33 \times t_d}$
$T_i = 3.33 \times 0.2\ \text{min}$
$T_i = 0.666\ \text{min}$

$\mathbf{rpm = 1 \div (3.33 \times t_d)}$
$\text{rpm} = 1 \div (3.33 \times 0.2)$
$\text{rpm} = 1 \div 0.666$
rpm = 1.5 repeats/min

Proportional, reset, and rate (PID):

$\mathbf{K_p = 1.2 \div (R_r \times t_d)}$
$K_p = 1.2 \div (2.4\%/\text{min} \times 0.2\ \text{min})$
$K_p = 1.2 \div 0.48$
$K_p = 2.5$

$\mathbf{PB = 100\% \div K_p}$
$PB = 100\% \div 2.5$
$PB = 40\%$

$\mathbf{T_i = 2.0 \times t_d}$
$T_i = 2.0 \times 0.2\ \text{min}$
$T_i = 0.4\ \text{min}$

$\mathbf{rpm = 1 \div (2.0 \times t_d)}$
$\text{rpm} = 1 \div (2.0 \times 0.2\ \text{min})$
$\text{rpm} = 1 \div 0.4$
rpm = 2.5 repeats/min

$\mathbf{T_d = 0.5 \times t_d}$
$T_d = 0.5 \times 0.2\ \text{min}$
$T_d = 0.1\ \text{min}$

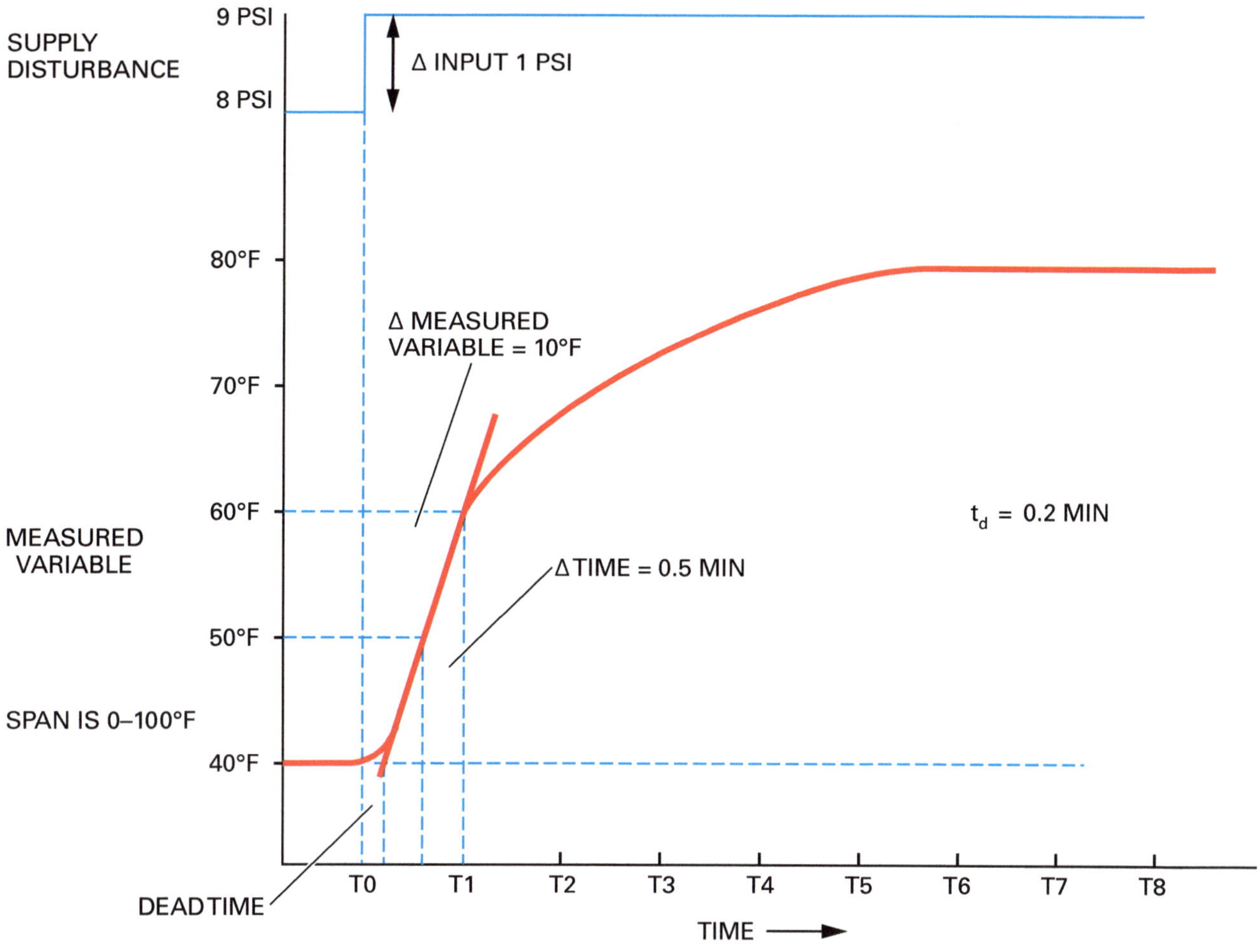

Figure 21 Test data for open-loop reaction rate method.

Both methods of open-loop control tuning yield a very close approximation of a one-quarter dampened wave. As with the closed-loop methods, a certain amount of trimming is required to achieve an exact one-quarter dampened wave. The major difference between the two methods for open-loop tuning is the information that is gathered to set the proportional adjustment of the controller.

Additional Resources

It is assumed that you will learn to perform these procedures under the direct supervision of your crew leader, foreman, or instructor. For independent study, refer to the recommended Additional Resources at the back of this module.

1.0.0 Section Review

1. If many inputs from one area are used, the controlling component (device) for a process control loop may be a _____.
   a. PLC
   b. PID
   c. PCS
   d. PIG

2. The three modes of PID loops are proportional, reset, and _____.
   a. transmit
   b. rate
   c. sensing
   d. control

3. Which of the following is a method for obtaining the values of process characteristics in order to determine the controller setting when tuning a loop?
   a. Testing method
   b. Derivative method
   c. Documented method
   d. Transduced method

4. The loop tuning method in which the gain (or proportional band) is adjusted until a response curve with a one-quarter decay ratio is obtained is the _____.
   a. notch method
   b. ultimate period method
   c. Z-N method
   d. dampened oscillation method

5. One type of open-loop tuning method is the _____.
   a. notch method
   b. ultimate period method
   c. reaction rate method
   d. dampened oscillation method

SECTION TWO

2.0.0 VERIFYING AND SETTING PROTECTION PARAMETERS

Objective

Explain how to verify and set protection parameters (CT26_0-17).

a. Identify prerequisites, competencies, and abnormal operating conditions (AOCs) associated with verifying and setting protection parameters.
b. Explain the general procedure for verifying and setting protection parameters.

Performance Task

1. Verify and set protection parameters for programmable controllers and/or other instrumentation control loops (CT26_0-17).
   - Identify potential abnormal operating conditions (AOCs) that may occur during performance of this CT, and know the appropriate actions to take in response to them.
   - Utilize the appropriate personal protective equipment according to relevant company procedures.
   - Identify the controlling device.
   - Notify operations control center and/or affected personnel before work begins.
   - Verify that the device setpoint or operating parameters agree with documentation or drawings.
   - Ensure proper isolation of the controlling device from the process system. Verify and document data "as found".
   - Implement changes to parameters and verify acceptance of parameters.
   - Return controller parameters to desired settings for permanent operation.
   - Verify accuracy and scaling for the analog input device.
   - If a loop controller or PLC is used for overpressure protection, verify activation of digital control output and/or operation of the end device at the established setpoint.
   - If pipeline pressure is maintained using a control valve or speed control, verify operation of analog output and control devices, such as control valve or throttle actuator.
   - Document system parameters "as left."
   - Notify operations control center and/or affected personnel.
   - Complete appropriate documentation as required by operator's procedures.

Trade Term

Abnormal operating conditions (AOCs): Conditions within a pipeline system, identified by the operator, that may indicate (1) a malfunction of a component or (2) a deviation from normal operations that may indicate a condition exceeding design limits or result in a hazard to persons, property, or the environment.

When pipeline E&I technicians are verifying and setting protection parameters for PLCs or other instrumentation control loops, they must perform the procedure without causing upset to the process system. Safety of personnel and of the system must always be the priority. Even when technicians use great care, an abnormal condition could still occur. All personnel need to be familiar with potential **abnormal operating conditions (AOCs)**, and they must know how to respond to these conditions.

2.1.0 Prerequisites, Competencies, and Abnormal Operating Conditions

Personnel must possess a variety of skills in order to successfully verify and set protection parameters. The various operations require a series of steps, which a pipeline maintenance technician must be competent to perform. Personnel must also be aware of the particular AOCs that may occur during the procedure and know the proper responses to them.

2.1.1 Prerequisites and Competencies

Personnel must possess the following basic knowledge and skills prior to verifying and setting protection parameters for PLCs and/or other instrument control loops:

- Ability to operate and properly use test equipment to perform the functions required in this task, such as multimeters, multi-function calibrators, and gauges (analog, digital, and digital pressure modules)

- Ability to operate and safely handle test media, such as pressure regulators, nitrogen sources, and process connections
- Ability to operate and safely handle electrical systems, in regard to voltages applied to the device, electrical requirements, and connection of test equipment to the device to be tested
- Sound knowledge of the control instrumentation for the process system, including knowing the alarm, control, indicating, and recording function of each instrument device in the control system

2.1.2 Abnormal Operating Conditions

Before verifying and setting protection parameters, personnel must be able to identify potential AOCs and know the proper response to each, should they occur. The following are some potential AOCs, and the appropriate action personnel should take in response to them (always follow applicable company/operator procedures regarding AOCs and responses):

- *An abnormal facility condition occurs during dynamic condition testing resulting from a failure to properly isolate the safety and control functions of the device from the process system* – Use bench condition testing and static condition testing if at all possible. Return to the correct condition, and call the operations control center.
- *A safety device is activated as a result of incorrect application of the test medium during testing, loss of control or regulation, or an overpressure condition on the device* – Correct the application of the test medium and call the control center. Follow any instructions from the control center.
- *Serious electrical shock through contact with energized circuits* – Call 911. Shut off the power if possible.
- *A component failure may cause an unexpected release of a fluid through a rupture disc, a loss of pilot pressure due to filtration, or a control valve failure* – If there is a component failure or any unexpected encounter with a hazardous liquid or carbon dioxide, notify local 911 emergency personnel if appropriate. Limit access to the area and make the proper company notifications. Try to identify the source of the commodity release and isolate the source of the release if it is possible and safe to do so.
- *Fire and/or explosion or the ignition of hazardous, flammable liquids or gases* – Notify local 911 emergency personnel if appropriate. Limit access to the area and make the proper company notifications.

2.2.0 Procedure

Before making any changes to protection parameters, pipeline E&I technicians must usually obtain multiple approvals or a single multi-level approval. Most companies refer to this as a Management of Change (MOC) procedure. In any case, changes to protection parameters must be within safe operating parameters of the pipeline.

The following steps outline the general procedure for verifying or setting protection parameters for programmable controllers and/or other instrumentation control loops:

CAUTION

This is a general procedure. These steps may vary depending on the situation; always follow applicable company/operator procedures when performing this task, including safety and emergency procedures.

Step 1 Identify potential abnormal operating conditions (AOCs) that may occur, and know the appropriate actions to take in response to them.

Step 2 Utilize the appropriate personal protective equipment according to relevant company procedures.

Step 3 Identify the controlling device.

Step 4 Notify the operations control center and/or affected personnel so that they are aware of the task activities and can ensure the safety of the pipeline and of personnel.

Step 5 Verify that the device setpoint or the operating parameters agree with the documentation or drawings. Make certain that setpoint changes are always controlled and verified to ensure safe operations.

Step 6 Make sure that the controlling device is properly isolated from the process system, because the interaction could create dangerous abnormal operating conditions (AOCs) with the process system.

Step 7 Verify and document the data as found to ensure that the values in place before a change are documented correctly and provide a historical record.

Step 8 Implement changes to the parameters.

CAUTION

Closely monitor the process system to detect losses of control or communications that could occur if total isolation from the process system does not exist.

Step 9 Verify acceptance of parameters to ensure that accurate modifications have been implemented.

Step 10 Return the controller parameters to the desired settings for permanent operation. Make sure that the parameters agree with the documentation and the desired system controls when the system is returned to normal.

Step 11 Verify the accuracy and scaling for the analog input device. This ensures that when pressure is applied to the input device or devices, the control scheme will operate to protect the integrity of the pipeline.

Step 12 If a loop controller or PLC is used for overpressure protection, verify activation of the digital control output and/or the operation of the end device at the established setpoint. This ensures that when tripped, the loop controller or PLC will operate to protect the integrity of the pipeline.

Step 13 If the pipeline pressure is maintained using a control valve or speed control, verify the operation of the analog output and the control devices, such as a control valve or a throttle actuator.

Step 14 Examine the control circuit and verify that it operates as intended.

NOTE

Analog outputs may be used to help control pressure on a pipeline. The control must operate as intended to protect the integrity of the pipeline.

Step 15 Document the system parameters as left to provide a means of conducting a review of the change or changes in order to track possible incorrect data entry and provide a historical record.

WARNING!

Incorrect data entry on PLC ladder logic can result in dangerous conditions. Correct input is critical!

Step 16 Notify the operations control center and/or affected personnel to verify the integrity of the system and return the system to normal operation.

Step 17 Identify potential abnormal operating conditions (AOCs) and indicate the appropriate action to be taken.

Additional Resources

It is assumed that you will learn to perform these procedures under the direct supervision of your crew leader, foreman, or instructor. For independent study, refer to the recommended Additional Resources at the back of this module.

2.0.0 Section Review

1. If incorrect application of the test medium causes a safety device to be activated during the setting of protection parameters for a control loop, personnel should correct the application of the test medium and _____.
   a. call the control center for further instruction
   b. wait for the loop to stabilize
   c. notify local 911 emergency personnel
   d. evacuate the work area

2. One of the last steps in verifying and setting the protection parameters for an instrumentation control loop is to _____.
   a. activate the sensing device
   b. isolate the controlling device
   c. apply the appropriate test medium
   d. document the system parameters as left

SECTION THREE

3.0.0 TROUBLESHOOTING LOOPS

Objective

Explain how to troubleshoot loops.

a. Explain how continuity checks are completed.
b. Describe various troubleshooting procedures.
c. Describe how to troubleshoot an oscillating process.

Troubleshooting is the process of finding the cause of a problem in order to correct it. This is a very important job because an entire process operation may depend on the troubleshooter's ability to solve the problem. The following sections provide information on performing continuity checks of electrical and pneumatic loops and also describe troubleshooting procedures.

3.1.0 Continuity Checks

Although continuity checks of electrical loops are easily performed, the locations of the faults are not always easily detected. Unlike pneumatic systems in which leakage makes noise, electricity is silent and invisible. Because of this, technicians must use special instruments to locate faults in electrical circuits.

Several methods of testing the continuity of electrical circuits are available. Some of the methods identify the actual location of the fault, while others detect the presence of a fault but not the actual location.

Before performing a continuity check on an electrical loop, determine the scope of the test. Using system diagrams, prints, or schematics, select an origin and a destination. At this point also determine the number of wires as well as the system used to label them.

> **NOTE**
> Always confirm that your test equipment is working before performing a continuity test. For example, to test an ohmmeter or electrical continuity tester, short the test probes together.

Once the origin and destination of the portion of the electrical circuit to be tested have been determined, disconnect the origin and destination points from the electrical circuit. During the performance of a continuity check, electrical current is sent through the circuit, originating from the multimeter battery. Even the voltage from a battery can damage some instruments, so it is essential to isolate the instruments from the test circuit.

The next step in the performance of a continuity check is to connect the multimeter (*Figure 22*) or similar test instrument to the wiring undergoing the test and check the continuity of the circuit.

3.1.1 Continuity Check of Pneumatic Loops

Many of the systems used in the area of instrumentation and control use pressurized air as an operating medium. Carefully calibrated controllers send air signals to actuators with a high level of precision. If a rupture or leak occurs in the air lines leading to the actuator, the signal may degrade, causing system operating inaccuracies.

By performing a continuity check on these air lines, technicians can verify not only the continuity of the lines but also their leak-tight integrity and freedom from blockages. If these three items can be verified, the technicians can be sure that the air signal sent by one component in the instrument loop will arrive at the intended receiver of that signal within the proper tolerances.

Figure 22 Multimeter.

Technicians may use a pneumatic bubbler system or a pressure calibrator to perform continuity and leak tests of pneumatic control lines.

3.2.0 Seven-Step Troubleshooting Procedure

Troubleshooting loops is an analytical process. Approaching troubleshooting in a logical step-by-step manner can save valuable time. The following troubleshooting considerations and practices form the basis for the seven-step troubleshooting procedure:

- Verify that there is really a problem. A problem is usually indicated by a change in equipment performance or product quality (i.e., recognizable symptoms). Do not simply accept a report that something is wrong: check for yourself.
- Identify and locate the cause of the problem. Problems are often caused by a change in the system. The better that you understand the system and how it is supposed to work, the easier it will be to compare normal to actual conditions and find the cause of the problem.

> **NOTE**
>
> Check documentation to see who was the last person to work on, test, or calibrate the system and when that work was performed. A change made at that time may be causing the problem that you are now troubleshooting.

- Dividing the instrument loop into smaller sections is a useful technique for locating the cause of the trouble. Knowing the different signal flow paths (*Figure 23*) is essential to locating the problem. Different paths can give false indications.
- Fix the problem at its source. It is very important to correct the cause of the fault, not just the effect or symptom. This involves replacing or repairing a part or instrument, or making an adjustment.
- Verify that the problem has been corrected. Do this by repeating the same check that originally indicated the problem. If the fault has been corrected, the system should operate properly.
- Follow up to prevent future problems. Determine the underlying cause of the trouble. Suggest a plan to your supervisor that will prevent future problems of this kind. This may include recommending the use of upgraded or improved parts, if applicable, or if the manufacturer has issued a component recall.

Successful troubleshooting requires step-by-step compliance to the fault-isolating procedure, plus attention to the philosophical concepts that contribute to logical and timely troubleshooting. Technicians may use an individual step of the procedure more than once as troubleshooting narrows down from functions to circuits to components.

The seven-step troubleshooting procedures involves the following steps:

Step 1 Symptom recognition

Step 2 Symptom definition

Step 3 Listing of probable faulty functions

Step 4 Localization of the faulty function(s)

Step 5 Localization of the trouble to the circuit

Step 6 Failure analysis

Step 7 Retesting

3.2.1 Symptom Recognition

The first step in troubleshooting is to verify that a problem actually exists, and to recognize the problem. (In some cases, it may simply be an error in setup, rather than an actual problem with the loop.) Start by checking obvious problem sources:

- Verify that the equipment is properly connected to the power source and is receiving power.
- Make sure that the machine is turned on and is set to the proper operating mode.
- If an alarm condition occurred, be sure that the condition has been cleared and that the equipment and the emergency stop devices have been reset.
- Check photoelectric eyes, proximity switches, and other sensors to be sure that they are properly aligned, are not blocked, and are not otherwise tripped.
- Be familiar with the equipment. Recognize the signs of impending failure.

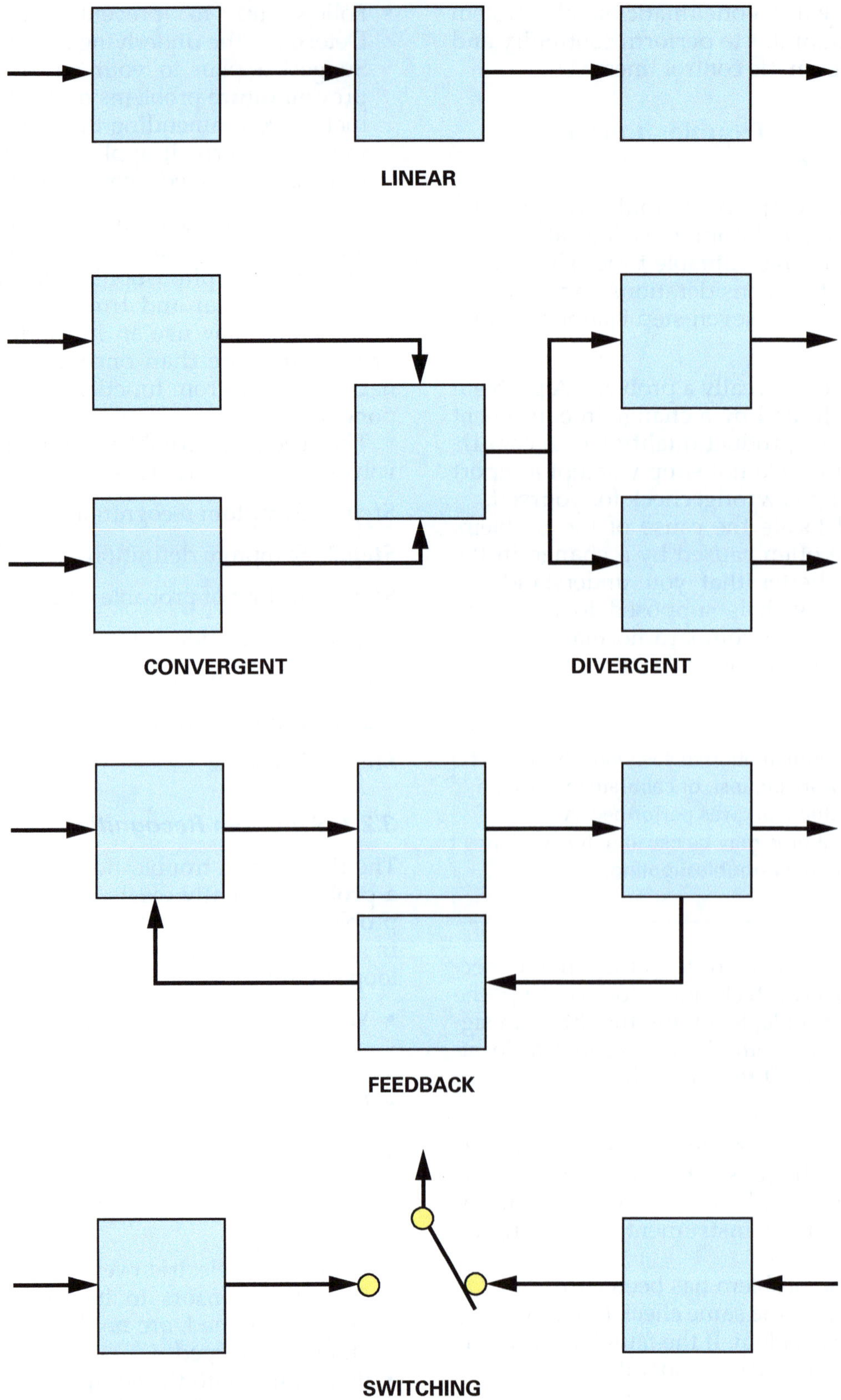

Figure 23 Common signal paths.

If the problem still exists after checking the obvious possibilities, follow standard troubleshooting procedures, and document each step so that another technician can follow the steps already taken and continue the process if necessary.

3.2.2 Symptom Definition

The object of symptom definition is to obtain as much information as possible about the problem found in Step 1. As the name implies, the symptom found in Step 1 is defined in Step 2. This step provides the information necessary to begin narrowing the problem down in a logical fashion. Observe the following considerations in the symptom definition step:

- Be aware that a large number of equipment faults can produce similar trouble symptoms. During this step, try to differentiate as much as possible between the characteristics of the symptoms.
- Start a troubleshooting log with as much background information as possible, and document each adjustment and its results. Also, consult any available historical records as they may document similar problems and their resolutions. Always utilize prior information and history to avoid "reinventing the wheel."
- Note how readings are affected by all modes of operation and switch lineups.
- Be sure to observe remote meters and other indicators. Pay special attention to how they are responding to the problem.
- Note whether out-of-specification readings are linear or occur only at set values.
- Note variations from specified readings, even if they are still within the tolerance band.
- Always note if an adjustment has no effect on the trouble symptom; this will help in eliminating troubleshooting steps later.
- Determine whether the trouble has developed slowly, or drifted, or if it is a sudden failure.

Perform control manipulation with care. Faulty voltages and currents exceeding maximum design specifications may occur during manipulation. This could have detrimental effects on both associated equipment and components within the failed equipment.

Do not try to determine the solution in one step. Troubleshooting should be a series of small, logical steps, each one intended to show a result leading to a faulty area.

Remember, troubleshooting can last two hours or two weeks. Be sure to write your troubleshooting log accordingly. Do not leave anything to memory.

3.2.3 Listing of Probable Faulty Functions

Once the problem is clearly defined, use a functional flowchart of the equipment to help narrow down the probable causes. If the engineering department cannot supply a functional flowchart, it may be possible to make a simple one with the help of the operator if the operation of the machine is not too complex. Developed processes may already have cause-and-effect matrices that describe proper equipment operation. If these exist, they can make troubleshooting easier.

This step is intended to narrow down the possible faulty functions based on the information obtained in the first two steps. The functional block diagram of the equipment, if available, and the troubleshooting log (Steps 1 and 2) are all that is needed for this step. Key points for this step include the following:

- Whenever possible, use a functional block diagram to avoid leaving out any possible functions.
- Write down all probable faulty functions, even if it is obvious that some of them are working correctly. Then write down your justification for concluding that they are functioning correctly.
- Be sure to include as functions such things as detectors, switches, cables, meters, point-to-point wiring, and plugs. Wiring is always a probable cause. Connections become loose over time due to heat/contraction and vibration. High-resistance joints caused by improperly torqued (loose) connections or crimps are very common. An easy way to find this type of fault is to tug gently on a termination or junction.
- Do not get locked in by what you expect the problem to be. Past troubleshooting experience and first hunches have a part in deciding possible faulty functions, but do not ignore hard evidence because of your gut feeling.
- Always ask yourself, "Would a failure of this function cause the symptoms I am seeing?"

3.2.4 Localization of Faulty Functions

Localizing faulty functions requires measuring an electrical value. Before this, no measurements should have been taken except for possibly checking calibration voltages if applicable.

This step narrows down the list generated in the previous step by checking inputs and outputs of functions. To systematically localize the faulty function, use logic, test point accessibility, ease of measurement, and bracketing techniques. Refer to *Figure 24* while keeping these points in mind.

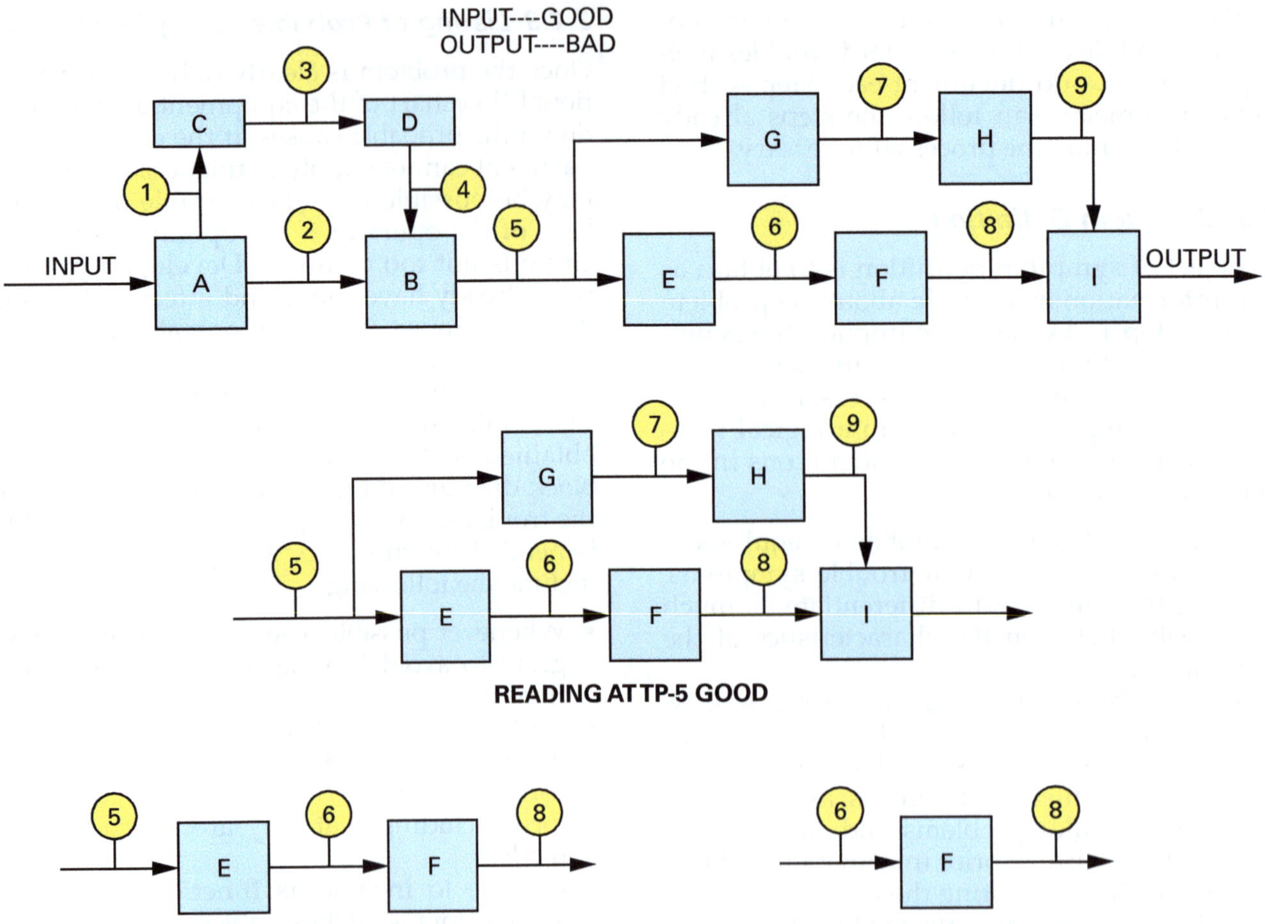

Figure 24 Bracketing technique.

Know approximately what value to expect before taking a reading, both for test equipment safety and to limit confusion about the results obtained. Check all inputs and all outputs of a function.

When trying to prove that a function is faulty by the input versus output method, note that with a good input it only takes one bad output to prove that the function is bad. Just because nine out of ten outputs are good, don't assume the entire function is good. Ten out of ten outputs must be correct.

If you obtain what appears to be an incorrect reading, verify both the reading and the test equipment setup. Vary the inputs (test signals) to a function to determine if it is inoperative with all inputs or only at a specified level. Note if a function is really working but at an incorrect level or setpoint.

> **NOTE**
> When noting results of measurements, don't use any general descriptive remarks. Always write down the exact reading.

If using a test signal to troubleshoot, use the test signal that generates the worst symptoms because it will yield the greatest circuit deviations from normal circuit parameters.

When taking a waveform, check the following three characteristics:

- Is it the correct shape?
- Is it the correct amplitude?
- Is it the correct frequency?

If possible, compare voltages and waveforms with the actual equipment's normal values (baseline data) rather than general data values provided in technical manuals.

When taking waveforms, if the waveform deviates from normal, consider whether the deviation could cause the symptoms that you have observed. If not, just note the deviation, and continue troubleshooting the original problem.

Do not be discouraged if several hours of troubleshooting lead to proving that a function is good. It is as important to prove things to be functioning correctly as it is to look for an incorrect function. Often, the first function that you choose to check out will not be the faulty one.

Every so often, take time to look over your notes to ensure that you are going in the right direction. Be sure that you have not lost sight of your original troubleshooting goal.

3.2.5 Localization of the Trouble to the Circuit

This step continues with isolating the fault once a faulty function has been found. The purpose of this step is to localize the fault to the circuit that is causing the symptoms, rather than to a component. This step requires knowledge of circuit operation, component characteristics, and how to best use the proper test equipment. The prints used for this step now are the schematic diagrams, which can be obtained from the engineering department. Focus on the following points:

- Before taking a reading, have in mind an assumption that you are trying to prove. This way, you are not always trying to justify a reading.
- Always expect the worst possible condition in respect to circuit parameters, for both personal safety and test equipment protection (that is, start with meters on high scales).
- Do not try to justify a reading that is not what you expected it to be. Do not use phrases like, "This voltage is low probably due to . . ." If you cannot state hard evidence to support your opinion, do not write down any justification. It will only mislead others.
- Only take measurements that will prove a condition or aid in drawing a logical conclusion. Unnecessary readings only tend to confuse your thinking and cause unnecessary wear and tear on the equipment.
- If you are unsure of a reading, compare it to a parameter of a circuit that is known to be operating correctly and/or check the circuit or component with another meter that is known to be in good working order. Technicians often use two meters in conjunction to pinpoint a problem in between the two meters.
- Be sure to differentiate between a low-voltage and a no-voltage condition.
- Be careful not to narrow down and isolate circuits too rapidly. Otherwise, you will be troubleshooting effects rather than causes.
- Be aware of common failure modes of basic circuit designs and of their distinct displays of correct operation.
- Always write down every reading taken. Even if it does not help you at the moment, it may help the person relieving you or it may help you when you resume work on the problem later.

3.2.6 Failure Analysis

Failure analysis requires technicians to locate and replace the failed component. The most important part of this step is to correct the cause of the failure and not just the component. Restore the circuit board to its original condition, and address the cause of the failure, rather than just repairing the component that failed as a result of the problem. Note the following key points:

- Move through a circuit in a logical manner, always taking readings that will narrow down the number of possible components.
- Check each component in flow order, and be sure to measure all the parameters of each component, including static condition. Continue even if the first parameter appears to be bad—a single parameter may not reveal the entire problem.
- Before replacing a component that appears to be faulty, analyze the input to the component. If the input is correct, then replace the component. If the input is not correct, back up and recheck the preceding component to find the problem.
- Always ask yourself, "Would a failure of this component cause the loss of this function?"
- Be familiar with component failure modes and rates.
- Always check the associated components of the failed unit thoroughly.
- Carefully perform visual inspections first. A great amount of information can be gained rapidly through a thorough visual inspection.
- In the case of failed electrical components, your sense of smell works well to detect that wire insulation or similar materials have burned or seriously overheated.
- Never replace a component until the exact cause of trouble is found and repaired.

NOTE

The wise pipeline E&I technician always keeps one truth in mind: a replacement device may itself be defective out of the box. Just because a critical spare is new, don't assume that it is working. Also, be aware that parts that have been sitting on the shelf for a long time may not have received firmware or IC updates to match the components that they're replacing. If the spare has configuration jumpers, these may not be properly set to match the current system configuration. Before using a spare, confirm that these issues have been addressed. Some technicians pre-test and configure replacement parts and label them "known good" before putting them on the shelf.

Documentation is imperative at this point, both to aid in troubleshooting this problem if it returns and to point out recurring design deficiencies.

3.2.7 Retesting

Retesting is the final step of the procedure. Although the equipment has been repaired and appears to be functioning properly, all functions must be checked and verified. The information obtained in this step can also provide some baseline information that can aid in future troubleshooting. One key point to remember is to fail safe. (If in doubt, do all checks that will ensure that the equipment is operating correctly.)

3.3.0 Troubleshooting Oscillating Processes

Troubleshooting a dynamic, or oscillating, process takes a good working knowledge of the system, as well as patience. Since the process is constantly changing, it is difficult to compare static readings. The basic procedure for troubleshooting this type of process is presented in the following sections using an example situation of an operator discovering an oscillating process showing up on the recorder controller.

3.3.1 Verifying That a Problem Exists

In many cases, trouble in a system is discovered by an operator. The operator may observe symptoms of a problem, or an alarm may indicate that something is wrong. When this type of trouble is discovered, the first step for the pipeline E&I technician is to verify that there is a problem in the system by asking the following questions:

- Do we have all the facts?
- How was the problem discovered?
- When did the trouble start?
- Is the trouble constant or intermittent?

Observe the system in operation to see the symptoms and to verify that the system is being operated properly.

3.3.2 Gathering Information

In addition to learning about the trouble, technicians must gather information about the system and how it operates. This information is necessary for determining whether the possible causes of the trouble actually relate to a malfunction in the system. There are three important sources of information: the operator, the panel graphic, and the loop diagram. A fourth possibility, if it exists, is a history/maintenance database, accessible by unit ID, which contains records of prior issues. If this service exists, be sure to consult it.

The operator can provide information about the system operation. The panel graphic, shown in *Figure 25,* is another source of information. While its primary function is to serve as a control center for the operator, the panel graphic also provides some information about the process.

The loop diagram or piping and instrumentation diagram (P&ID) provides detailed information about the components and design of the loop. For our example, the P&ID in *Figure 26* provides the following information:

- The flow element is an orifice.
- The flow transmitter senses the differential pressure across the orifice and sends a pneumatic signal to the flow recording controller.
- The recording controller sends a pneumatic signal to the flow control valve.
- The flow control valve responds to the controller signal to control the flow rate of the process fluid.

3.3.3 Identifying Possible Causes

Once you have determined the symptoms of the problem and are acquainted with how the system should work, begin identifying the possible cause of the problem. In this example, the symptom is an oscillating process in a flow control loop. There are several possible causes for this type of problem, including the following:

- The transmitter may be malfunctioning and sending improper signals.
- A faulty controller could be causing the oscillations. This condition can occur if the controller is not tuned properly.
- The control valve may be sticking, or the actuator may not be operating properly.
- While orifice plates are reliable for the most part, they may become damaged. It's also possible for the lines on either side of an orifice plate to become plugged with debris or frozen water. Problems of this type are numerous, although relatively easy to fix.
- The process itself may be oscillating due to a change in process fluid characteristics.

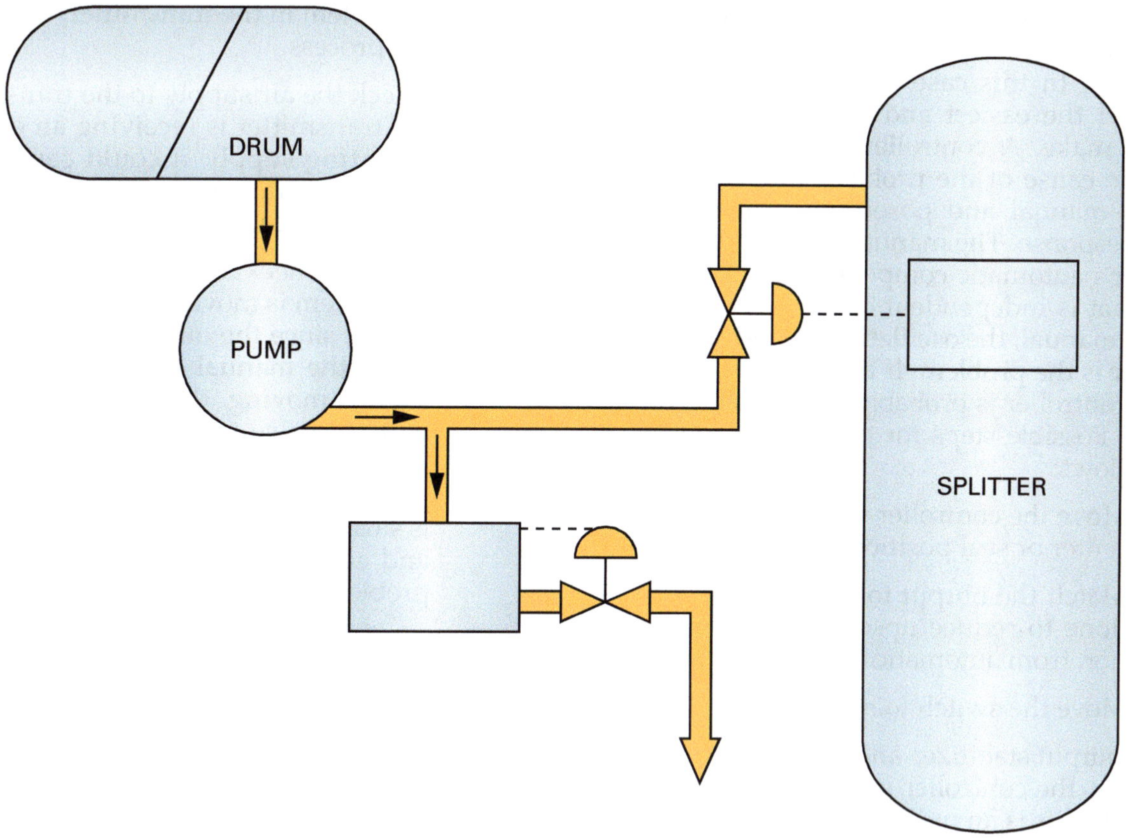

Figure 25 Panel graphic.

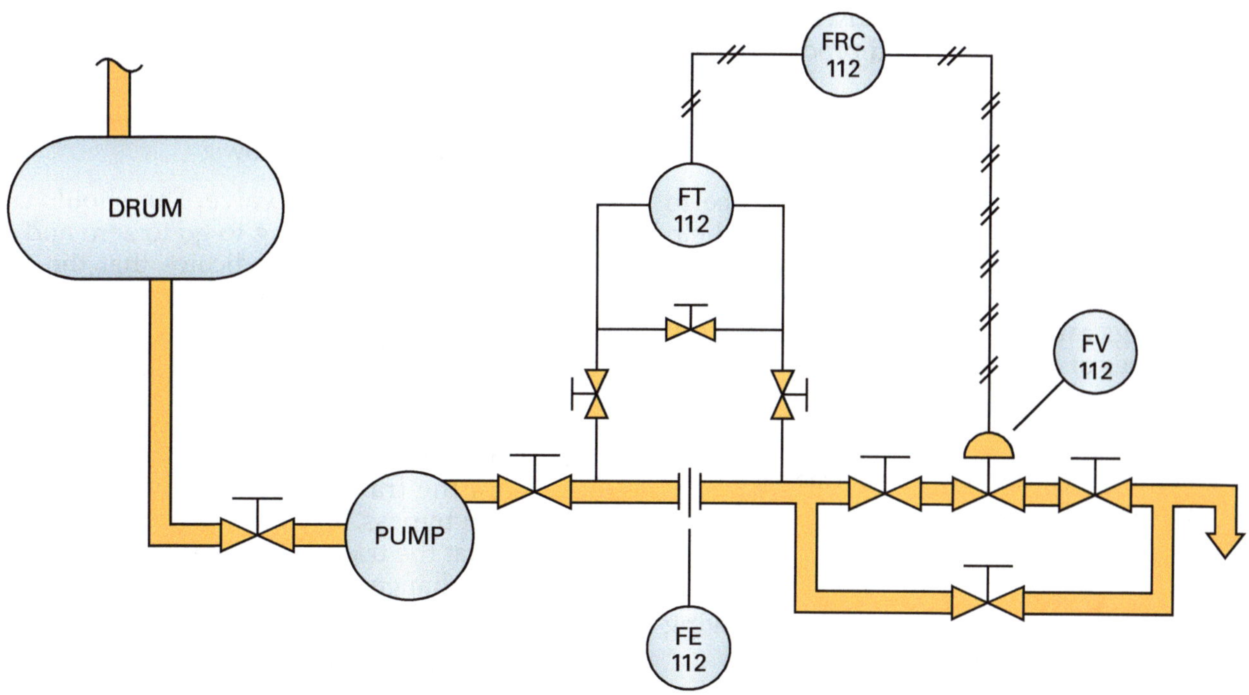

Figure 26 Piping and instrumentation diagram.

In this example, the cause of the problem could be in the controller, the transmitter, the valve, or the process. In this case, controller operation is considered the easiest and most efficient initial check to make. A controller can often be ruled out as the cause of the problem simply by shifting it to manual and observing controller and process response. The manual mode bypasses the controller's automatic components to produce an output that is independent of these components. When in manual, the oscillation should stop if the controller is the problem. If the oscillation continues, the controller is probably not the cause of the problem. Possible steps for checking a controller are as follows:

Step 1 Move the controller transfer switch to the center or seal position.

Step 2 Match the output to the setpoint. This is done to reduce upset during the transition from automatic to manual.

Step 3 Move the switch to manual.

If the output stabilizes and the input continues to oscillate, the controller is receiving an oscillating signal. This is an indication that the controller is probably not the cause of the problem.

If the controller is not the cause of the problem, continue the process of identifying the cause by checking out the other possible causes. In this example, the valve, transmitter, and process remain to be checked.

3.3.4 Checking Instruments in the Field

Before going to the field, it is good practice to check the P&ID again. This information can help determine the most efficient way to approach the problem. Making visual checks and observing gauge readings can frequently identify or rule out possible causes of a problem.

> **NOTE**
> Be sure to check that the receiver-type gauges at transmitters are working correctly.

As applied to our example, use the following steps to check instruments in the field:

Step 1 Visually check the output gauge on the transmitter. In this case, the gauge is oscillating. This indicates that the transmitter output is still oscillating. Oscillation in the transmitter output could be due to a problem in the transmitter, the valve, or the process.

Step 2 Check the air supply to the transmitter. If the transmitter is receiving an erratically pulsating supply, it could cause oscillation in the transmitter output. In this case, the gauge reading is relatively steady.

Step 3 Visually check the valve. Look to see if the valve stem is moving. The stem should be steady, since the output from the controller in the manual mode is steady. If the stem is moving, there may be a problem with the valve actuator or positioner. Look for air supply problems, as well as erratic positioner outputs. In this case, the stem is steady, which helps to rule out the valve and actuator as the possible cause of the problem.

After performing these steps, the controller and the valve can be ruled out as the cause of the oscillation.

3.3.5 Checking the Transmitter

Check transmitter operation by performing a zero check. This is done by equalizing the pressure on both sides of the differential pressure transmitter. With zero differential pressure, the transmitter output should go to a steady zero. If the output drops but continues to oscillate, the transmitter is probably the cause of the problem. Possible steps for performing a zero check on a differential pressure transmitter are as follows:

Step 1 Open the bypass valve. This should cause the output reading to go to zero and stop oscillating. This indicates that the transmitter is responding to a change in differential pressure.

Step 2 Close the valve on the high-pressure side of the transmitter.

Step 3 Close the valve on the low-pressure side of the transmitter. With the transmitter on bypass, a stable zero output indicates that the transmitter is at least responding to changes in differential pressure.

Step 4 Once its position has been verified, put the transmitter back into service.

Step 5 Open the low-pressure valve.

Step 6 Open the high-pressure valve.

Step 7 Close the bypass valve.

In this example, as differential pressure was restored, the output began to rise, and the oscillation appeared again. At this point, it seems that the cause of the problem is in the process itself. There are some possible instrument problems, such as a loose control valve plug or a transmitter that does not malfunction until a differential pressure is sensed. Checking these problems, however, is a time-consuming job. It is most efficient to make a quick check of the process first.

3.3.6 Locating the Problem

When the cause of a problem appears to be in the process, the instrumentation diagram can provide information about the equipment and its function in the process loop. As shown in *Figure 27*, there is a drum with a pipe going from the bottom through a suction valve to a pump. From the pump, the pipe goes to a discharge valve and to the orifice, or primary flow element.

The cause of the problem could be in the process piping, one of the valves, or the pump. Checking these possibilities first may avoid the costly time involved in removing and disassembling instruments.

In this example, the operator found that the suction valve between the tank and the pump was partially closed. The pump was trying to pump more fluid than the valve would allow.

A partially blocked valve can cause abnormally low pressures to occur in the pump suction as it attempts to draw in liquid. These low pressures can cause vapor bubbles to form and subsequently collapse at the pump discharge. This process is called cavitation. Even if the valve is fully open, cavitation may still occur if the pump is not receiving sufficient net positive suction head (NPSH) or there is insufficient tank level. A fouled or improperly sized strainer or filter prior to the pump can cause a similar lack of NPSH. Again, knowing the specific process that you're working with is basic for effective troubleshooting.

Cavitation leads to pulsating flow and pump pressure. When the pump cavitates, this causes oscillation in the pump discharge and pump damage if the condition persists. The oscillation is accurately measured by the instrument system, but the action of the controller cannot correct the oscillation.

Because the cause of the problem in this example is in the process and not in the instrumentation, the operator, rather than the instrument technician, makes the necessary fix. The operator opens the valve that is partially closed. With the valve open, the process stabilizes. This stabilization is an indication that the problem has been solved.

Once the repair or change has been made, the troubleshooter should verify that the problem has been corrected. Verification of the fix usually involves making the same check performed earlier to verify that the problem existed. In our example, this check involves the following steps:

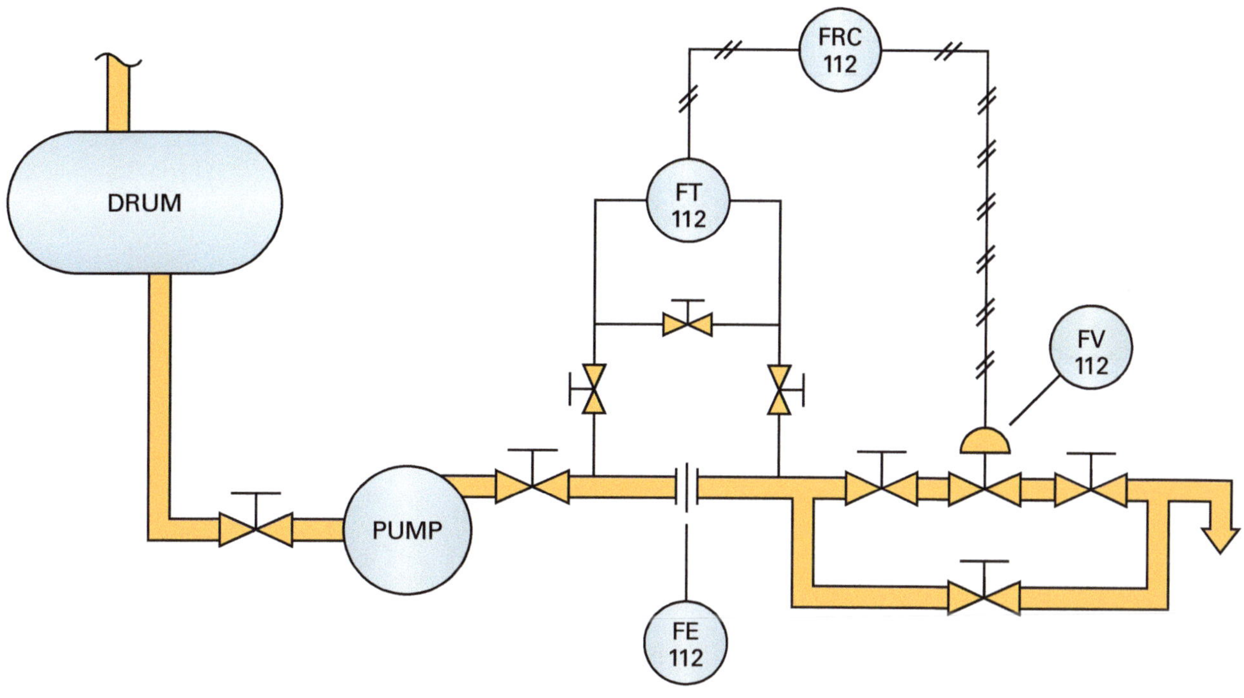

Figure 27 Process diagram.

Step 1 Move the controller transfer switch to the seal position.

Step 2 Match the output to the setpoint.

Step 3 Move the switch to the automatic position.

Step 4 Observe the input and the output. They should be tracking without oscillation. If this is the case, the problem has been corrected.

3.3.7 Using a Troubleshooting Flowchart

The information gathered from this example problem and the method used to find the failure can be applied to a flowchart, as shown in *Figure 28*. Some key points about troubleshooting flowcharts are the following:

- Diamonds represent decision-making points.
- Squares represent facts or solutions.
- Circles represent references to another source.

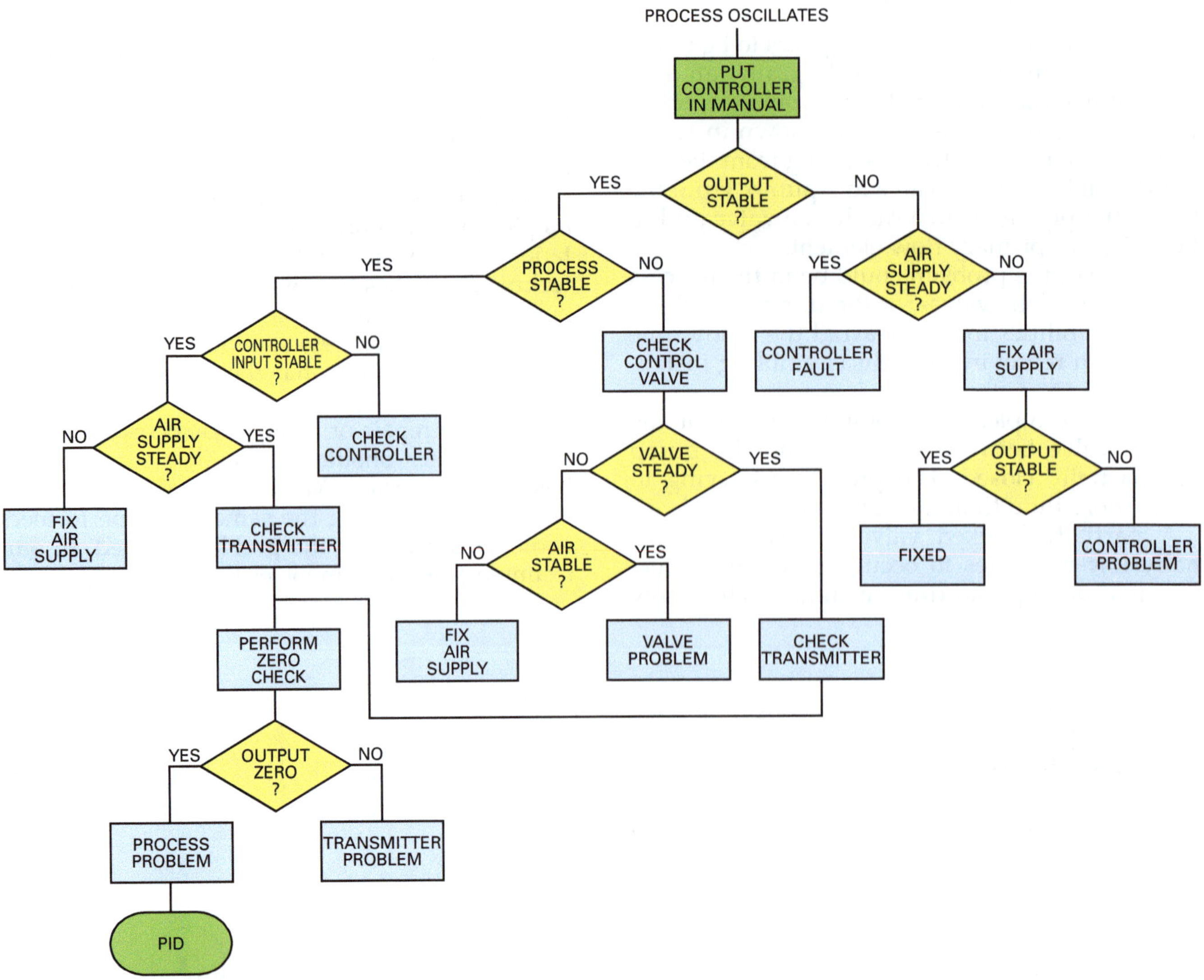

Figure 28 Flowchart.

Additional Resources

It is assumed that you will learn to perform these procedures under the direct supervision of your crew leader, foreman, or instructor. For independent study, refer to the recommended Additional Resources at the back of this module.

3.0.0 Section Review

1. Performing a continuity test on the air lines in pneumatic loops enables a technician to verify the continuity of the lines and _____.
   a. the probable location of any pneumatic faults
   b. their airtightness and freedom from blockages
   c. the scope of the problem
   d. alarm system functionality

2. To localize a faulty function, systematically use the following techniques: logic, test point accessibility, ease of measurement, and _____.
   a. bracketing
   b. reading
   c. testing
   d. verification

3. When troubleshooting a loop in an oscillating process, technicians can find detailed information about the components and design of the loop in a(n) _____.
   a. MOC procedure
   b. panel graphic
   c. troubleshooting flowchart
   d. P&ID

Summary

To maintain process controllers safely and efficiently, pipeline E&I technicians must be familiar with control and PID loops, including the four basic loop components and how each component operates. Technicians must have a good understanding of the function of PID loops, including proportional band, reset, and rate.

Other important skills needed for maintaining loops are the ability to verify and set protection parameters, to recognize and respond to abnormal operating conditions (AOCs), and to follow proper procedures.

Troubleshooting a loop requires testing the performance of the loop and making continuity checks. Technicians must be familiar with the seven basic steps involved in recognizing and defining the problem, locating and identifying the faulty circuit or component, and analyzing the cause of the failure. In addition, troubleshooting oscillating processes requires specific skills, such as knowing which information to obtain, what the likely causes are, and how to check the various components.

Once a technician verifies that a loop is functioning correctly, he or she must tune the loop. To do this, the technician must use the appropriate method—open-loop or closed-loop—depending on the type of loop and its function in the process.

Review Questions

1. In a control loop, almost any device that can accurately determine the condition of a process variable such as speed, pressure, temperature, and other variables can be used as a _____.
   a. transmitting component
   b. sensing component
   c. controlling component
   d. final control component

2. In which type of loop does no comparison take place between the actual and desired values?
   a. PID
   b. open
   c. closed
   d. control

3. The first mode of control in a PID loop is the range of values of the process variable that requires _____.
   a. reset control
   b. integral control
   c. proportional band control
   d. derivative control

4. In a PID loop, the interval by which derivative action advances the time of proportional action is called _____.
   a. proportional band
   b. reset (integral)
   c. gain
   d. derivative time

5. In tuning loops, the most difficult problem in dealing with process control is the setting of the adjustments that are located on the _____.
   a. sensing component
   b. transmitting component
   c. controller
   d. final control component

6. One of the simplest methods of tuning a controller is by using the frequency response of the controller to determine the point when the gain is less than 1 at the critical frequency. This is called the _____.
   a. ultimate period method
   b. dampened oscillation method
   c. time constant method
   d. notch method

7. The closed-loop tuning method that requires placing the process into continuous oscillation only once is the _____.
   a. ultimate period method
   b. dampened oscillation method
   c. time constant method
   d. notch method

8. The class of tuning method that requires that only one disturbance be imposed on the process is the _____.
   a. dampened oscillation method
   b. open-loop method
   c. closed-loop method
   d. notch method

9. The information required from the process reaction curve for the time constant method includes the time constant, the process gain, and the _____.
   a. span width
   b. Z-N coefficient
   c. reaction rate
   d. deadtime

10. Before implementing changes to the protection parameters for instrumentation control loops make sure that the controlling device is _____.
    a. insulated with a test medium
    b. isolated from the process system
    c. back-linked to the operations control center
    d. modified for dampened oscillation cycling

11. Once the origin and destination of the portion of an electrical circuit to be continuity tested have been determined, the origin and destination points must be _____.
    a. labeled with color-coded electrical tape
    b. disconnected from the electrical circuit
    c. cleaned with a non-corrosive cleaning solution
    d. tested by applying an appropriate test medium

12. In the seven-step method, after troubleshooting and repairing a loop in a process, the last step is to _____.
    a. retest the loop
    b. restart the process
    c. restore the power
    d. check for abnormal conditions

13. In the seven-step troubleshooting procedure, the step for localizing faulty functions requires _____.
    a. resetting photoelectric eyes
    b. applying a test medium
    c. measuring an electrical value
    d. replacing possibly faulty components

14. Before checking instruments in the field when troubleshooting an oscillating process, it is a good practice to _____.
    a. review the as-left SCADA values
    b. set all controllers to manual mode
    c. check the P&ID again
    d. call the operations control center

15. Transmitter operation can be checked by equalizing the pressure on both sides of the differential pressure transmitter, which is called performing a(n) _____.
    a. zero check
    b. delta P test
    c. oscillating test
    d. pressurization check

Trade Terms Quiz

Fill in the blank with the correct term that you learned from your study of this module.

1. The ________ is an open-loop tuning method that is based on the time constant, process gain, and deadtime from the process reaction curve.
2. The frequency at which a 360-degree phase shift around the control loop exists is the ________.
3. Equations developed in the early 1940s by J.G. Ziegler and N.B. Nichols that are still used today as the basis of some of the most common methods of loop tuning are the ________.
4. ________ is the acronym for proportional-plus-integral-plus-derivative control.
5. An open-loop tuning method based on the slope of the maximum rate of rise line (the reaction rate) from the process reaction curve and the deadtime is called the ________.
6. A closed-loop tuning method that requires the determination of the ultimate gain or proportional band and the ultimate period is called the ________.
7. A closed-loop tuning method that uses the frequency response of a controller to determine the point at which the loop gain is less than one at the critical frequency is called the ________.
8. ________ is the maximum allowable value of gain or proportional band at which a system is stable.
9. The mode of control that uses a range of values of the process variable, usually expressed as a percentage, is known as ________.
10. The difference between the point of stabilization and the setpoint is referred to as ________.
11. The ________ is a closed-loop tuning method that is a modification of the ultimate period method; it is used for processes that cannot tolerate continuous oscillations.
12. The acceptable amount by which a process variable can deviate from its setpoint, such as 68 degrees ±3, is called ________.
13. Any physical conditions that need to be controlled, such as pressure, speed, level, temperature, or another measurable quantity, are called ________.
14. Conditions within a pipeline system that may indicate (1) a malfunction of a component or (2) a condition exceeding design limits or resulting in a hazard to persons, property, or the environment are called ________.
15. The mode of control that adjusts the output signal to bring the process back to the setpoint and eliminate offset is known as ________.
16. The period of response during which either the gain or proportional band is set at the ultimate value is called the ________.
17. The mode of control based on how fast the process variable is changing is called ________.
18. The delay interval between the point when the controlled variable changes and the point when the controller begins to respond to the change is called ________.
19. The condition in which a process variable is stabilized at an acceptable value is called ________.
20. The ________ is a computer-based system used remotely by a controller in a control room that displays information about a pipeline facility.

Trade Terms

Abnormal operating conditions (AOCs)
Critical frequency
Dampened oscillation method
Deadtime
Notch method
Offset
PID
Process variables
Proportional band (gain)
Rate (derivative)
Reaction rate method
Reset (integral)
Steady-state
Supervisory Control and Data Acquisition (SCADA)
Time constant method
Tolerance
Ultimate gain
Ultimate period
Ultimate period method
Ziegler-Nichols equations

Trade Terms Introduced in This Module

Abnormal operating conditions (AOCs): Conditions within a pipeline system, identified by the operator, that may indicate (1) a malfunction of a component or (2) a deviation from normal operations that may indicate a condition exceeding design limits or result in a hazard to persons, property, or the environment.

Critical frequency: The frequency at which a 360-degree phase shift around the control loop exists.

Dampened oscillation method: A closed-loop tuning method that is a modification of the ultimate period method; it is used for processes that cannot tolerate continuous oscillations.

Deadtime: The delay interval between the point when the controlled variable changes and the point when the controller begins to respond to the change.

Notch method: A closed-loop tuning method that uses the frequency response of a controller to determine the point at which the loop gain is less than one at the critical frequency.

Offset: The difference between the point of stabilization and the setpoint.

PID: A method of control made up of proportional band, integral, and derivative control. (It is an acronym for proportional + integral + derivative control.)

Process variables: Any physical conditions of a process that need to be controlled, such as pressure, speed, level, temperature, or another measurable quantity.

Proportional band (gain): A method of control based on the range of values of a process variable that requires proportional (gain) control; usually shown as a percentage of the full-scale range of the controller.

Rate (derivative): A method of control based on derivative, or rate of change, action.

Reaction rate method: An open-loop tuning method based on the slope of the maximum rate of rise line (the reaction rate) from the process reaction curve and the deadtime.

Reset (integral): A method of integral control based on reset rate in repeats per minute required in proportion to the size and duration of an error, or offset.

Steady-state: The condition in which a process variable is stabilized at an acceptable value.

Supervisory Control and Data Acquisition (SCADA): A computer-based system used remotely by a controller in a control room that displays information about a pipeline facility, and which may have the ability to send commands back to the facility.

Time-constant method: An open-loop tuning method that is based on the time constant, process gain, and deadtime from the process reaction curve.

Tolerance: The acceptable amount by which a process variable can deviate from its setpoint, such as 68 degrees ±3.

Ultimate gain: The maximum allowable value of gain or proportional band at which a system is stable.

Ultimate period: The period of response during which either the gain or proportional band is set at the ultimate value.

Ultimate period method: A closed-loop tuning method that requires the determination of the ultimate gain or proportional band and the ultimate period.

Ziegler-Nichols equations: Equations developed in the early 1940s by J.G. Ziegler and N.B. Nichols that are still used today as the basis of some of the most common methods of loop tuning.

Additional Resources

The following recommended resources can provide additional helpful information related to the requirements for this covered task:

API Recommended Practice 1161, Third Edition, © 2014 American Petroleum Institute. API Publishing Services, 1220 L Street, NW, Washington, DC 20005. Distributed by Thompson Reuters (Scientific) LLC, **www.techstreet.com**.

NCCER Module AOCCC-17, *Abnormal Operating Conditions Control Center*.

NCCER Module AOCFG-17, *Abnormal Operating Conditions Field and Gas*.

NCCER Module 12405-16, *Tuning Loops*.

NCCER Module 12204-16, *Process Control Theory*.

Instrumentation Level Three, NCCER. Third Edition. 2016. New York, NY: Pearson Education, Inc.

Instrumentation Level Four, NCCER. Third Edition. 2016. New York, NY: Pearson Education, Inc.

Instrument Engineers Handbook, Volume 1: Process Measurement and Analysis, Béla G. Lipták (editor). Fourth Edition. 2003. Boca Raton, FL: CRC Press.

Instrument Engineers' Handbook, Volume 2: Process Control and Optimization, Béla G. Lípták (editor). Fourth Edition. 2005. Boca Raton, FL: CRC Press.

Instrumentation and Process Control, Terry Bartelt. 2007. Clifton Park, NT: Cengage Learning.

Instrumentation Reference Book, Walt Boyes (editor). Fourth Edition. 2009. Oxford, UK: Butterworth-Heinemann.

Measurement and Control Basics, Thomas A. Hughes. Fifth Edition. 2014. Research Triangle Park, NC: The International Society of Automation.

Measurement and Instrumentation Principles, Alan S. Morris. Third Edition. 2001. Oxford, UK: Butterworth-Heinemann.

Purdy's Instrument Handbook, Ralph Dewey. 2015 Edition. Deer Park, TX: Glen Enterprises.

The Condensed Handbook of Measurement and Control, N.E. Battikha. Third Edition, 2006. Research Triangle Park, NC: The International Society of Automation.

The following websites offer resources for products and training:

Dwyer Instruments, Inc., **www.dwyer-inst.com**

Emerson Electric Co., **www.emerson.com**

The International Society of Automation, **www.isa.org**

OMEGA Engineering Inc., **www.omega.com**

Figure Credits

Biwater International Limited, Module Opener

© rCarner/Shutterstock.com, SA01

Greenlee / A Textron Company, Figure 22

Section Review Answer Key

Answer	Section Reference	Objective
Section One		
1. a	1.1.3	1a
2. b	1.2.0	1b
3. a	1.3.0	1c
4. d	1.4.3	1d
5. c	1.5.2	1e
Section Two		
1. a	2.1.2	2a
2. d	2.2.0	2b
Section Three		
1. b	3.1.1	3a
2. a	3.2.4	3b
3. d	3.3.2	3c

NCCER CURRICULA — USER UPDATE

NCCER makes every effort to keep its textbooks up-to-date and free of technical errors. We appreciate your help in this process. If you find an error, a typographical mistake, or an inaccuracy in NCCER's curricula, please fill out this form (or a photocopy), or complete the online form at **www.nccer.org/olf**. Be sure to include the exact module ID number, page number, a detailed description, and your recommended correction. Your input will be brought to the attention of the Authoring Team. Thank you for your assistance.

Instructors – If you have an idea for improving this textbook, or have found that additional materials were necessary to teach this module effectively, please let us know so that we may present your suggestions to the Authoring Team.

NCCER Product Development and Revision

13614 Progress Blvd., Alachua, FL 32615

Email: curriculum@nccer.org

Online: www.nccer.org/olf

❑ Trainee Guide ❑ Lesson Plans ❑ Exam ❑ PowerPoints Other ______________________

Craft / Level: Copyright Date:

Module ID Number / Title:

Section Number(s):

Description:

Recommended Correction:

Your Name:

Address:

Email: Phone:

CT30_0-17

Test Overfill Protective Devices

Overview

When working with storage tanks of various kinds, it's vital to avoid overfilling them. An overflow not only results in lost product, but it could also cause a number of other serious problems. If the material is hazardous, it could endanger personnel, facilities, and the environment. Hazardous material cleanup is both time-consuming and costly. Spills of all kinds can damage equipment, resulting in expensive repairs. Preventing overfill situations relies on a group of technologies known as overfill protective devices (OPDs), which are kept in-service by pipeline E&I technicians. This module examines the different types of OPDs and surveys the procedures used to test them.

Trainees with successful module completions may be eligible for credentialing through the NCCER Registry. To learn more, go to **www.nccer.org** or contact us at 1.888.622.3720. Our website has information on the latest product releases and training, as well as online versions of our *Cornerstone* magazine and Pearson's product catalog.

Your feedback is welcome. You may email your comments to **curriculum@nccer.org**, send general comments and inquiries to **info@nccer.org**, or fill in the User Update form at the back of this module.

This information is general in nature and intended for training purposes only. Actual performance of activities described in this manual requires compliance with all applicable operating, service, maintenance, and safety procedures under the direction of qualified personnel. References in this manual to patented or proprietary devices do not constitute a recommendation of their use.

CT30_0-17
Test Overfill Protective Devices

Objectives

When you have completed this module, you should be able to do the following:

1. Identify and describe the various types of overfill protective devices.
   a. Identify and describe mechanical OPDs.
   b. Identify and describe non-contact OPDs.
2. Describe the procedures used to test overfill protective devices (CT30_0-17).
   a. Identify the prerequisites, competencies, and abnormal operating conditions (AOCs) associated with testing OPDs.
   b. Describe the procedures used to test OPDs.

Performance Tasks

Under the supervision of the instructor, you should be able to do the following:

1. Test overfill protective devices (CT30_0-17).
   - Identify potential abnormal operating conditions that may occur during performance of this CT, and know the appropriate actions to take in response to them.
   - Utilize the appropriate personal protective equipment according to relevant company procedures.
   - Notify operations control center prior to performing any test or working on the system.
   - Manually trip the overfill protective device alarm.
   - Verify the operations control center has received the alarms and they have cleared appropriately.
   - Verify that the alarms were received locally and that the alarms cleared appropriately.
   - Verify operation of any devices, such as valves, that might operate as part of a relief flow system.
   - Demonstrate the ability to request equipment repairs or replacement.
   - Notify operations control center and/or affected personnel.
   - Complete appropriate documentation as required by operator's procedures.

Trade Terms

Abnormal operating conditions (AOCs)
Automatic overfill protection system (AOPS)
Critical high (CH)
High-high (HH)
Maximum working (MW)
Still pipe
Tank gauge

Industry Recognized Credentials

If you are training through an NCCER-accredited sponsor, you may be eligible for credentials from NCCER's Registry. The ID number for this module is CT30_0-17. Note that this module may have been used in other NCCER curricula and may apply to other level completions. Contact NCCER's Registry at 1.888.622.3720 or go to **www.nccer.org** for more information.

Contents

1.0.0 Overfill Protective Devices 1

- 1.1.0 Mechanical OPDs 1
  - 1.1.1 Mechanical OPD Operation 2
  - 1.1.2 Mechanical OPD problems 2
  - 1.1.3 Level Switches 2
- 1.2.0 Non-Cont wwact OPDs 5
  - 1.2.1 Radar, Sonar, and Laser OPDs 5

2.0.0 Section TwoTesting Overfill Protective Devices 8

- 2.1.0 Prerequisites, Competencies, and Abnormal Operating Conditions 8
  - 2.1.1 Prerequisites and Competencies 8
  - 2.1.2 Abnormal Operating Conditions 8
- 2.2.0 Procedure 9

Figures

Figure 1 A mechanical OPD 2

Figure 2 Level switch 3

Figure 3 Level switch operation 3

Figure 4 Mounted level switch 4

Figure 5 Tubing connections 4

Figure 6 External cage assembly for mounting a level switch 5

Figure 7 A radar OPD uses reflection timing to determine the liquid's height 6

Figure 8 Radar OPD antennas 6

SECTION ONE

1.0.0 OVERFILL PROTECTIVE DEVICES

Objective

Identify and describe the various types of overfill protective devices.

a. Identify and describe mechanical OPDs.
b. Identify and describe non-contact OPDs.

Trade Terms

Automatic overfill protection system (AOPS): Refers to an automatic system designed to prevent overfilling. AOPS also refers to the level at which this system will be activated.

Critical high (CH): The level at which a tank is close to overflowing and at which an emergency response system will be activated.

High-high (HH): The level above the normal high level. An alarm will be activated if the liquid level reaches this point.

Maximum working (MW): The normal high level in a tank and the lowest of the high levels. Reaching this height does not trigger an alarm.

Still pipe: A perforated pipe that runs from the top to the bottom of a tank. Still pipes are designed to provide a non-turbulent place from which to take measurements.

Tank gauge: Refers to one of several technologies designed to report the liquid level in a storage tank. Tank gauges are used for everyday tank operations, such as custody transfer management.

Overfill protective devices (OPDs) serve to prevent overflow of storage tanks. They come in many different forms and with varying degrees of sophistication. Simple OPDs are purely mechanical and may do nothing more than shut off flow to the tank. More advanced technologies have no moving parts and may be part of an automatic control system that can handle overfill situations with far more sophistication. OPDs are separate from other tank level monitoring systems, such as a **tank gauge**, although they may use some of the same technologies.

In a typical tank system there are several defined high points (high liquid levels in the tank) that can be detected. Historically, these have gone by a number of different names and have had varied definitions. The following four levels are defined by API *Standard 2350* 4th Edition.

The **maximum working (MW)** height is the lowest high level and refers to the normal maximum height to which liquid should rise. There is, however, a significant cushion of space above this point to allow for variation. Normally, no alarms will be raised if the liquid rises to the maximum working level, although many control systems will report the event as information.

The next level is **high-high (HH)**. If the liquid reaches this point, it will trigger an alarm and the flow should be terminated within a specific period of time. Above the high-high level is the **automatic overfill protection system (AOPS)** point. An AOPS is an automatic system that takes corrective action without operator intervention. Only tanks equipped with an AOPS will sense this level. The final level is the **critical high (CH)** point. At this level, the tank is about to overflow. If this level is detected, emergency actions must be initiated at once to prevent a spill.

Overfill protection is no longer simply synonymous with mechanical level switches. Meeting two key standards for overfill protection, API *Standard2350* and *IED 61511*, may require the use of an AOPS. In such a system, certified overfill protection sensors measure level continuously and independently from the automatic tank gauges. When a shutdown condition occurs, the AOPS automatically initiates a shutdown of incoming flow to a tank without any manual intervention.

An AOPS can reduce risks, but it can also increase risks if it is not applied and designed properly. Particular care is required when applying AOPS to any marine or pipeline operation. Thorough engineering analysis is critical to prevent the potential for pipeline rupture. For instance, an improperly designed AOPS could close off a flowing tank receipt delivering from a pipeline, which could cause the pipeline to rupture.

The pipeline operator and the terminal operator must work together to determine if and how an AOPS will be used. They need to devise an arrangement that maximizes benefits yet also meets all of the relevant standards for overfill protection.

1.1.0 Mechanical OPDs

Mechanical OPDs are the simplest type of protective device. They either restrict flow or cut it off entirely, depending on their design. Like all mechanical devices, however, they can wear out or malfunction due to problems with their moving parts. The following sections discuss the most common varieties of mechanical OPDs.

1.1.1 Mechanical OPD Operation

Most mechanical OPDs are integrated into the tank's drop tube along with some type of valve. Liquid flows through them, usually encountering some kind of valve. Almost all mechanical OPDs rely on a float to sense the liquid level in the tank. When the liquid is below the desired high level point, the float is at its lowest position. This causes the valve in the drop tube to be fully open, permitting unrestricted filling.

As the liquid level rises and reaches the bottom of the float, the float will begin to rise with the liquid. Many mechanical OPDs start restricting the flow at this point. If an operator is manually filling the tank, the change in flow will be detected and the operation can be halted. Since the valve is still partially open, however, hoses can be drained into the tank.

If filling continues, the float keeps rising and eventually closes the valve completely, halting the fill operation. Some mechanical OPDs might also include contacts that can trigger an alarm or provide status information to a control system.

Mechanical OPDs (*Figure 1*) vary primarily in their float style and valve type. Some floats are hinged and rise at an angle. Others rise vertically along a pair of guide rods. Valves are often of a piston construction, although there are other styles. Ball-type floats are used in some OPDs and can function both as the level-sensing element and as the valve itself. These are more common in vent pipes.

Figure 1 A mechanical OPD.

1.1.2 Mechanical OPD problems

Since mechanical OPDs rely on moving parts, they are prone to a number of failure points. One common problem is sticking in the float linkage mechanism. This can prevent the float from rising with the liquid, causing failure to close the valve. It's also possible for the control mechanism to stick in the closed position. In this case, when the liquid level drops, the float will be stuck in its highest position, rather than falling with the liquid and opening the valve.

Many mechanical OPDs come with a manual test mechanism that allows pipeline E&I technicians to confirm their operation. In many cases, this is nothing more than a cord or chain connected to the float. Pulling up on the cord tests the OPD's shutoff behavior, while releasing it confirms that the device re-opens the valve. Actuating the manual mechanism is usually a part of OPD testing procedures.

1.1.3 Level Switches

Level switches are used in many different pipeline applications. One of these is as an OPD. An example of a level switch (*Figure 2*) is the Magnetrol Tuffy® liquid level control with a pneumatic switch. These switches are designed for horizontal mounting in a tank or vessel through threaded or flanged pipe connections.

Switching action is obtained through the use of a magnet attached to a float and a switching mechanism. A non-magnetic, pressure-tight barrier separates these two basic component assemblies. The float moves the magnet upward and into the field of the switch mechanism's magnet. As a result, the magnets repel, causing the pivot cam and actuating pin to seat the ball firmly, which shuts off the air supply. Level switch operation is shown *Figure 3*.

Before the switch in a tank or vessel is assembled, the threaded or flanged mounting nozzle should be checked for the following:

- The nozzle length and diameter must be the correct size to allow for switch actuation at design levels (*Figure 4*).
- The horizontal alignment of the nozzle must be checked. Since the switch is mounted differently for high-or low-level operation, it must be mounted correctly for system requirements:

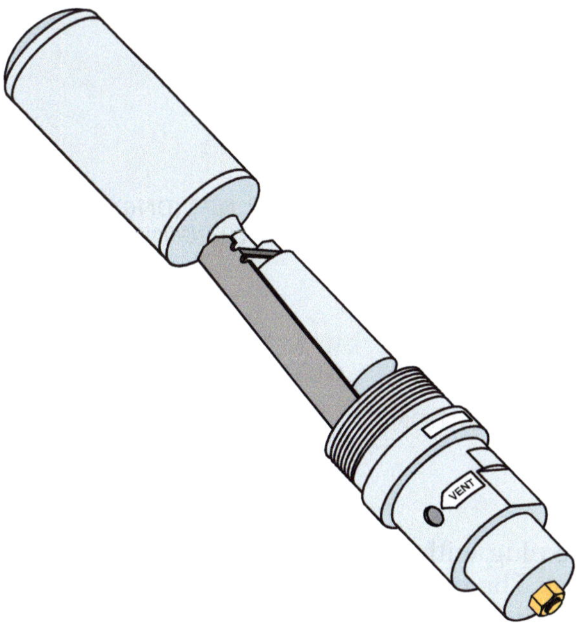

Figure 2 Level switch.

- For high-level operation, the switch must have the Hi Level label on top and the wrench flats horizontally level. In this position, with the float down, no air flows. With the float in the up position, air flows into the outlet. This would be the normal mounting when a level switch is acting as an OPD.
- For low-level operation, the switch must have the Low Level label on top and the wrench flats horizontally level. In this position, with the float up, no air flows. With the float in the down position, air will flow into the outlet.

The inlet air line is connected to the $\frac{1}{8}$ inch national pipe thread (NPT) (DN6) fitting labeled INLET on the end of the valve. Inlet air pressure should be in the range of 10–100 psig (69–689 kPa). Outlet air is connected to the $\frac{1}{8}$ inch NPT (DN6) connection on the side of the valve labeled OUTLET. These tubing connections are shown in *Figure 5*.

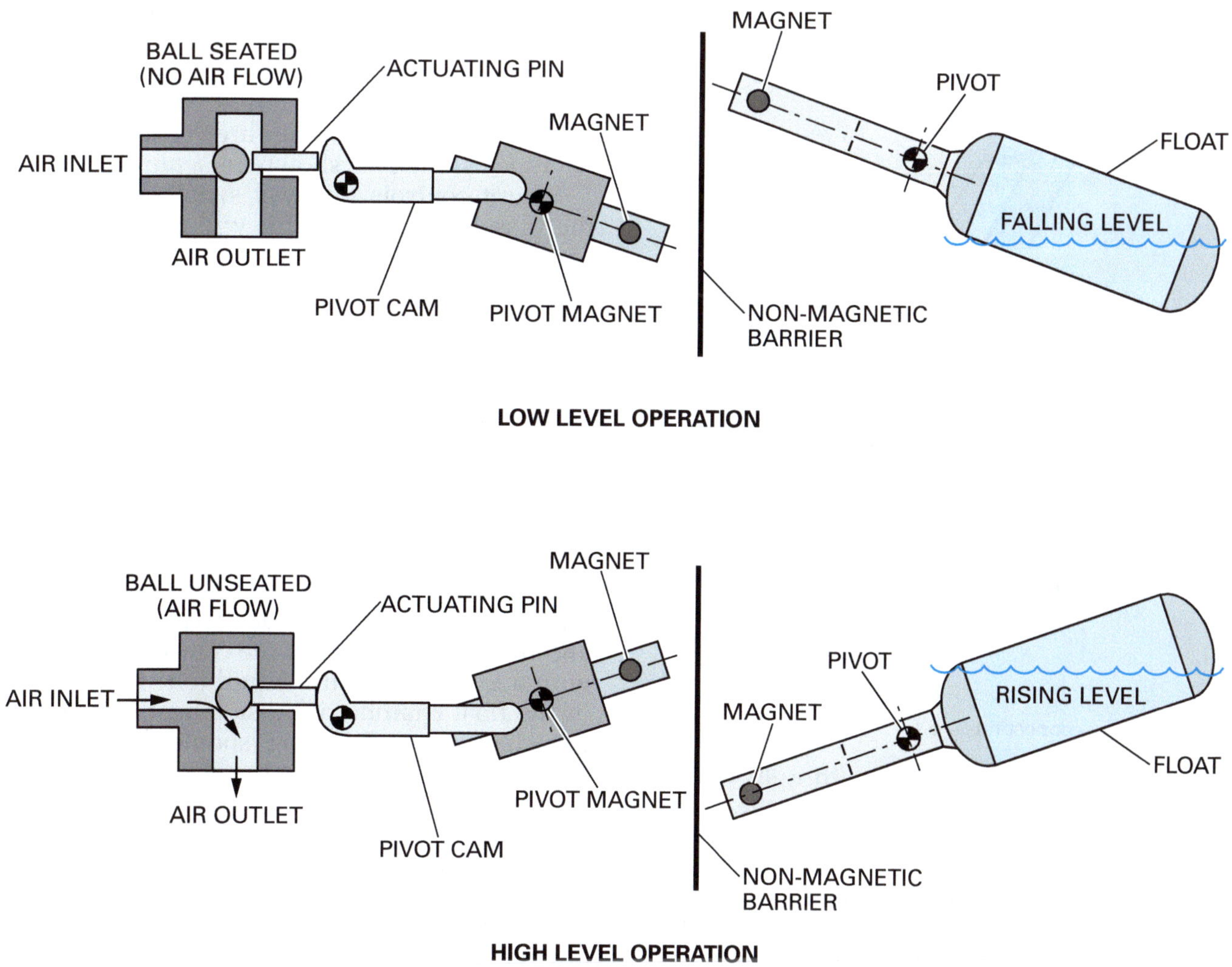

Figure 3 Level switch operation.

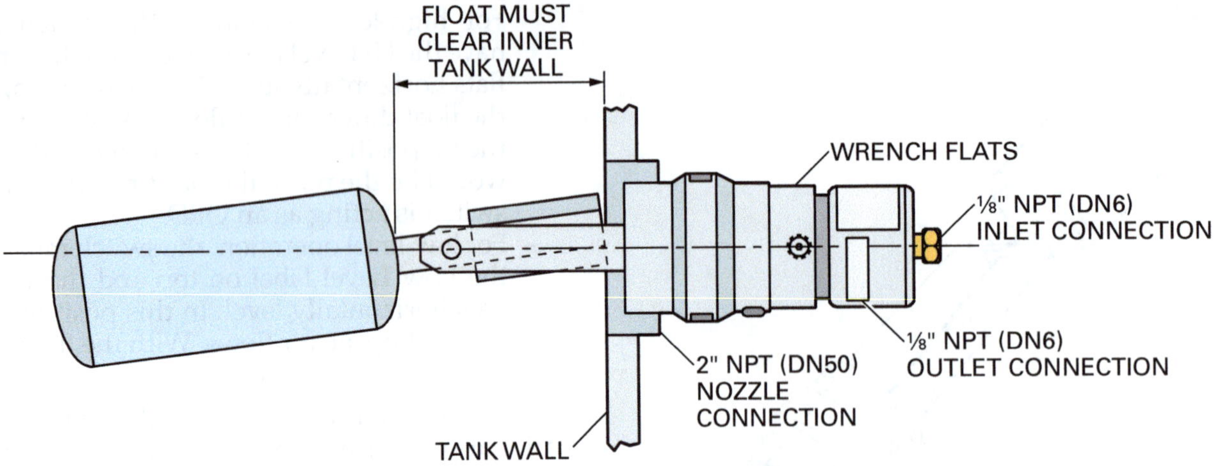

Figure 4 Mounted level switch.

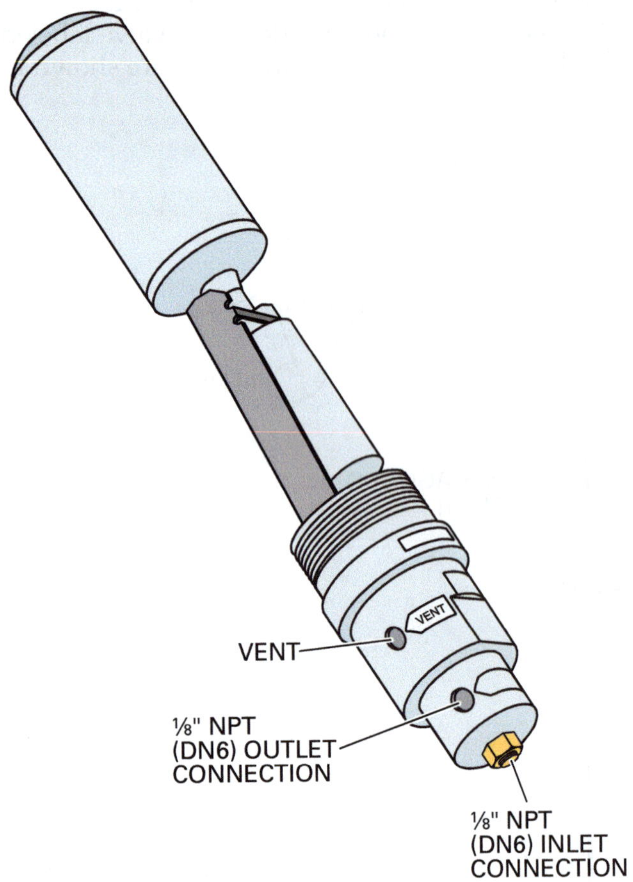

Figure 5 Tubing connections.

Inlet air should be filtered with a 40 micron (0.04 mm) or finer filter. This will extend the times between cleaning of the mechanism. If required, a vent line may be connected to the opening labeled VENT on the side of the housing. A vent line takes any vented air away from the device location. If a vent line is not required, the pipe plug with a small hole in it should be left in place. The opening in the plug allows vented air to be discharged at the valve location.

The switch can be mounted externally using the external cage assembly. The switch actuates within a $^3/_4$ inch (19 mm) distance above the cage center line for the rising level, and within a $^3/_4$ inch (19 mm) distance below the center line for the falling level. The external cage must be installed so that the tappings are vertical. An external cage assembly is shown in *Figure 6*.

Mount controls for level switches as close to the vessel as possible. This will result in a more responsive and accurate level change in the cage. Liquid in a long pipeline may be cooler and denser than liquid in a vessel, causing the level indication in the control to be lower than the actual level in the vessel.

Use pipe of sufficient strength to support the control. If necessary, provide a stand or hanger to help support its weight. All piping should be straight and free of low spots or pockets. This ensures that the lower liquid line will drain towards the vessel and the upper vapor line will drain toward the control. Shutoff valves are recommended for installation between the vessel and the cage assembly. If the control is to be used with a low-temperature liquid (one that will boil in the float chamber if outside heat is absorbed), the chamber and piping should be insulated. Otherwise, the boiling in the chamber will cause false level indications. However, the switch mechanism housing should not be insulated.

Level switch repair is typically limited to individual component replacement of the switch. Follow manufacturer's instructions to perform any repair.

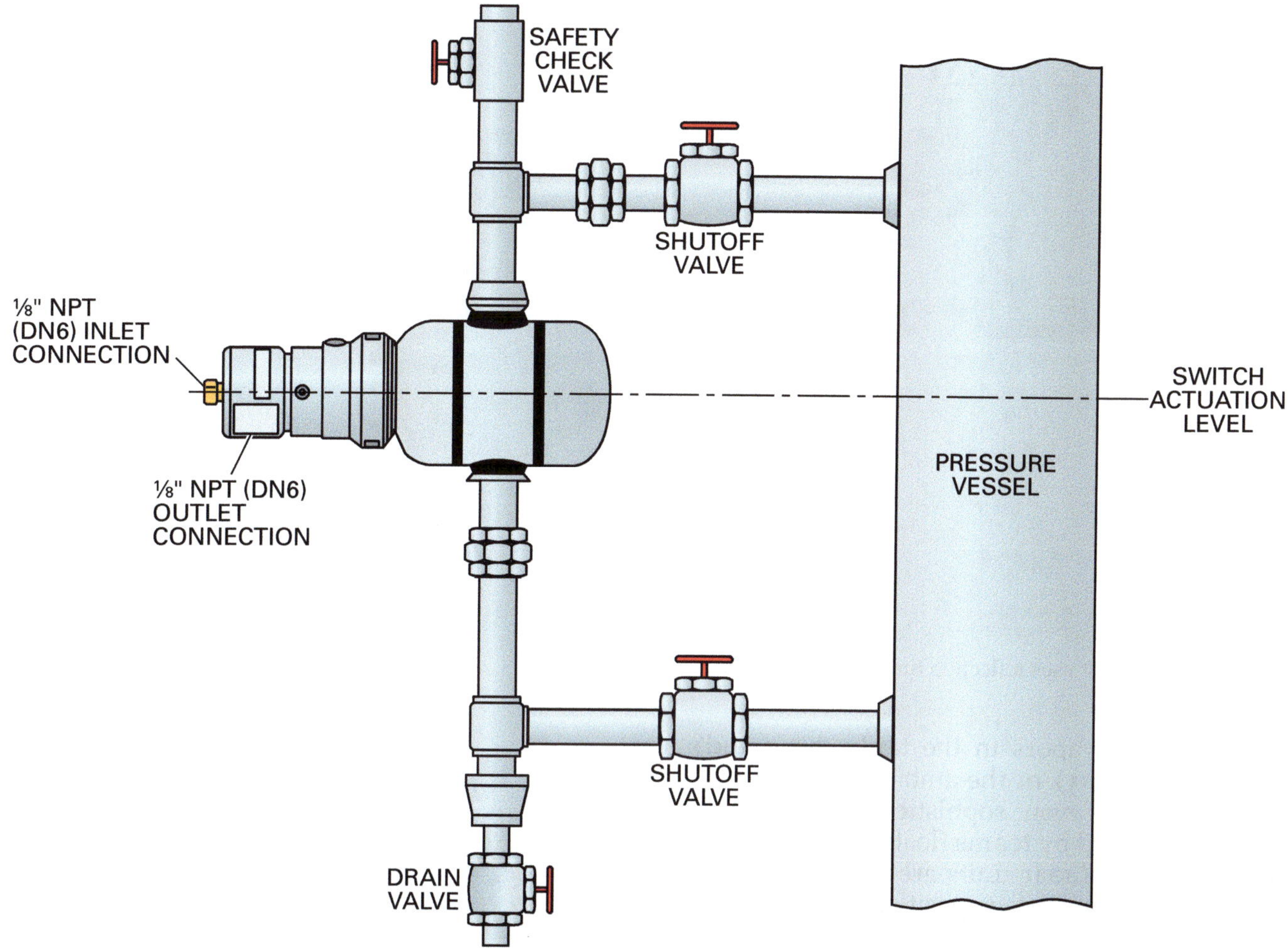

Figure 6 External cage assembly for mounting a level switch.

NOTE

Certain liquids contain particulates or salts that can accumulate in level measuring devices. The accumulated material can result in inaccurate measurements or even make the measuring device cease to function. Consider using non-contact level measuring options for liquids that have these characteristics.

1.2.0 Non-Contact OPDs

In recent years, many operations have been switching over to OPDs that are non-mechanical. These devices rely on methods of sensing the liquid level without actually coming into contact with the liquid itself. Most of these technologies rely on the principle of reflection to determine the liquid surface's location. Since they have no moving parts, they are less prone to wear or malfunctions. However, they are usually much more complex and expensive than the simple mechanical OPDs.

All non-contact OPDs are electronic, generating a signal that indicates when the liquid level has reached one of the several defined high levels. These signals may trigger alarms or valves directly. Alternatively, they may communicate with a controller that makes decisions about the appropriate actions to take.

1.2.1 Radar, Sonar, and Laser OPDs

Non-contact OPDs generally work by aiming a signal at the liquid's surface and timing how long it takes the reflection to return. Using the speed at which the signal travels, the OPD performs a simple calculation to determine the distance to the liquid's surface. There are three kinds of OPD that use this method: radar, sonar, and laser.

A radar OPD (*Figure 7*) uses microwave radio frequency energy as the signal. Since RF energy travels at the speed of light, timing and calculations are based on this value. The same principle is used for radars of all kinds. A radar OPD is

Figure 7 A radar OPD uses reflection timing to determine the liquid's height.

not affected by vapors in the tank, the liquid's thickness (viscosity), or the ambient temperature. Some units are even sophisticated enough to avoid being fooled by foams floating on top of the liquid's surface. Radar units are extremely accurate, reporting the surface position to within just a few millimeters.

Typical radar OPDs usually have interchangeable antennas (*Figure 8*) that alter the shape of the microwave beam. Cone/horn and parabolic antennas are designed to be used in larger tanks with plenty of space around the antenna. Array and rod antennas are meant to be used in a still pipe, in which the space is much more limited. It's important to select and install the best antenna for the individual situation. Each manufacturer provides complete information to guide the selection process.

The sonar OPD works in much the same way as the radar version except that it uses high-frequency sound waves instead of radio waves. Like the radar OPD, it has no moving parts and relies on measuring reflection time to determine the position of the liquid's surface. Unlike the radar OPD, however, sonar OPD accuracy can be affected by temperature and vapors since both can alter the local speed of sound. These factors must be taken into account in order to obtain accurate results. Some sonar OPDs include temperature sensors to help compensate for this component.

Figure 8 Radar OPD antennas.

Finally, laser OPDs perform essentially the same function as radar and sonar units. Instead of RF or sound energy, they use laser light, which bounces off the liquid's surface. Like radar OPDs, laser units are not affected by temperature or the atmosphere in the tank.

All reflection OPDs must go through a setup process designed to eliminate false readings due to reflections from tank seams and welds. Once this process has been completed, however, the system requires relatively little maintenance or adjustment. Setup often involves using manufacturer-supplied software and an attached notebook computer.

Additional Resources

It is assumed that you will learn to perform these procedures under the direct supervision of your crew leader, foreman, or instructor. For independent study, refer to the recommended Additional Resources at the back of this module.

1.0.0 Section Review

1. A mechanical OPD sense liquid level with a(n) _____.
   a. DP flow sensor
   b. acoustic transducer
   c. float
   d. valve

2. A reflection-type OPD requires initial setup to compensate for _____.
   a. electrical interference
   b. seams and welds
   c. viscosity
   d. tank flow rate

SECTION TWO

2.0.0 TESTING OVERFILL PROTECTIVE DEVICES

Objective

Describe the procedures used to test overfill protective devices (CT30_0-17).

a. Identify the prerequisites, competencies, and abnormal operating conditions (AOCs) associated with testing OPDs.
b. Describe the procedures used to test OPDs.

Performance Task

Test overfill protective devices (CT30_0-17).

- Identify potential abnormal operating conditions that may occur during performance of this CT, and know the appropriate actions to take in response to them.
- Utilize the appropriate personal protective equipment according to relevant company procedures.
- Notify operations control center prior to performing any test or working on the system.
- Manually trip the overfill protective device alarm.
- Verify the operations control center has received the alarms and they have cleared appropriately.
- Verify that the alarms were received locally and that the alarms cleared appropriately.
- Verify operation of any devices, such as valves, that might operate as part of a relief flow system.
- Demonstrate the ability to request equipment repairs or replacement.
- Notify operations control center and/or affected personnel.
- Complete appropriate documentation as required by operator's procedures.

Trade Terms

Abnormal operating conditions (AOCs): Conditions within a pipeline system, identified by the operator, that may indicate (1) a malfunction of a component or (2) a deviation from normal operations that may indicate a condition exceeding design limits or result in a hazard to persons, property, or the environment.

As with other instrumentation, OPDs require regular inspection and testing to keep them functioning properly. Since the consequences of a spill can be very serious, this task should never be neglected. Each OPD has its own test procedure. Always consult the manufacturer's documentation for guidelines.

2.1.0 Prerequisites, Competencies, and Abnormal Operating Conditions

Personnel must possess a variety of skills in order to successfully test OPDs. The various operations require a series of steps, which a pipeline E&I technician must be competent to perform. Personnel must also be aware of the particular abnormal operating conditions (AOCs) that may occur during the procedure and know the proper responses to them.

2.1.1 Prerequisites and Competencies

Personnel must possess the following basic knowledge and skills prior to testing OPDs:

- Familiarity with tank types as well as their filling and draining systems
- Ability to communicate with the control center using appropriate protocols and technologies
- Knowledge of company procedure for testing OPDs
- Familiarity with the control system, its alarms, and their behaviors and consequences

2.1.2 Abnormal Operating Conditions

Personnel testing OPDs must be able to identify abnormal operating conditions (AOCs) and know the proper response to each, should they occur. The following are some potential AOCs, and the appropriate action the E&I technician should take in response to them:

- *The tank's liquid level is inappropriately high or low* – The pipeline E&I technician must alert appropriate personnel to take the requisite corrective action.
- *During testing, inappropriate actions or a shutdown occurs* – The pipeline E&I technician must alert appropriate personnel about the event and the status of the OPD as required by company procedure.
- *An OPD is inoperable* – The pipeline E&I technician must alert appropriate personnel about the situation and the status of the OPD as required by company procedure.

2.2.0 Procedure

The following steps outline the general procedure for testing overfill protective devices:

> **NOTE**
>
> Obtain and check device history before inspecting or calibrating an OPD. This will identify things that may have gone wrong previously with the device, and it can save a lot of time if there is documented history of malfunction (and how it was resolved).

> **NOTE**
>
> Always follow operator procedures related to lone worker status, grounding, and fall protection.

Step 1 Identify potential abnormal operating conditions that may occur, and know the appropriate actions to take in response to them.

Step 2 Utilize the appropriate personal protective equipment according to relevant company procedures.

Step 3 Before beginning the process, notify the control center, local personnel, and any other personnel who might be affected by the test procedure. The control center may have to provide feedback about the OPD alarm response during the process. It may also have to override automatic shutdowns or safety systems to protect the process from interruption. Since an OPD trip indicates a potentially serious condition, it's important that a test does not trigger any undesirable outcomes. These events are often known as Critical Interlock and Trips (CIATs). They usually require higher authorization to override. Pipeline E&I technicians should be aware of these rules and know what they're allowed to do.

Step 4 Determine the device's number and nameplate data in order to verify that it is the correct unit to test and that any data coming from the unit will be properly identified as such.

Step 5 Confirm that you understand the system that the OPD affects. Be aware of any actions that may result from testing the OPD. Confirm that the control center is aware of any unexpected actions that could result from the test.

> **NOTE**
>
> Company policy may require using a handline or hand gauge to check level height, even for overfill protective devices that are equipped with mechanical gauges.

> **NOTE**
>
> Accuracy is affected if there has been recent flow into or out of the tank, especially high flow. Normally, to ensure accuracy, wait four hours after flow into or out of a tank before checking the level height. As little as a $^1/_4$-inch (6 mm) inaccuracy in level on a 100M-barrel tank represents a considerable volume.

Step 6 Alert appropriate personnel that you are about to trigger the OPD. Using appropriate measures, activate the OPD to simulate an overfill condition.

> **CAUTION**
>
> In some terminals with multiple-receipt tanks, alarms are wired fail safe and daisy-chained together. Triggering an OPD on one tank will cause all the tanks to be considered in an alarm condition. As a result, automated valves close, possibly activating relief flow systems or overpressure relief devices. Automated sequences may include shut down of incoming product, closure of automated isolation valves, and operation of relief flow devices to prevent exceeding the maximum allowable operating pressure (MAOP).

Step 7 Confirm that the correct actions occurred and that the appropriate personnel received the OPD alert. If the OPD triggered any automated actions, be sure that the control center is aware of them.

Step 8 Return the OPD to its normal condition and verify that all alarms and other events have been reset and returned to normal. Confirm that any automatic responses have also returned to a normal condition.

Step 9 Notify the control center, local personnel, and any affected personnel that the test is complete and that the unit has been returned to normal operation.

Step 10 Document the results of the test. Follow company procedure.

Additional Resources

It is assumed that you will learn to perform these procedures under the direct supervision of your crew leader, foreman, or instructor. For independent study, refer to the recommended Additional Resources at the back of this module.

2.0.0 Section Review

1. If an OPD is not working, what is the first thing that the pipeline E&I technician should do?
   a. Alert appropriate personnel.
   b. Repair or replace the OPD.
   c. Enter an override into the control system.
   d. Request that the process be halted as a precaution.

2. Why is it especially important to have the control center fully involved in testing an OPD?
   a. The control center has the test instrument calibration records.
   b. An OPD is a delicate device so the control center will want to be sure that the test is properly handled.
   c. An OPD alert is a potentially serious event and automatic actions could occur as a result of the test.
   d. The control center supervises all tests.

Summary

The consequences of overfilling a storage tank can be very serious. Damage, injury, environmental contamination, and financial loss are all possibilities when product escapes its storage tank. Overfill protection devices help to prevent overflows and spills by shutting down the fill process if the tank's level rises above an appropriate level. Keeping these devices working reliably is the job of the pipeline E&I technician. This module surveyed the different types of OPDs and provided the basic procedure for testing them.

Review Questions

1. Which of the four defined tank high levels is considered normal and will not trigger an alarm?
   a. Maximum working (MW)
   b. High-high (HH)
   c. Automatic overfill protection system (AOPS)
   d. Critical high (CH)

2. In their first stage of activation, many mechanical OPDs _____.
   a. halt the fill flow
   b. restrict the fill flow
   c. initiate an emergency response
   d. open the valve

3. What does a laser OPD bounce off the surface of the liquid?
   a. Microwaves
   b. Sound
   c. Light
   d. Magnetic flux

4. Pipeline E&I technicians preparing to test an OPD should be familiar with _____.
   a. the local speed of sound
   b. the tank diameter
   c. digital multimeters
   d. the control system

5. If personnel who are testing an OPD find that the tank level is inappropriately high or low, they should __.
   a. Adjust the floating roof.
   b. Open or close the valve.
   c. Call 911.
   d. Notify appropriate personnel.

6. Once you have triggered an OPD for testing purposes, what should you do first?
   a. Confirm that the correct actions occurred.
   b. Confirm that the entire process shuts down.
   c. Scan the fieldbus.
   d. Document what you have just done.

Trade Terms Quiz

Fill in the blank with the correct term that you learned from your study of this module.

1. _________ refers to an automatic system designed to prevent overfilling. Also refers to the level at which this system will be activated.
2. A(n) _________ is a perforated pipe that runs from the top to the bottom of a tank. It is designed to provide a non-turbulent place from which to take measurements.
3. Conditions within a pipeline system, identified by the operator, that may indicate (1) a malfunction of a component or (2) a deviation from normal operations that may indicate a condition exceeding design limits or result in a hazard to persons, property, or the environment are called _________.
4. The _________ is the level above the normal high level. An alarm will be activated if the liquid level reaches this point.
5. _________ refers to one of several technologies designed to report the liquid level in a storage tank. It is used for everyday tank operations, such as custody transfer management.
6. The level at which a tank is close to overflowing and at which an emergency response system will be activated is called _________.
7. _________ is the normal high level in a tank and the lowest of the high levels. Reaching this height does not trigger an alarm.

Trade Terms

Abnormal operating conditions (AOCs)
Automatic overfill protection system (AOPS)
Critical high (CH)
High-high (HH)
Maximum working (MW)
Still pipe
Tank gauge

Trade Terms Introduced in This Module

Abnormal operating conditions (AOCs): Conditions within a pipeline system, identified by the operator, that may indicate (1) a malfunction of a component or (2) a deviation from normal operations that may indicate a condition exceeding design limits or result in a hazard to persons, property, or the environment.

Automatic overfill protection system (AOPS): Refers to an automatic system designed to prevent overfilling. Also refers to the level at which this system will be activated.

Critical high (CH): The level at which a tank is close to overflowing and at which an emergency response system will be activated.

High-high-(HH): The level above the normal high level. An alarm will be activated if the liquid level reaches this point.

Maximum working (MW): The normal high level in a tank and the lowest of the high levels. Reaching this height does not trigger an alarm.

Still pipe: A perforated pipe that runs from the top to the bottom of a tank. Still pipes are designed to provide a non-turbulent place from which to take measurements.

Tank gauge: Refers to one of several technologies designed to report the liquid level in a storage tank. Tank gauges are used for everyday tank operations, such as custody transfer management.

Additional Resources

The following recommended resources can provide additional helpful information related to the requirements for this covered task:

API Recommended Practice 1161, Third Edition, © 2014 American Petroleum Institute. API Publishing Services, 1220 L Street, NW, Washington, DC 20005. Distributed by Thompson Reuters (Scientific) LLC, **www.techstreet.com**.

NCCER Module AOCCC-17, *Abnormal Operating Conditions Control Center*

NCCER Module AOCFG-17, *Abnormal Operating Conditions Field and Gas*

Instrumentation Level One, NCCER. Third Edition. 2014. New York, NY: Pearson Education, Inc.

Instrumentation Level Two, NCCER. Third Edition. 2016. New York, NY: Pearson Education, Inc.

Instrumentation Level Three, NCCER. Third Edition. 2016. New York, NY: Pearson Education, Inc.

Instrumentation Level Four, NCCER. Third Edition. 2016. New York, NY: Pearson Education, Inc.

Instrument Engineers Handbook, Volume 1: Process Measurement and Analysis, Béla G. Lipták (editor). Fourth Edition. 2003. Boca Raton, FL: CRC Press.

Instrument Engineers' Handbook, Volume 2: Process Control and Optimization, Béla G. Lípták (editor). Fourth Edition. 2005. Boca Raton, FL: CRC Press.

Instrumentation, Franklyn W. Kirk, Thomas A. Weedon, and Philip Kirk. Fifth Edition. 2010. Orland Park, IL: American Technical Publishers.

Instrumentation and Process Control, Terry Bartelt. 2007. Clifton Park, NT: Cengage Learning.

Instrumentation Reference Book, Walt Boyes (editor). Fourth Edition. 2009. Oxford, UK: Butterworth-Heinemann.

Measurement and Control Basics, Thomas A. Hughes. Fifth Edition. 2014. Research Triangle Park, NC: The International Society of Automation.

Purdy's Instrument Handbook, Ralph Dewey. 2015. Deer Park, TX: Glen Enterprises.

The Condensed Handbook of Measurement and Control, N.E. Battikha. Third Edition. 2006. Research Triangle Park, NC: The International Society of Automation.

NCCER Module 60102-02, *Basic Pipeline Hydraulics and Equipment*

NCCER Module 64202-02, *Use of Meters and Test Equipment*

NCCER Module CT44_4-17, *Inspection, Testing, Corrective and Preventative Maintenance of Tank Gauging for Hazardous Liquid Leak Detection*

NCCER Module CT63_1-17, *Start-up of a Liquid Pipeline (Field)*

NCCER Module CT63_2-17, *Shutdown of a Liquid Pipeline (Field)*

NCCER Module CT63_4-17, *Locally Operate Valves on a Liquid Pipeline System*

NCCER Module CT64_1-17, *Start-up of a Liquid Pipeline (Control Center)*

NCCER Module CT64_2-17, *Shutdown of a Liquid Pipeline (Control Center)*

NCCER Module CT64_4-17, *Remotely Operate Valves on a Liquid Pipeline System*

The following websites offer resources for products and training:

Emerson Electric Co., **www.emerson.com**

Morrison Bros. Co., **www.morbros.com**

OPW, A Dover Company, **www.opwglobal.com**

The International Society of Automation, **www.isa.org**

Figure Credits

© Supakitmod/Dreamstime.com, Module Opener
Morrison Bros. Co., Figure 1
Emerson Process Management, Figures 7 and 8

Section Review Answer Key

Answer	Section Reference	Objective
Section One		
1. c	1.1.1	1a
2. b	1.2.1	1b
Section Two		
1. a	2.1.2	2a
2. c	2.2.0	2b

NCCER CURRICULA — USER UPDATE

NCCER makes every effort to keep its textbooks up-to-date and free of technical errors. We appreciate your help in this process. If you find an error, a typographical mistake, or an inaccuracy in NCCER's curricula, please fill out this form (or a photocopy), or complete the online form at **www.nccer.org/olf**. Be sure to include the exact module ID number, page number, a detailed description, and your recommended correction. Your input will be brought to the attention of the Authoring Team. Thank you for your assistance.

Instructors – If you have an idea for improving this textbook, or have found that additional materials were necessary to teach this module effectively, please let us know so that we may present your suggestions to the Authoring Team.

NCCER Product Development and Revision

13614 Progress Blvd., Alachua, FL 32615

Email: curriculum@nccer.org
Online: www.nccer.org/olf

❑ Trainee Guide ❑ Lesson Plans ❑ Exam ❑ PowerPoints Other ____________

Craft / Level: Copyright Date:

Module ID Number / Title:

Section Number(s):

Description:

Recommended Correction:

Your Name:

Address:

Email: Phone:

NCCER CURRICULA — USER UPDATE

NCCER makes every effort to keep its textbooks up-to-date and free of technical errors. We appreciate your help in this process. If you find an error, a typographical mistake, or an inaccuracy in NCCER's curricula, please fill out this form (or a photocopy), or complete the online form at www.nccer.org/olf. Be sure to include the exact module ID number, page number, a detailed description, and your recommended correction. Your input will be brought to the attention of the Authoring Team. Thank you for your assistance.

Instructors – If you have an idea for improving this textbook, or have found that additional materials were necessary to teach this module effectively, please let us know so that we may present your suggestions to the Authoring Team.

NCCER Product Development and Revision
13614 Progress Blvd., Alachua, FL 32615
Email: curriculum@nccer.org
Online: www.nccer.org/olf

☐ Trainee Guide ☐ Lesson Plans ☐ Exam ☐ PowerPoints Other ____________

Module ID Number / Title:

Section Number(s):

Description:

Recommended Correction:

Your Name:

Address:

Phone:

CT31_0-17

Inspect and Calibrate Overfill Protective Devices (Liquid)

Overview

When working with storage tanks of various kinds, it's vital to avoid overfilling them. Not only would an overflow result in lost product, but it could also cause a number of more serious problems. If the material is hazardous, it could endanger personnel, facilities, and the environment. Hazardous material cleanup is both time-consuming and costly. Furthermore, spills of all kinds can damage equipment, resulting in expensive repairs. Preventing overfill relies on a group of technologies known as overfill protective devices (OPDs). Keeping them working is the task of a pipeline E&I technician. This module examines the different types of OPDs and surveys the procedures used to inspect and calibrate them.

Trainees with successful module completions may be eligible for credentialing through the NCCER Registry. To learn more, go to **www.nccer.org** or contact us at 1.888.622.3720. Our website has information on the latest product releases and training, as well as online versions of our *Cornerstone* magazine and Pearson's product catalog.

Your feedback is welcome. You may email your comments to **curriculum@nccer.org**, send general comments and inquiries to **info@nccer.org**, or fill in the User Update form at the back of this module.

This information is general in nature and intended for training purposes only. Actual performance of activities described in this manual requires compliance with all applicable operating, service, maintenance, and safety procedures under the direction of qualified personnel. References in this manual to patented or proprietary devices do not constitute a recommendation of their use.

CT31_0-17
Inspect and Calibrate Overfill Protective Devices (Liquid)

When you have completed this module, you will be able to do the following:

1. Identify and describe the various types of overfill protective devices.
   a. Identify and describe mechanical OPDs.
   b. Identify and describe non-contact OPDs.
2. Describe the procedures used to inspect and calibrate overfill protective devices (CT31_0-17).
   a. Identify the prerequisites, competencies, and abnormal operating conditions (AOCs) associated with inspecting and calibrating OPDs.
   b. Describe the procedures used to inspect and calibrate OPDs.

Performance Tasks

Under the supervision of the instructor, you should be able to do the following:

1. Inspect and calibrate overfill protective devices (CT31_0-17).
   - Identify potential abnormal operating conditions that may occur during performance of this CT, and know the appropriate actions to take in response to them.
   - Utilize the appropriate personal protective equipment according to relevant company procedures.
   - Notify operations control center prior to performing any test or working on system.
   - Inspect OPD for corrosion or damage. Clean, repair, or replace per the manufacturer's instructions.
   - Test OPD switch.
   - Verify operation of flow control devices and flow switches on relief tanks for operability and/or damage.
   - Demonstrate ability to calibrate OPD switch per manufacturer's instructions.
   - Notify operations control center that inspection, testing and calibration is complete and control is returned to SCADA system.
   - Complete appropriate documentation as required by operator's procedures.

Trade Terms

Abnormal operating conditions (AOCs)
Automatic overfill protection system (AOPS)
Critical high (CH)
High-high (HH)
Maximum working (MW)
Still pipe
Tank gauge

Industry Recognized Credentials

If you are training through an NCCER-accredited sponsor, you may be eligible for credentials from NCCER's Registry. The ID number for this module is CT31_0-17. Note that this module may have been used in other NCCER curricula and may apply to other level completions. Contact NCCER's Registry at 1.888.622.3720 or go to **www.nccer.org** for more information.

Contents

1.0.0 Overfill Protective Devices ... 1
1.1.0 Mechanical OPDs ... 1
1.1.1 Mechanical OPD Operation ... 2
1.1.2 Mechanical OPD problems ... 2
1.1.3 Level Switches ... 2
1.2.0 Non-Contact OPDs ... 5
1.2.1 Radar, Sonar, and Laser OPDs ... 5
2.0.0 Inspecting And Calibrating Overfill Protective Devices ... 8
2.1.0 Prerequisites, Competencies, and Abnormal Operating Conditions ... 8
2.1.1 Prerequisites and Competencies ... 8
2.1.2 Abnormal Operating Conditions ... 9
2.2.0 Procedure ... 9

Figures

Figure 1 A mechanical OPD. ... 2
Figure 2 Level switch. ... 3
Figure 3 Level switch operation. ... 3
Figure 4 Mounted level switch. ... 4
Figure 5 Tubing connections. ... 4
Figure 6 External cage used for mounting a level switch. ... 5
Figure 7 A radar OPD uses reflection timing to determine the liquid's height. ... 6
Figure 8 Radar OPD antennas. ... 6

SECTION ONE

1.0.0 OVERFILL PROTECTIVE DEVICES

Objectives

Identify and describe the various types of overfill protective devices.

a. Identify and describe mechanical OPDs.
b. Identify and describe non-contact OPDs.

Trade Terms

Automatic overfill protection system (AOPS): Refers to an automatic system designed to prevent overfilling. Also refers to the level at which this system will be activated.

Critical high (CH): The level at which a tank is close to overflowing and at which an emergency response system will be activated.

High-high (HH): The level above the normal high level. An alarm will be activated if the liquid level reaches this point.

Maximum working (MW): The normal high level in a tank and the lowest of the high levels. Reaching this height does not trigger an alarm.

Still pipe: A perforated pipe that runs from the top to the bottom of a tank. Still pipes are designed to provide a non-turbulent place from which to take measurements.

Tank gauge: Refers to one of several technologies designed to report the liquid level in a storage tank. Tank gauges are used for everyday tank operations, such as custody transfer management.

Overfill protective (or prevention) devices come in many different forms and with varying degrees of sophistication. Simple OPDs are purely mechanical and may do nothing more than shut off the flow. More advanced technologies have no moving parts and may be part of an automatic control system that can handle overfill situations with far more sophistication. OPDs are separate from other tank level monitoring systems, such as a tank gauge, although they may use some of the same technologies.

In a typical tank system there are actually several defined high points that can be detected. Historically, these have gone by a number of different names and have had varied definitions. The following four levels are defined by *API 2350* 4th Edition.

The maximum working (MW) height is the lowest high level and refers to the normal maximum height to which liquid should rise. There is, however, a significant cushion of space above this point to allow for variation. Normally, no alarms will be raised if the liquid rises to the maximum working level, although many control systems will report the event as information.

The next level is high-high (HH). If the liquid reaches this point, it will trigger an alarm and the flow should be terminated within a specific period of time. Above the high-high level is the automatic overfill protection system (AOPS) point. An AOPS is an automatic system that takes corrective action without operator intervention. Only tanks equipped with an AOPS will sense this level. The final level is the critical high (CH) point. If the liquid reaches this level, the tank is about to overflow. Should this level be sensed, emergency actions must be initiated at once to prevent a spill.

Overfill prevention is no longer simply synonymous with mechanical level switches. Meeting two key standards for overfill protection, *API 2350* and *IED 61511*, may require the use of an AOPS. In such a system, certified overfill prevention sensors measure level continuously and independently from the automatic tank gauges. When a shutdown condition occurs, the AOPS automatically initiates a shutdown of incoming flow to a tank without any manual intervention.

An AOPS can reduce risks, but it can also increase risks if it is not applied and designed properly. Particular care is required when applying AOPS to any marine or pipeline operation. Thorough engineering analysis is critical to prevent the potential for pipeline rupture. For instance, an improperly designed AOPS could close off a flowing tank receipt delivering from a pipeline, which could cause the pipeline to rupture.

The pipeline operator and the terminal operator must work together to determine if and how an AOPS will be used. They need to devise an arrangement that maximizes benefits yet also meets all of the relevant standards for overfill protection.

1.1.0 Mechanical OPDs

Mechanical OPDs are the simplest type of protective device. They either restrict flow or cut it off entirely, depending on their design. Like all mechanical devices, however, they can wear out or malfunction due to problems with their moving parts. The following material surveys the most common types of mechanical OPDs.

1.1.1 Mechanical OPD Operation

Most mechanical OPDs are integrated into the tank's drop tube. Liquid flows through them, usually encountering some kind of valve. Almost all mechanical OPDs rely on a float to sense the liquid level in the tank. When the liquid is below the desired high level point, the float is at its lowest position. This causes the valve in the drop tube to be fully open, permitting unrestricted filling.

As the liquid level rises to the bottom of the float, it begins to rise in response. Many mechanical OPDs start restricting the flow at this point. If an operator is manually filling the tank, the change in flow will be detected and the operation can be halted. Since the valve is still partially open, however, hoses can be drained into the tank.

If filling continues, the float keeps rising and eventually closes the valve completely, halting the fill operation. Some mechanical OPDs might also include contacts that can trigger an alarm or provide status information to a control system.

Mechanical OPDs (*Figure 1*) vary primarily in their float style and valve type. Some floats are hinged and rise at an angle. Others rise vertically along a pair of guide rods. Valves are often of a piston construction, although there are other styles. Ball-type floats are used in some OPDs and can function both as the level-sensing element and as the valve itself. These are more common in vent pipes.

Figure 1 A mechanical OPD.

1.1.2 Mechanical OPD problems

Since mechanical OPDs rely on moving parts, they have a number of failure points. A common problem is sticking in the float linkage mechanism. If this happens, the float may not be able to rise with the liquid and will fail to close the valve. It's also possible for the control mechanism to stick in the closed position. In this case, when the liquid level drops, the float will be stuck in its highest position and won't fall with the liquid and open the valve.

Many mechanical OPDs come with a manual test mechanism that allows pipeline E&I technicians to confirm their operation. In many cases, this is nothing more than a cord or chain connected to the float. Pulling up on the cord tests the OPD's shutoff behavior, while releasing it confirms that the device re-opens the valve. Actuating the manual mechanism is usually a part of OPD testing procedures.

1.1.3 Level Switches

Levels switches are used in many different pipeline applications. One of these is as an OPD. An example of a level switch (*Figure 2*) is the Magnetrol Tuffy® liquid level control with a pneumatic switch. These switches are designed for horizontal mounting in a tank or vessel through threaded or flanged pipe connections.

Switching action is obtained through the use of a magnet attached to a float and a switching mechanism. A non-magnetic, pressure-tight barrier separates these two basic component assemblies. The float moves the magnet upward and into the field of the switch mechanism magnet. As a result, the magnets repel, causing the pivot cam and actuating pin to seat the ball firmly, which shuts off the air supply. Level switch operation is shown in *Figure 3*.

Before assembling the switch in a tank or vessel, check the threaded or flanged mounting nozzle for the following:

- Correct sizing of the nozzle length and diameter to allow for switch actuation at design levels (*Figure 4*).
- The horizontal alignment of the nozzle. Since the switch is mounted differently for high- or low-level operation, it must be mounted correctly for system requirements:

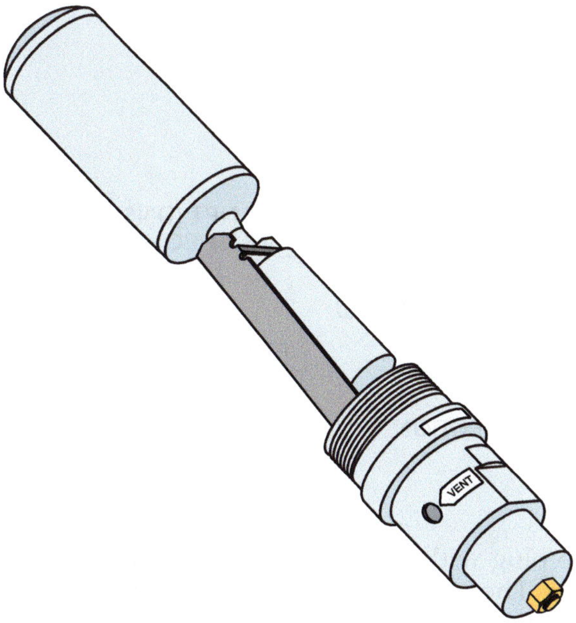

Figure 2 Level switch.

- For high-level operation, the switch must have the Hi Level label on top and the wrench flats horizontally level. In this position, with the float down, no air flows. With the float in the up position, air flows into the outlet. This would be the normal mounting when a level switch is acting as an OPD.
- For low-level operation, the switch must have the Low Level label on top and the wrench flats horizontally level. In this position, with the float up, no air flows. With the float in the down position, air will flow into the outlet.

The inlet air line is connected to the $\frac{1}{8}$ inch national pipe thread (NPT) (DN6) fitting labeled INLET on the end of the valve. Inlet air pressure should be in the range of 10–100 psig (69–689 kPa). Outlet air is connected to the $\frac{1}{8}$ inch NPT (DN6) connection on the side of the valve labeled OUTLET. *Figure 5* shows these tubing connections.

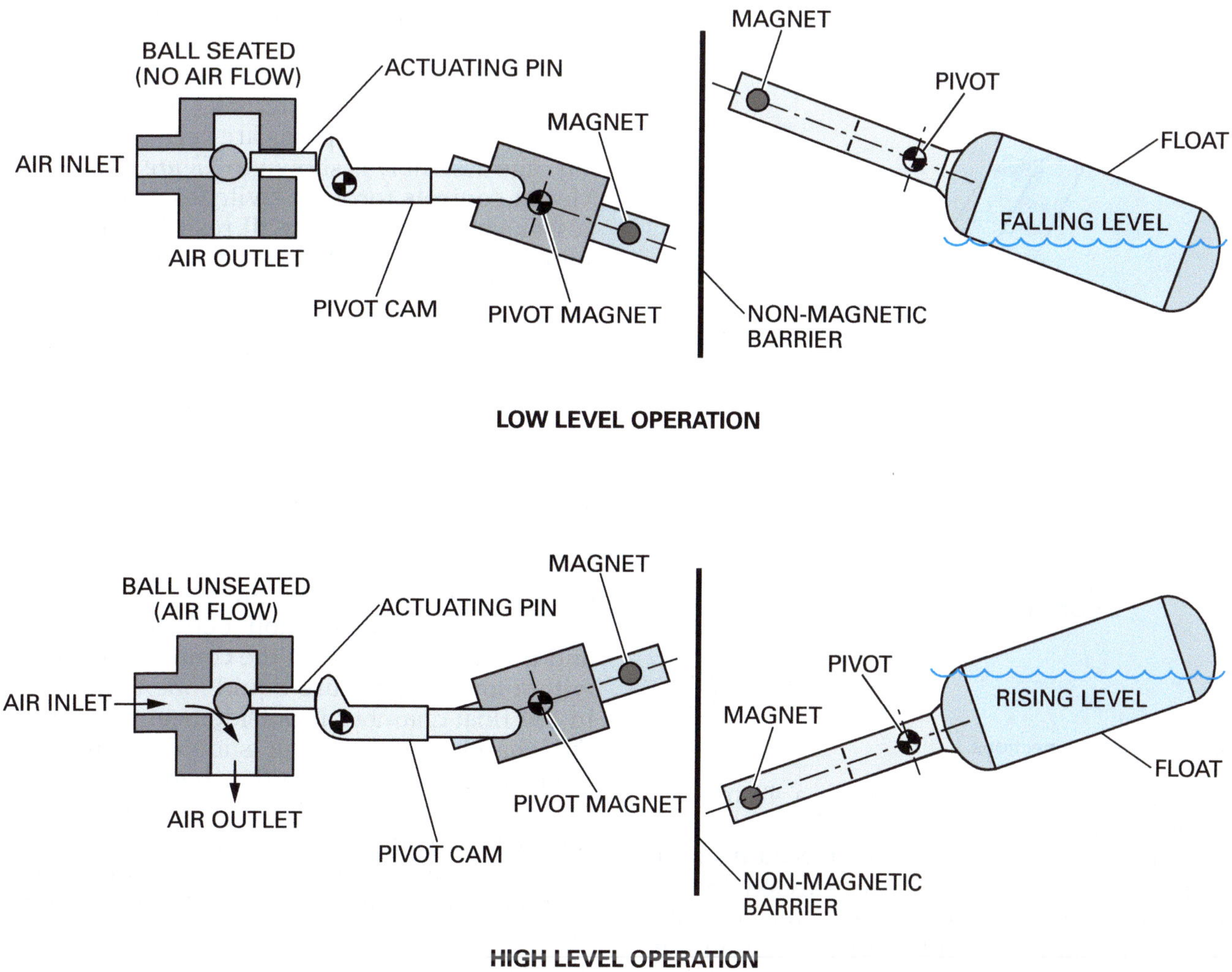

Figure 3 Level switch operation.

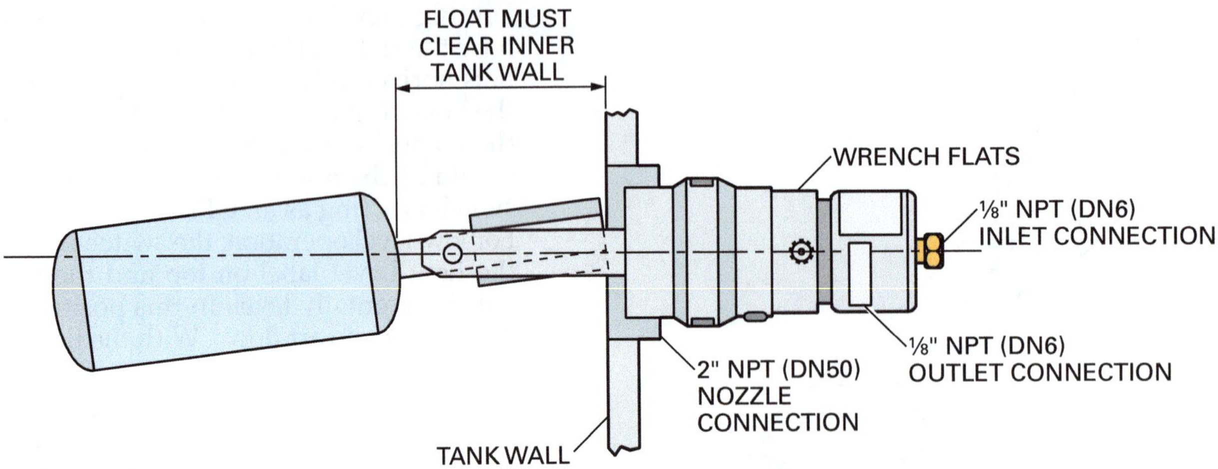

Figure 4 Mounted level switch.

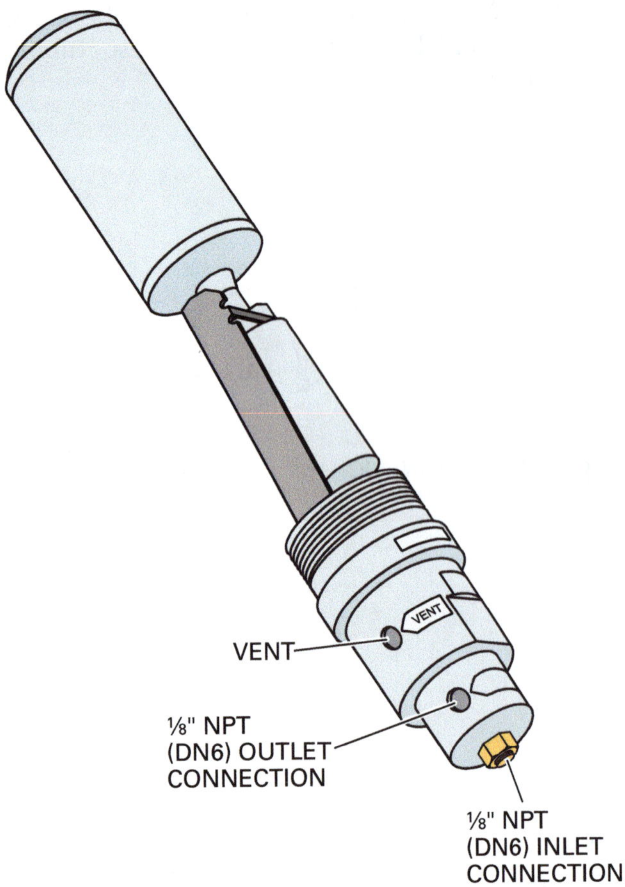

Figure 5 Tubing connections.

Inlet air should be filtered with a 40 micron (0.04 mm) or finer filter. This will extend the times between cleaning of the mechanism. If required, a vent line may be connected to the opening labeled VENT on the side of the housing. A vent line takes any vented air away from the device location. If a vent line is not required, the pipe plug with a small hole in it should be left in place. The opening in the plug allows vented air to be discharged at the valve location.

The switch can be mounted externally using the external cage assembly. The switch actuates within a $^3/_4$ inch (19 mm) distance above the cage center line for the rising level, and within a $^3/_4$ inch (19 mm) distance below the center line for the falling level. The external cage must be installed so that the tappings are vertical. The external cage assembly is shown in *Figure 6*.
Mount controls for level switches as close to the vessel as possible. This will result in a more responsive and accurate level change in the cage. Liquid in a long pipeline may be cooler and denser than liquid in a vessel, causing the level indication in the control to be lower than the actual level in the vessel.

Use pipe of sufficient strength to support the control. If necessary, provide a stand or hanger to help support its weight. All piping should be straight and free of low spots or pockets. This ensures that the lower liquid line will drain towards the vessel and the upper vapor line will drain toward the control. Shutoff valves are recommended for installation between the vessel and the cage assembly. If the control is to be used with a low-temperature liquid (one that will boil in the float chamber if outside heat is absorbed), the chamber and piping should be insulated. Otherwise, the boiling in the chamber will cause false level indications. Do not, however, insulate the switch mechanism housing.

Level switch repair is typically limited to individual component replacement of the switch. Follow manufacturer's instructions to perform any repair.

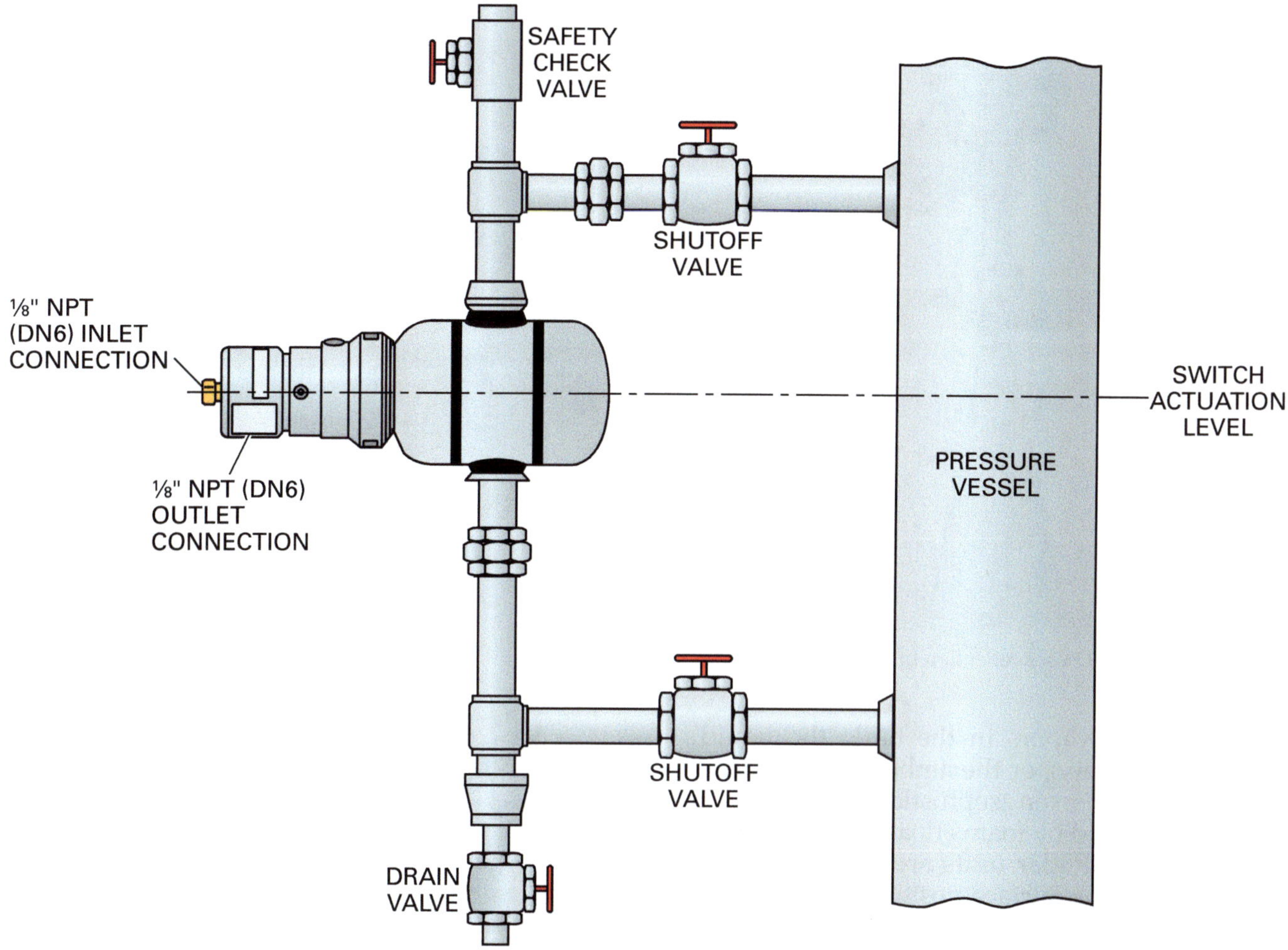

Figure 6 External cage used for mounting a level switch.

> **NOTE**
> Certain liquids contain particulates or salts that can accumulate in level measuring devices. The accumulated material can result in inaccurate measurements or even make the measuring device cease to function. Consider using non-contact level measuring options for liquids that have these characteristics.

1.2.0 Non-Contact OPDs

In recent years, many operations have been switching over to OPDs that are non-mechanical. These devices rely on methods of sensing the liquid level without actually coming into contact with the liquid itself. Most of these technologies rely on the principle of reflection to determine the liquid surface's location. They have the advantage of no moving parts to wear out or malfunction, although they are usually much more complex and expensive than the simple mechanical OPDs.

All non-contact OPDs are electronic, generating a signal that indicates when the liquid level has reached one of the several defined high levels. These signals may trigger alarms or valves directly. Alternatively, they may communicate with a controller that makes decisions about the appropriate actions to take.

1.2.1 Radar, Sonar, and Laser OPDs

Non-contact OPDs generally work by aiming a signal at the liquid's surface and timing how long it takes the reflection to return. Using the speed at which the signal travels, the OPD performs a simple calculation to determine the distance to the liquid's surface. There are three kinds of OPD that use this method: radar, sonar, and laser.

A radar OPD (*Figure 7*) uses microwave radio frequency energy as the signal. Since RF energy travels at the speed of light, timing and calculations are based on this value. The same principle is used for radars of all kinds. A radar OPD is

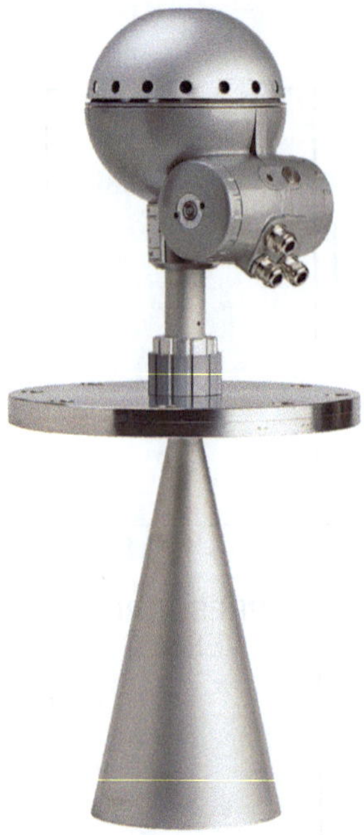

Figure 7 A radar OPD uses reflection timing to determine the liquid's height.

Figure 8 Radar OPD antennas.

not affected by vapors in the tank, the liquid's thickness (viscosity), or the ambient temperature. Some units are even sophisticated enough to avoid being fooled by foams floating on top of the liquid's surface. Radar units are extremely accurate, reporting the surface position to within just a few millimeters.

Typical radar OPDs usually have interchangeable antennas (*Figure 8*) that alter the shape of the microwave beam. Cone/horn and parabolic antennas are designed to be used in larger tanks with plenty of space around the antenna. Array and rod antennas are meant to be used in a still pipe, in which the space is much more limited. It's important to select and install the best antenna for the individual situation. Each manufacturer provides complete information to guide the selection process.

The sonar OPD works in much the same way as the radar version except that it uses high-frequency sound waves instead of radio waves. Like the radar OPD, it has no moving parts and relies on measuring reflection time to determine the position of the liquid's surface. Unlike the radar OPD, however, sonar OPD accuracy can be affected by temperature and vapors since both can alter the local speed of sound. These factors must be taken into account in order to obtain accurate results. Some sonar OPDs include temperature sensors to help compensate for this component.

Finally, laser OPDs perform essentially the same function as radar and sonar units. Instead of RF or sound energy, they use laser light, which bounces off the liquid's surface. Like radar OPDs, laser units are not affected by temperature or the atmosphere in the tank.

All reflection OPDs must go through a setup process designed to eliminate false readings due to reflections from tank seams and welds. Once this process has been completed, however, the system requires relatively little maintenance or adjustment. Setup often involves using manufacturer-supplied software and an attached notebook computer.

Additional Resources

It is assumed that you will learn to perform these procedures under the direct supervision of your crew leader, foreman, or instructor. For independent study, refer to the recommended Additional Resources at the back of this module.

1.0.0 Section Review

1. A mechanical OPD senses liquid level with a(n) _____.
   a. DP flow sensor
   b. acoustic transducer
   c. float
   d. valve

2. A reflection-type OPD requires initial setup to compensate for _____.
   a. electrical interference
   b. seams and welds
   c. viscosity
   d. tank flow rate

SECTION TWO

2.0.0 INSPECTING AND CALIBRATING OVERFILL PROTECTIVE DEVICES

Objectives

Describe the procedures used to inspect and calibrate overfill protective devices (CT31_0-17).

a. Identify the prerequisites, competencies, and abnormal operating conditions (AOCs) associated with inspecting and calibrating OPDs.
b. Describe the procedures used to inspect and calibrate OPDs.

Performance Task

Inspect and calibrate overfill protective devices (CT31_0-17).

- Identify potential abnormal operating conditions that may occur during performance of this CT, and know the appropriate actions to take in response to them.
- Utilize the appropriate personal protective equipment according to relevant company procedures.
- Notify operations control center prior to performing any test or working on system.
- Inspect OPD for corrosion or damage. Clean, repair, or replace per the manufacturer's instructions.
- Test OPD switch.
- Verify operation of flow control devices and flow switches on relief tanks for operability and/or damage.
- Demonstrate ability to calibrate OPD switch per manufacturer's instructions.
- Notify operations control center that inspection, testing and calibration is complete and control is returned to SCADA system.
- Complete appropriate documentation as required by operator's procedures.

Trade Term

Abnormal operating conditions (AOCs): Conditions within a pipeline system, identified by the operator, that may indicate (1) a malfunction of a component or (2) a deviation from normal operations that may indicate a condition exceeding design limits or result in a hazard to persons, property, or the environment.

As with other instrumentation, OPDs require regular inspection and calibration to keep them functioning properly. Since the consequences of a spill can be very serious, this task should never be neglected. Each OPD has its own calibration procedure. Consult the manufacturer's documentation for guidelines.

2.1.0 Prerequisites, Competencies, and Abnormal Operating Conditions

Personnel must possess a variety of skills in order to successfully inspect and calibrate OPDs. The various operations require a series of steps, which a pipeline E&I technician must be competent to perform. Personnel must also be aware of the particular **abnormal operating conditions (AOCs)** that may occur during the procedure and know the proper responses to them.

2.1.1 Prerequisites and Competencies

Personnel must possess the following basic knowledge and skills prior to inspecting and calibrating OPDs:

- Familiarity with the various OPD technologies and types
- Familiarity with tank types as well as their filling and draining systems
- Skills with test and calibration equipment, including multimeters and test media
- Knowledge about the specific OPD in question, including its setpoint
- Familiarity with the control system, its alarms, and their behaviors and consequences

2.1.2 Abnormal Operating Conditions

Personnel inspecting and calibrating OPDs must be able to identify abnormal operating conditions (AOCs) and know the proper response to each, should they occur. The following are some potential abnormal operating conditions (AOCs), and the appropriate action the E&I technician should in response to them:

- *The tank's liquid level is inappropriately high or low* – The pipeline E&I technician must alert appropriate personnel to the current level status.
- *An OPD is inoperable* – The pipeline E&I technician must alert appropriate personnel about the status of the OPD as required by company procedure.
- *During inspection and calibration, inappropriate device activation or a shutdown occurs* – The pipeline E&I technician must alert appropriate personnel about the event and the status of the OPD as required by company procedure.
- *The pipeline E&I technician discovers damage to the tank roof* – The pipeline E&I technician must alert appropriate personnel to take the required corrective action.
- *The pipeline E&I technician discovers free product or other inappropriate material on the tank roof* – The pipeline E&I technician must alert appropriate personnel to take the required corrective action.

2.2.0 Procedure

The following steps outline the general procedures for inspection and calibration of overfill protective devices:

> **NOTE**
>
> Obtain and check device history before inspecting or calibrating an OPD. This will identify things that may have gone wrong previously with the device, and it can save a lot of time if there is documented history of malfunction (and how it was resolved).

> **NOTE**
>
> Always follow operator procedures related to lone worker status, grounding, and fall protection.

Step 1 Identify potential abnormal operating conditions that may occur, and know the appropriate actions to take in response to them.

Step 2 Utilize the appropriate personal protective equipment according to relevant company procedures.

Step 3 Before beginning the process, notify the control center, local personnel, and any other personnel who might be affected by the procedure. The control center may have to provide feedback about the OPD's behavior, its ID, or its output data. It may also have to override automatic shutdowns or safety systems to protect the process from interruption. These events are often known as Critical Interlock and Trips (CIATs). They usually require higher authorization to override. Pipeline E&I technicians should be aware of these rules and know what they're allowed to do.

Step 4 Check the appropriate company documentation to confirm the correct OPD setpoint.

Step 5 Identify and obtain the correct calibration equipment. Consult company documentation and the manufacturer's material for guidelines. Confirm that the test equipment itself has been certified, calibrated, and verified. Some technicians will not use test equipment until it has been tested on a known source, even if its calibration is current and properly documented.

Step 6 Determine the device's number and nameplate data in order to verify that it is the correct unit to test and that any data coming from the unit will be properly identified as such.

Step 7 Visually inspect the device and any associated equipment. Check for mechanical condition, corrosion, leakage, electrical connection problems, or problems with associated electronics. Notify appropriate personnel if maintenance is required.

Step 8 Check and confirm that the OPD's setpoint is correct. Using this information, determine if the device requires calibration.

> **NOTE**
>
> Current or recent flow(s) into or out of the tank can cause inaccurate level measurements. In order to avoid inaccuracies, be sure to allow the required settling time before taking readings!

> **NOTE**
>
> If the liquid measurements are incorrect, notify the operations control center or appropriate personnel that the equipment is not working properly.

Step 9 Based on your conclusions from Step 8, calibrate the device if necessary, following the manufacturer's guidelines and company procedure. Repeat this step as often as required to establish a repeatable calibration.

Step 10 Follow the appropriate procedures and test the OPD.

> **NOTE**
>
> Testing overfill protection devices is a separate covered task, and is defined in NCCER Module CT30_0-17, *Test Overfill Protective Devices (Liquid)*.

> **CAUTION**
>
> In some terminals with multiple-receipt tanks, alarms are wired fail safe and daisy-chained together. Triggering an OPD on one tank will cause all the tanks to be considered in an alarm condition. As a result, automated valves close, possibly activating relief flow systems or overpressure relief devices. Automated sequences may include shut down of incoming product, closure of automated isolation valves, and operation of relief flow devices to prevent exceeding the maximum allowable operating pressure (MAOP).

Step 11 Verify operation of flow control devices and flow switches on relief tanks for operability and/or damage. (Ensure proper operation of flow relief equipment to shut down or divert flow in the event of high-level condition.)

Step 12 Return the OPD to its normal condition and verify that all alarms and other events have been reset and returned to normal. Confirm that any automatic responses have also returned to a normal condition.

Step 13 Notify the control center, local personnel, and any affected personnel that the procedure is complete and that the unit has been returned to normal operation.

Step 14 Document the results of the inspection and calibration. Follow company procedure.

Additional Resources

It is assumed that you will learn to perform these procedures under the direct supervision of your crew leader, foreman, or instructor. For independent study, refer to the recommended Additional Resources at the back of this module.

2.0.0 Section Review

1. If an OPD is not working, what is the first thing that the pipeline E&I technician should do?
   a. Alert appropriate personnel.
   b. Repair or replace the OPD.
   c. Enter an override into the control system.
   d. Request that the process be halted as a precaution.

2. What is an authoritative source for guidance on calibrating a specific OPD?
   a. The control center historian
   b. The manufacturer's documentation
   c. The technician who last calibrated the OPD
   d. The process operator

Summary

Storage tanks should never be allowed to overflow since the consequences could be very serious. Damage, injury, environmental contamination, and financial loss are all possibilities when product escapes its storage tank. Overfill protection devices help to prevent spills by shutting down the fill process if the tank's level rises above an appropriate level. Keeping these devices working reliably is the job of the pipeline E&I technician. This module surveyed the different types of OPDs and provided the basic procedure for inspecting and calibrating them.

Review Questions

1. Which of the four defined tank high levels is considered normal and will not trigger an alarm?
   a. Maximum working (MW)
   b. High-high (HH)
   c. Automatic overfill protection system (AOPS)
   d. Critical high (CH)

2. In their first stage of activation, many mechanical OPDs _____.
   a. halt the fill flow
   b. restrict the fill flow
   c. initiate an emergency response
   d. open the valve

3. What does a laser OPD bounce off the surface of the liquid?
   a. Microwaves
   b. Sound
   c. Light
   d. Magnetic flux

4. Pipeline E&I technicians preparing to calibrate an OPD should be familiar with _____.
   a. the local speed of sound
   b. the tank diameter
   c. fieldbus debugging procedures
   d. the specific OPD in question

5. Which of the following is an abnormal operating condition that could occur during the calibration of an OPD?
   a. A floating roof collapses.
   b. Tank containment fails, releasing hazardous vapors.
   c. The OPD overheats and catches fire.
   d. The tank's liquid level is inappropriately high or low.

6. Once you have completed sufficient calibration runs on an OPD, what is the next step?
   a. Restart the process.
   b. Confirm that all valves have re-opened.
   c. Test the OPD.
   d. Scan the fieldbus.

Trade Terms Quiz

Fill in the blank with the correct term that you learned from your study of this module.

1. The ________ is the level above the normal high level. An alarm will be activated if the liquid level reaches this point.
2. ________ refers to one of several technologies designed to report the liquid level in a storage tank. It is used for everyday tank operations, such as custody transfer management.
3. The level at which a tank is close to overflowing and at which an emergency response system will be activated is called ________.
4. A(n) ________ is a perforated pipe that runs from the top to the bottom of a tank. It is designed to provide a non-turbulent place from which to take measurements.
5. Conditions within a pipeline system, identified by the operator, that may indicate (1) a malfunction of a component or (2) a deviation from normal operations that may indicate a condition exceeding design limits or result in a hazard to persons, property, or the environment are called ________.
6. ________ is the normal high level in a tank and the lowest of the high levels. Reaching this height does not trigger an alarm.
7. ________ refers to an automatic system designed to prevent overfilling. Also refers to the level at which this system will be activated.

Trade Terms

Abnormal operating conditions (AOCs)
Automatic overfill protection system (AOPS)
Critical high (CH)
High-high (HH)
Maximum working (MW)
Still pipe
Tank gauge

Trade Terms Introduced in This Module

Abnormal operating conditions (AOCs): Conditions within a pipeline system, identified by the operator, that may indicate (1) a malfunction of a component or (2) a deviation from normal operations that may indicate a condition exceeding design limits or result in a hazard to persons, property, or the environment.

Automatic overfill protection system (AOPS): Refers to an automatic system designed to prevent overfilling. Also refers to the level at which this system will be activated.

Critical high (CH): The level at which a tank is close to overflowing and at which an emergency response system will be activated.

High-high (HH): The level above the normal high level. An alarm will be activated if the liquid level reaches this point.

Maximum working (MW): The normal high level in a tank and the lowest of the high levels. Reaching this height does not trigger an alarm.

Still pipe: A perforated pipe that runs from the top to the bottom of a tank. Still pipes are designed to provide a non-turbulent place from which to take measurements.

Tank gauge: Refers to one of several technologies designed to report the liquid level in a storage tank. Tank gauges are used for everyday tank operations, such as custody transfer management.

Additional Resources

The following recommended resources can provide additional helpful information related to the requirements for this covered task:

API Recommended Practice 1161, Third Edition, © 2014 American Petroleum Institute. API Publishing Services, 1220 L Street, NW, Washington, DC 20005. Distributed by Thompson Reuters (Scientific) LLC, **www.techstreet.com**.

NCCER Module AOCCC-17, *Abnormal Operating Conditions Control Center.*

NCCER Module AOCFG-17, *Abnormal Operating Conditions Field and Gas.*

Instrumentation Level One, NCCER. Third Edition. 2014. New York, NY: Pearson Education, Inc.

Instrumentation Level Two, NCCER. Third Edition. 2016. New York, NY: Pearson Education, Inc.

Instrumentation Level Three, NCCER. Third Edition. 2016. New York, NY: Pearson Education, Inc.

Instrumentation Level Four, NCCER. Third Edition. 2016. New York, NY: Pearson Education, Inc.

Instrument Engineers Handbook, Volume 1: Process Measurement and Analysis. Béla G. Lipták (editor). Fourth Edition. 2003. Boca Raton, FL: CRC Press.

Instrument Engineers' Handbook, Volume 2: Process Control and Optimization. Béla G. Lípták (editor). Fourth Edition. 2005. Boca Raton, FL: CRC Press.

Instrumentation. Franklyn W. Kirk, Thomas A. Weedon, and Philip Kirk. Fifth Edition. 2010. Orland Park, IL: American Technical Publishers.

Instrumentation and Process Control. Terry Bartelt. 2007. Clifton Park, NT: Cengage Learning.

Instrumentation Reference Book. Walt Boyes (editor). Fourth Edition. 2009. Oxford, UK: Butterworth-Heinemann.

Measurement and Control Basics. Thomas A. Hughes. Fifth Edition. 2014. Research Triangle Park, NC: The International Society of Automation.

Purdy's Instrument Handbook. Ralph Dewey. 2015 Edition. Deer Park, TX: Glen Enterprises.

The Condensed Handbook of Measurement and Control. N.E. Battikha. Third Edition, 2006. Research Triangle Park, NC: The International Society of Automation.

NCCER Module 60102-02, *Basic Pipeline Hydraulics and Equipment*

NCCER Module 64202-02, *Use of Meters and Test Equipment*

NCCER Module CT30_0-17, *Test Overfill Protective Devices (Liquid)*

NCCER Module CT44_4-17, *Inspection, Testing, Corrective and Preventative Maintenance of Tank Gauging for Hazardous Liquid Leak Detection*

NCCER Module CT63_1-17, *Start-up of a Liquid Pipeline (Field)*

NCCER Module CT63_2-17, *Shutdown of a Liquid Pipeline (Field)*

NCCER Module CT63_4-17, *Locally Operate Valves on a Liquid Pipeline System*

NCCER Module CT64_1-17, *Start-up of a Liquid Pipeline (Control Center)*

NCCER Module CT64_2-17, *Shutdown of a Liquid Pipeline (Control Center)*

NCCER Module CT64_4-17, *Remotely Operate Valves on a Liquid Pipeline System*

The following websites offer resources for products and training:

Emerson Electric Co., **www.emerson.com**

Morrison Bros. Co., **www.morbros.com**

OPW, A Dover Company., **www.opwglobal.com**

The International Society of Automation, **www.isa.org**

Figure Credits

© Supakitmod/Dreamstime.com, Module Opener
Morrison Bros. Co., Figure 1
Emerson Process Management, Figures 7 and 8

Section Review Answer Key

Answer	Section Reference	Objective
Section One		
1. c	1.1.1	1a
2. b	1.2.1	1b
Section Two		
1. a	2.1.2	2a
2. b	2.2.0	2b

NCCER CURRICULA — USER UPDATE

NCCER makes every effort to keep its textbooks up-to-date and free of technical errors. We appreciate your help in this process. If you find an error, a typographical mistake, or an inaccuracy in NCCER's curricula, please fill out this form (or a photocopy), or complete the online form at **www.nccer.org/olf**. Be sure to include the exact module ID number, page number, a detailed description, and your recommended correction. Your input will be brought to the attention of the Authoring Team. Thank you for your assistance.

Instructors – If you have an idea for improving this textbook, or have found that additional materials were necessary to teach this module effectively, please let us know so that we may present your suggestions to the Authoring Team.

NCCER Product Development and Revision

13614 Progress Blvd., Alachua, FL 32615

Email: curriculum@nccer.org

Online: www.nccer.org/olf

❑ Trainee Guide ❑ Lesson Plans ❑ Exam ❑ PowerPoints Other ____________________

Craft / Level: Copyright Date:

Module ID Number / Title:

Section Number(s):

Description:

Recommended Correction:

Your Name:

Address:

Email: Phone:

CT44_3-17

Inspect, Test, and Maintain Flow Computer for Hazardous Liquid Leak Detection

Overview

Many different technologies must work together to maintain a pipeline's integrity. From an instrumentation perspective, leak detection and pipeline performance center upon the computational pipeline management (CPM) system. CPM depends upon a steady stream of data, and flow computers play a key role in supplying that information. This module examines flow computers and technologies they interact with, as well as procedures followed for their use and maintenance.

Trainees with successful module completions may be eligible for credentialing through the NCCER Registry. To learn more, go to **www.nccer.org** or contact us at 1.888.622.3720. Our website has information on the latest product releases and training, as well as online versions of our *Cornerstone* magazine and Pearson's product catalog.

Your feedback is welcome. You may email your comments to **curriculum@nccer.org**, send general comments and inquiries to **info@nccer.org**, or fill in the User Update form at the back of this module.

This information is general in nature and intended for training purposes only. Actual performance of activities described in this manual requires compliance with all applicable operating, service, maintenance, and safety procedures under the direction of qualified personnel. References in this manual to patented or proprietary devices do not constitute a recommendation of their use.

CT44_3-17
Inspect, Test, and Maintain Flow Computer for Hazardous Liquid Leak Detection

Objectives

When you have completed this module, you will be able to do the following:

1. Describe CPM, flow computers, and the technologies used with them.
   a. Describe computational pipeline management systems.
   b. Describe flow computers and list their features and capabilities.
   c. Identify the kinds of digital technologies that flow computers interact with.
2. Describe the procedures used to inspect, test, and maintain flow computers for hazardous liquid leak detection (CT44_3-17).
   a. Identify the prerequisites, competencies, and abnormal operating conditions (AOCs) associated with inspecting, testing, and maintaining flow computers.
   b. Describe the procedures used to inspect, test, and maintain flow computers for hazardous liquid leak detection.

Performance Tasks

Under the supervision of the instructor, you should be able to do the following:

1. Inspect, test, and maintain a flow computer for hazardous liquid leak detection (CT44_3-17).
   - Identify potential abnormal operating conditions (AOCs) that may occur during performance of this CT, and know the appropriate actions to take in response to them.
   - Utilize the appropriate personal protective equipment according to relevant company procedures.
   - Determine the flow computer's manufacturer and model identifier, preferably from its faceplate.
   - Examine the flow computer and identify everything connected to it.
   - Before beginning any diagnostic or maintenance work, particularly actions that may affect the pipeline operation, obtain permission from the control center.
   - Visually inspect the flow computer for physical damage and proper electrical and I/O connections.
   - If possible, using the flow computer's front panel and controls, check I/O parameters and the status of each connected device.
   - Check each fieldbus, serial data line, and network connection to confirm that it is working correctly.
   - Correct any problems detected, following the appropriate procedure as specified by the manufacturer. After corrective action, re-check the flow computer and its associated devices to confirm that everything is operating correctly.
   - Notify the control center before bringing any equipment back on line.
   - Complete appropriate documentation as required by operator's procedures.

Trade Terms

Abnormal operating conditions (AOCs)
Ethernet
Fieldbus
Highway Addressable Remote Transducer (HART)
Human-machine interface (HMI)
Modbus
Standard cubic feet (SCF)
Supervisory Control and Data Acquisition (SCADA)

Industry Recognized Credentials

If you are training through an NCCER-accredited sponsor, you may be eligible for credentials from NCCER's Registry. The ID number for this module is CT44_3-17. Note that this module may have been used in other NCCER curricula and may apply to other level completions. Contact NCCER's Registry at 888.622.3720 or go to **www.nccer.org** for more information.

Contents

1.0.0 CPM And Flow Computers 1
1.1.0 CPM Function and Application 1
1.1.1 CPM Trending 1
1.2.0 Measuring Flow 3
1.2.1 Flow Computers 3
1.3.0 Digital Systems Used with Flow Computers 3
1.3.1 Remote Terminal Units (RTUs) 3
1.3.2 Programmable Logic Controllers (PLCs) 4
1.3.3 Troubleshooting PLCs and RTUs 4
2.0.0 Working With Flow Computers 6
2.1.0 Prerequisites, Competencies, and Abnormal Operating Conditions 6
2.1.1 Prerequisites and Competencies 6
2.1.2 Abnormal Operating Conditions 7
2.2.0 Inspecting, Testing, and Maintaining Flow Computers 7
2.2.1 Procedure 7

Figures

Figure 1 Negative-pressure wave screen 2
Figure 2 Flow computers 3
Figure 3 A remote terminal unit 4
Figure 4 A PLC 4

SECTION ONE

1.0.0 CPM AND FLOW COMPUTERS

Objective

Describe CPM, flow computers, and the technologies used with them.

a. Describe computational pipeline management systems.
b. Describe flow computers and list their features and capabilities.
c. Identify the kinds of digital technologies that flow computers interact with.

Trade Terms

Ethernet: An extremely popular, flexible wired networking standard used in corporate and industrial settings.

Fieldbus: A simple digital communications network and protocol set designed to connect instruments, controllers, and other intelligent devices together. There are several different fieldbus standards.

Highway Addressable Remote Transducer (HART): A fieldbus protocol used to connect smart instruments to controllers and other digital technologies. HART has the advantage because it can co-exist with traditional analog 4–20 mA instrument signals on the same wire.

Human-machine interface (HMI): A station attached to a PLC or SCADA system that allows users to monitor and/or interact with the device and the process that it's controlling. HMIs, also known as a man-machine interface (MMI), can be very simple or highly sophisticated;.

Modbus: A popular fieldbus standard used in the pipeline industry. Modbus comes in several versions, including standard Modbus, Modbus RTU, and Modbus TCP.

Standard cubic feet (SCF): A measure of quantity used in the natural gas industry. It refers to one cubic foot of product at a specified standard temperature and pressure.

Supervisory Control and Data Acquisition (SCADA): An industrial control system that acquires data from a process, organizes it through supervisory computers, and makes it available in various forms so operators can monitor and control the process.

Computational pipeline monitoring (CPM) is a computerized leak detection program required by 49 *CFR* 195.444. In a CPM, the computers carry on a continuous program of monitoring pressure, flow rate, line volume balance, and variations between points. The information provided by this system can be used for a number of purposes. Leak detection is a key feature, but other kinds of data can be gathered as well. Flow computers are a major component in any CPM system, and their proper upkeep is crucial to keeping the system working properly and providing useful data.

1.1.0 CPM Function and Application

CPMs use line balance calculations to compare the volume received into a system with the volume being delivered. These calculations are adjusted for product retained in relief tanks, product removed for other uses, and any other factors affecting the final total. If there is a discrepancy, the system will detect it and alert operations personnel. Personnel must then investigate alarms as quickly as possible. CPM systems can detect deviations of less than 1 percent.

Negative-pressure wave monitors employ detectors in combination with the system to display the exact locations of line breaks. When a line break occurs, there is a sudden drop in pressure, followed immediately by a return to pressure. These sudden changes result in a wave that travels in both directions through the pipe at the speed of sound, growing smaller as it travels. With sensors located every few miles, the location of the release is calculated and known within seconds (*Figure 1*). It is then displayed in graphical form on a screen in the control center.

A further expansion of computational leak detection is accomplished by additional modules added to the program. These modules gather the information that **Supervisory Control and Data Acquisition (SCADA)** systems use and, together with the necessary interface and software, create a model of the hydraulic state of the pipeline. Along with this model, the computer program gives operations personnel the capability of controlling the pipeline. All information used and received is displayed on a **human-machine interface (HMI)** in real time.

SCADA may be fitted with a transient volume balance leak detection system that enables the computer to generate graphical displays of the

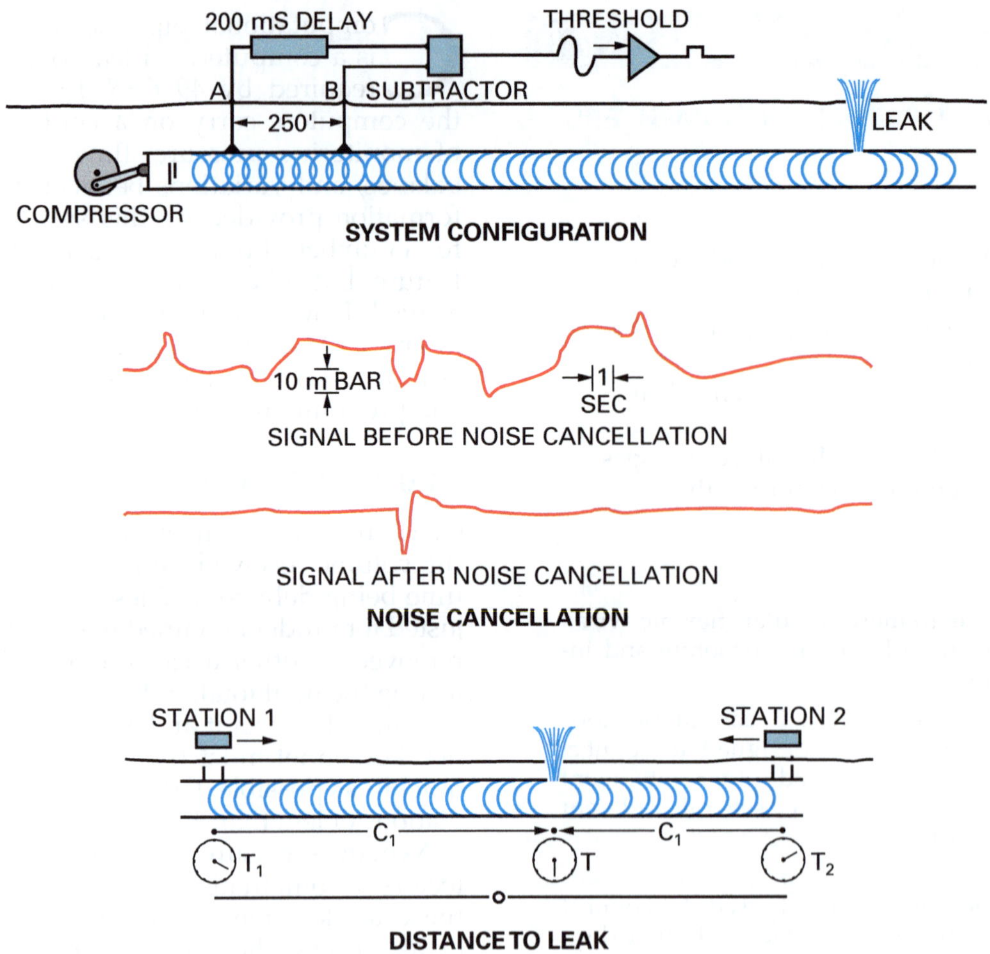

Figure 1 Negative-pressure wave screen.

hydraulic profile of the pipeline. This program operates by using complete time-dependent equations that calculate the flow of product through the line at designated points. SCADA uses complex differential equations that include calculations for head loss due to friction, line pack, and compression. Such influences often generate deviations in flow and surge-like effects. SCADA performs these complex calculations within a realistic time frame so that they show a useful profile to operations personnel. Results are shown on the hydraulic profile and updated at regular intervals (every few seconds).

Using a SCADA system to provide an accurate hydraulic model allows operations personnel to better schedule the starting and stopping of main-line pumps. Operations personnel are able to better control operating head, pressure, and relief and generally keep the pipeline in a more even steady state condition.

1.1.1 CPM Trending

Trending is the computer operation that demonstrates the ongoing conditions of the pipeline and illustrates any changes that are occurring during a specified period. As previously mentioned, computational pipeline monitoring is used to detect possible leaks. Through calculations of pipeline pressure, flow, line balance, and line volume balance, some computers can detect a deviation as small as 0.00028 percent of the total flow. Computer programs demonstrate the trend of change on the screens by showing hydraulic profiles, line volume calculations, and pressure readings at the various points. A graphical display is available so that the entire pipeline can be observed at one time. The computer also calculates and demonstrates the rate of change taking place so that operations personnel can observe a problem that might be developing.

Operating parameters are defined and programmed into the computer. As information is collected from the sensors being polled, the computer performs various tests on each parameter. First, the electrical signal is checked to make certain that it falls into the permissible electrical signal range. A signal is posted on the computer screen and printout to show any out-of-range condition. The computer compares suction pressure against discharge pressure and against preset limits. The computer compares the present pressure reading with the previous base value. Setpoints transmitted are compared with the value entered by operations personnel. Flow rate is scanned and compared to the previous values. Any unexplained deviation in these calculations reveals the possibility of a leak on the pipeline. Alarms are sounded for each of the individual deviations mentioned, and operations personnel should begin investigation of the alarm's cause as quickly as possible.

This aspect of pipeline operation is extremely important. Although it is of enormous scale, it must be maintained meticulously. Input devices must be inspected and calibrated on a scheduled basis to ensure the accuracy of the CPM system. Inspection, testing, and repair of these systems involves work by pipeline E&I technicians, computer specialists, and mechanical personnel.

1.2.0 Measuring Flow

Flow rates are extremely important to the CPM system. The flow rate of a pipeline is the measure of how fast product is moving or how much volume is passing through the pipeline in specified periods of time. Products flowing through petroleum pipelines are normally measured in barrels per hour (bph) or barrels per day (bpd). With pipelines carrying gases, such as natural gas, flow rates are measured in **standard cubic feet (SCF)** per day. MSCF is the abbreviation for 1,000 standard cubic feet per day, while MMSCF is the abbreviation for 1,000,000 standard cubic feet per day.

Knowing pipeline flow is useful for a number of reasons. Most obvious is the fact that it is essential for accurate transfers of all types. Changes in flow rate can also alert control operators to the presence of leaks. Regardless of how the data is used, flow measurements typically come from flow meters, instruments that detect flow rate in a pipeline. This raw information feeds into digital control devices, such flow computers and other related technology.

1.2.1 Flow Computers

Flow computers (*Figure 2*) are used on the pipeline to calculate and record the volume that flows through the pipeline flow meters. A flow computer is a compact industrial computer with programming that combines the data received from sensors on the pipeline to calculate the amount of volume in a given time. The units total and record all flow as calculated. Adjustments for changes in gravity, pressure, temperature, and velocity are included in the calculations.

A typical flow computer is totally self-contained with an embedded processor, memory, I/O, and communications capabilities. Most have a small LCD display and a number of simple controls, such as a keypad or buttons, for field adjustments. Modern flow computers are usually extremely flexible in their communications capabilities and can talk to a wide variety of other pipeline technologies. Recent model flow computers can interact with serial communications devices as well as with any common industrial **fieldbus**, such as a **Highway Addressable Remote Transducer (HART)** or **Modbus**. Most also support **Ethernet** for communications with larger networks.

Although these units are usually small, their programs are quite sophisticated. The manufacturer's manual and recommendations must be followed when setting up these units. Some have diagnostic and reprogramming capabilities. However, most problems need to be referred to a specialist.

1.3.0 Digital Systems Used with Flow Computers

Flow computers are usually part of a much larger pipeline monitoring system that includes other digital components. Having a strong overall understanding of the technology that makes up the CPM system will make it easier to solve problems when they develop. The following is a summary of some of the technologies typically seen alongside flow computers.

1.3.1 Remote Terminal Units (RTUs)

RTUs are a key part of SCADA systems and may do some data processing. Most record and file data until SCADA polls the unit and asks for the

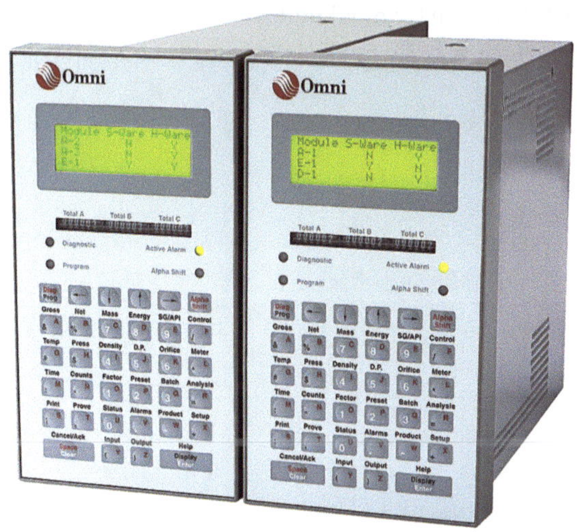

Figure 2 Flow computers.

information. The RTU will then upload the data to the SCADA supervisory computer for processing. RTUs vary in their size and capability. Many are battery powered. Smaller ones are stand-alone units that may run for over a year on a single set of batteries, only transferring data when polled by the host computer. Larger ones are able to power up at the appropriate time, collect data and measurements, report, and then shut down. RTUs come with a variety of communications capabilities. Most recent models can interact with popular fieldbuses as well as Ethernet networks. *Figure 3* shows a typical RTU.

1.3.2 Programmable Logic Controllers (PLCs)

A PLC (*Figure 4*) is a small industrial computer containing a program that operates in a loop with inputs and outputs. Units are fitted with interface devices so that they are able scan inputs, gather data, and modify outputs based on the inputs received. PLCs have extremely flexible communications capabilities, making it possible for them to talk to almost anything. Most recent models have a built-in Ethernet port. Communications modules may be added to allow interaction with popular fieldbuses as well as various proprietary PLC communications networks.

Figure 3 A remote terminal unit.

The various electronic devices that are connected to RTUs and PLCs to transmit information to the control center include density measuring instruments, flow meters, water detectors, and temperature and pressure sensors. Most operate as independent units that report results to the operations control center; these results are displayed as electronic readings. All the information is combined through the calculations of computer programs to give accurate readings and displays of every device on the pipeline. In order to calculate volume accurately, the computer must have an accurate measurement of flow rate, temperature, pressure, API gravity, S&W level, and elapsed time.

1.3.3 Troubleshooting PLCs and RTUs

Maintaining and troubleshooting PLCs and RTUs is complex and requires personnel who are technically trained in computer repair and programming to accomplish the task. Companies employ technicians who have these specialized skills to accomplish these tasks efficiently. Another point that must be understood is that many of these tasks cannot be accomplished by one person. The specialist works in the control center or other computer site, while another person, possibly a pipeline E&I technician, occupies a station at the RTU or PLC to assist, operate switches, and observe reactions.

Figure 4 A PLC.

Additional Resources

It is assumed that you will learn to perform these procedures under the direct supervision of your crew leader, foreman, or instructor. For independent study, refer to the recommended Additional Resources at the back of this module.

1.0.0 Section Review

1. One way that a CPM system can detect a leak is by monitoring _____.
   a. temperature
   b. pressure
   c. pipeline chirp
   d. pipeline flexing
2. Modern flow computers communicate with sensors using a(n) _____.
   a. APU
   b. SCADA linkup
   c. fieldbus
   d. UDP
3. Which of the following is an example of a computer-like device that collects data at the pipeline level as part of a SCADA system?
   a. RTU
   b. HMI
   c. MMI
   d. HART

SECTION TWO

2.0.0 WORKING WITH FLOW COMPUTERS

Objective

Describe the procedures used to inspect, test, and maintain flow computers for hazardous liquid leak detection (CT44_3-17).

a. Identify the prerequisites, competencies, and abnormal operating conditions (AOCs) associated with inspecting, testing, and maintaining flow computers.
b. Describe the procedures used to inspect, test, and maintain flow computers for hazardous liquid leak detection.

Performance Task

1. Inspect, test, and maintain a flow computer for hazardous liquid leak detection (CT44_3-17).
   - Identify potential abnormal operating conditions (AOCs) that may occur during performance of this CT, and know the appropriate actions to take in response to them.
   - Utilize the appropriate personal protective equipment according to relevant company procedures.
   - Determine the flow computer's manufacturer and model identifier, preferably from its faceplate.
   - Examine the flow computer and identify everything connected to it.
   - Before beginning any diagnostic or maintenance work, particularly actions that may affect the pipeline operation, obtain permission from the control center.
   - Visually inspect the flow computer for physical damage and proper electrical and I/O connections.
   - If possible, using the flow computer's front panel and controls, check I/O parameters and the status of each connected device.
   - Check each fieldbus, serial data line, and network connection to confirm that it is working correctly.
   - Correct any problems detected, following the appropriate procedure as specified by the manufacturer. After corrective action, re-check the flow computer and its associated devices to confirm that everything is operating correctly.
   - Notify the control center before bringing any equipment back on line.
   - Complete appropriate documentation as required by operator's procedures.

Trade Term

Abnormal operating conditions (AOCs): Conditions within a pipeline system, identified by the operator, that may indicate (1) a malfunction of a component or (2) a deviation from normal operations that may indicate a condition exceeding design limits or result in a hazard to persons, property, or the environment.

Because pipeline instrumentation systems can be quite complex, there are many opportunities for a failure. While the flow computer itself may not be the actual failure point, it often reveals the failure by outputting abnormal information or simply by not communicating at all. The following material explores the proper procedure for inspecting, testing, and maintaining flow computers.

2.1.0 Prerequisites, Competencies, and Abnormal Operating Conditions

Personnel must possess a variety of skills in order to successfully inspect, test, and maintain flow computers. The various operations require a number of steps, which a pipeline E&I technician must be competent to perform. Personnel must also be aware of the particular abnormal operating conditions (AOCs) that may occur during the procedure and know the proper responses to them.

2.1.1 Prerequisites and Competencies

Personnel must possess the following basic knowledge and skills prior to inspecting, testing, and maintaining flow computers:

- Flow computer configuration/programming
- Digital and analog I/O

- Digital communications systems
- Electrical systems knowledge and safety
- Understanding the particular pipeline system in question as well as its associated technologies
- Pressure sensors
- Digital technologies including PLCs, DCSs, SCADA, and HMIs.

Personnel must also understand the company's safety policies and procedures, including the proper use of personal protective equipment (PPE).

> **WARNING!**
> Always adhere to applicable safety policies and procedures, and utilize the appropriate PPE when working on or near a pipeline.

2.1.2 Abnormal Operating Conditions

Personnel inspecting, testing, and maintaining flow computers must be able to identify potential abnormal operating conditions and know the proper response to each, should they occur. Always follow applicable company/operator procedures relating to AOCs and appropriate responses.

One potential AOC that may be encountered when working with a flow computer is the discovery that it is not functioning or is not communicating with the HMI or SCADA system. If this happens, alert the control center or other appropriate personnel to take the requisite corrective action.

2.2.0 Inspecting, Testing, and Maintaining Flow Computers

Dealing with a flow computer problem requires systematic troubleshooting, beginning with the most likely causes of the problem and working through each logical possibility. The next section provides a general procedure for inspecting, testing, and maintaining flow computers.

2.2.1 Procedure

The troubleshooting of a flow computer typically begins when the system receiving its data (SCADA, PLC, etc.) indicates that the flow computer has stopped transmitting or is sending inappropriate data. When this happens, the flow computer must be inspected to determine its condition. Many have LCD displays, making it easy to immediately determine whether the unit is operating or not. Once that has been determined, continue diagnosing the problem from there. The following steps outline the general procedure for the inspection, testing, and maintaining of a flow computer:

Step 1 The first step in flow computer troubleshooting is to determine the flow computer's manufacturer and model identifier, preferably from its faceplate. Don't rely on company records for this information since a flow computer may have been changed to another model but not documented. Once you've identified the model, obtain the appropriate manual.

Step 2 Examine the flow computer and identify everything connected to it. This includes instrumentation devices such as flow meters, transmitters, and other sensors. Also included are fieldbuses, networks, and devices, such as PLCs or SCADA systems.

Step 3 Before beginning any diagnostic or maintenance work, particularly actions that may affect the pipeline operation, obtain permission from the control center. If certain devices go offline, automatic shutdowns may occur, Company procedures must be followed at all times.

Step 4 Visually inspect the flow computer for physical damage and proper electrical and I/O connections. Confirm that all cables are secure. If the flow computer has an LCD display, check its operational status and note any messages or error codes.

Step 5 If possible, using the flow computer's front panel and controls, check I/O parameters and the status of each connected device. Also check the flow computer's configuration parameters for correctness. You will need to consult the flow computer's manual and company documentation to know how to do this step since each installation and model is different.

Step 6 Check each fieldbus, serial data line, and network connection to confirm that it is working correctly. This may be done through front-panel controls, status indicators, or other tools. Again, consult the manual and company documentation for guidance.

Step 7 Correct any problems detected, following the appropriate procedure as specified by the manufacturer. After corrective action, re-check the flow computer and its associated devices to confirm that everything is operating correctly.

Step 8 Notify the control center before bringing any equipment back on line. Also notify control personnel that the problem has been corrected.

CAUTION

Many companies do not allow changes to limits and operating parameters without a Management of Change (MOC). Before making any change, be sure that you understand your company's policy and have confirmed that what you're doing is permitted. Some companies allow pipeline E&I technicians to make minor parameter adjustments without an MOC. In these cases, allowable variation usually has been defined and documented so the technician knows what degree of adjustment is permissible.

Step 9 Document all work performed on the system, following company procedure for guidelines. Changes to the flow computer's parameters or programming must be recorded.

Additional Resources

It is assumed that you will learn to perform these procedures under the direct supervision of your crew leader, foreman, or instructor. For independent study, refer to the recommended Additional Resources at the back of this module.

2.0.0 Section Review

1. If a pipeline E&I technician encounters the abnormal operating condition of a flow computer that has stopped communicating, what is the proper response?
   a. Isolate the flow computer from the SCADA system.
   b. Check for error codes on the flow computer's display.
   c. Halt the flow computer.
   d. Notify the control center.

2. When troubleshooting a flow computer, what is the first thing that an E&I technician should do?
   a. Reboot the flow computer and check for error codes.
   b. Check the fieldbus and network connections for activity.
   c. Identify the flow computer manufacturer and model number.
   d. Shut down the flow computer.

Summary

Accuracy of measurement is a basic requirement in the transportation of petroleum. One system that's particularly critical is the computational pipeline monitoring (CPM) system, as this program monitors hundreds of miles of pipeline and can detect very small releases almost instantaneously. This equipment must be maintained in the most meticulous manner for the protection of personnel, the public, and the environment. Understanding the various measuring technologies that help make up the CPM system is crucial to maintaining it in top operating condition. Since flow computers are a key component in a CPM system, handling flow computer problems promptly and efficiently is essential.

Review Questions

1. A CPM system can detect deviations between input and output of less than _____.
   a. 1 percent
   b. 1.5 percent
   c. 5 percent
   d. 10 percent

2. Human operators usually communicate with SCADA systems through _____.
   a. PLCs
   b. RTUs
   c. DCSs
   d. HMIs

3. Natural gas flow is measured in _____.
   a. BPDs
   b. SCFs
   c. BPHs
   d. PSIs

4. HART is an example of a(n) _____.
   a. Modbus
   b. RTU
   c. fieldbus
   d. human-machine interface

5. What kind of industrial computer can talk to almost anything if it's fitted with the right communications module?
   a. HMI
   b. PLC
   c. RTD
   d. SCADA

6. Which of the following flow computer skills should a pipeline E&I technician possess in order to be able to deal with flow computer problems?
   a. Configuration/programming
   b. Flow meter disassembly
   c. Orifice plate inspection
   d. DP cell testing

7. When you've identified the manufacturer and model of a problem flow computer, what is the next correct action?
   a. Shut down the flow computer.
   b. Check the supervisory computer history log.
   c. Obtain the proper manual.
   d. Perform a network scan for broken links.

8. In order to prevent pipeline problems, from whom should you get permission to work on a malfunctioning flow computer?
   a. The control center
   b. The head of maintenance
   c. The plant manager
   d. Permission isn't needed since it is an emergency

Trade Terms Quiz

Fill in the blank with the correct term that you learned from your study of this module.

1. ________ is a popular fieldbus standard used in the pipeline industry. It comes in several versions, including standard, RTU, and TCP.
2. ________ is a fieldbus protocol used to connect smart instruments to controllers and other digital technologies. It has the advantage that it can co-exist with traditional analog 4–20 mA instrument signals on the same wire.
3. An extremely popular and flexible wired networking standard used in corporate and industrial settings is called ________.
4. A measure of quantity used in the natural gas industry that refers to one cubic foot of product at a specified standard temperature and pressure is called ________.
5. ________ is a simple digital communications network and protocol set designed to connect instruments, controllers, and other intelligent devices together. It has several different standards.
6. The ________ is an industrial control system that acquires data from a process, organizes it through supervisory computers, and makes it available in various forms so operators can monitor and control the process.
7. Conditions within a pipeline system that may indicate (1) a malfunction of a component or (2) a condition exceeding design limits or resulting in a hazard to persons, property, or the environment are called ________.
8. The ________ is a station attached to a PLC or SCADA system that allows users to monitor and/or interact with the device and the process that it's controlling. It can be very simple or highly sophisticated.

Trade Terms

Abnormal operating conditions (AOCs)
Ethernet
Fieldbus
Highway Addressable Remote Transducer (HART)
Human-machine interface (HMI)
Modbus
Standard cubic feet (SCF)
Supervisory Control and Data Acquisition (SCADA)

Trade Terms Introduced in This Module

Abnormal operating conditions (AOCs): Conditions within a pipeline system, identified by the operator, that may indicate (1) a malfunction of a component or (2) a deviation from normal operations that may indicate a condition exceeding design limits or result in a hazard to persons, property, or the environment.

Ethernet: An extremely popular and flexible wired networking standard used in corporate and industrial settings.

Fieldbus: A simple digital communications network and protocol set designed to connect instruments, controllers, and other intelligent devices together. There are several different fieldbus standards.

Highway Addressable Remote Transducer (HART): A fieldbus protocol used to connect smart instruments to controllers and other digital technologies. HART has the advantage that it can co-exist with traditional analog 4–20 mA instrument signals on the same wire.

Human-machine interface (HMI): A station attached to a PLC or SCADA system that allows users to monitor and/or interact with the device and the process that it's controlling. HMIs can be very simple or highly sophisticated. Also known as a man-machine interface (MMI).

Modbus: A popular fieldbus standard used in the pipeline industry. Modbus comes in several versions, including standard Modbus, Modbus RTU, and Modbus TCP.

Standard cubic feet (SCF): A measure of quantity used in the natural gas industry. It refers to one cubic foot of product at a specified standard temperature and pressure.

Supervisory Control and Data Acquisition (SCADA): An industrial control system that acquires data from a process, organizes it through supervisory computers, and makes it available in various forms so operators can monitor and control the process.

Additional Resources

The following recommended resources can provide additional helpful information related to the requirements for this covered task:

API Recommended Practice 1161, Third Edition, © 2014 American Petroleum Institute. API Publishing Services, 1220 L Street, NW, Washington, DC 20005. Distributed by Thompson Reuters (Scientific) LLC, **www.techstreet.com**.

The Condensed Handbook of Measurement and Control, N. E. Battikha. Third Edition. 2006. Research Triangle Park, NC: The International Society of Automation.

NCCER Module 12406-16, *Programmable Logic Controllers*, from *Instrumentation Level Four*

NCCER Module 12407-16, *Distributed Control Systems*, from *Instrumentation Level Four*

NCCER Module 26102-14, *Electrical Safety*, from *Electrical Level One*

NCCER Module 60102-02, *Pipeline E&I Safety*

NCCER Module 64205-16, *Supervisory Control Systems*

The following websites offer resources for products and training:

ABB Group, **new.abb.com**

The International Society of Automation, **www.isa.org**

OMNI Flow Computers, Inc., **www.omniflow.com**

Yokogawa Corporation of America, **www.yokogawa.com**

Emerson Electric Co., **www.emerson.com**

Modbus® is a registered trademark of the Modbus Organization, Inc., **www.modbus.org**.

HART®, WirelessHART®, and HART-IP™ are registered trademarks of the FieldComm Group™, **en.hartcomm.org**.

Figure Credits

© Kodym/Dreamstime.com, Module Opener

OMNI Flow Computers, Inc., Figure 2

Yokogawa Corporation of America, Figure 3

Automationdirect.com, Figure 4

Section Review Answer Key

Answer	Section Reference	Objective
Section One		
1. b	1.1.0	1a
2. c	1.2.1	1b
3. a	1.3.1	1c
Section Two		
1. d	2.1.2	2a

NCCER CURRICULA — USER UPDATE

NCCER makes every effort to keep its textbooks up-to-date and free of technical errors. We appreciate your help in this process. If you find an error, a typographical mistake, or an inaccuracy in NCCER's curricula, please fill out this form (or a photocopy), or complete the online form at **www.nccer.org/olf**. Be sure to include the exact module ID number, page number, a detailed description, and your recommended correction. Your input will be brought to the attention of the Authoring Team. Thank you for your assistance.

Instructors – If you have an idea for improving this textbook, or have found that additional materials were necessary to teach this module effectively, please let us know so that we may present your suggestions to the Authoring Team.

NCCER Product Development and Revision

13614 Progress Blvd., Alachua, FL 32615

Email: curriculum@nccer.org

Online: www.nccer.org/olf

❏ Trainee Guide ❏ Lesson Plans ❏ Exam ❏ PowerPoints Other ____________

Craft / Level: Copyright Date:

Module ID Number / Title:

Section Number(s):

Description:

Recommended Correction:

Your Name:

Address:

Email: Phone:

CT44_4-17

Inspection, Testing, and Corrective and Preventative Maintenance of Tank Gauging for Hazardous Liquid Leak Detection

Overview

Determining how much product is in a storage tank is a fundamental operation in the pipeline industry. Product is transferred from the pipeline to tanks and then back again. Accuracy is essential in this process as errors can be very costly. For this reason, the technologies that measure the amount of product in a tank must be in good working condition and provide reliable results. This module explores the different methodologies used for this task and discusses the procedures used for inspecting, testing, and maintaining the equipment associated with it.

Trainees with successful module completions may be eligible for credentialing through the NCCER Registry. To learn more, go to **www.nccer.org** or contact us at 1.888.622.3720. Our website has information on the latest product releases and training, as well as online versions of our *Cornerstone* magazine and Pearson's product catalog.

Your feedback is welcome. You may email your comments to **curriculum@nccer.org**, send general comments and inquiries to **info@nccer.org**, or fill in the User Update form at the back of this module.

This information is general in nature and intended for training purposes only. Actual performance of activities described in this manual requires compliance with all applicable operating, service, maintenance, and safety procedures under the direction of qualified personnel. References in this manual to patented or proprietary devices do not constitute a recommendation of their use.

CT44_4-17
Inspection, Testing, and Corrective and Preventative Maintenance of Tank Gauging for Hazardous Liquid Leak Detection

Objectives

When you have completed this module, you will be able to do the following:

1. Identify and describe custody transfer and the various types of tank gauging devices associated with it as well as how they are serviced.
   a. Describe custody transfer and its relationship to tank gauging.
   b. Identify and describe various types of tank gauging devices.
   c. Describe how tank gauging devices are serviced.
2. Describe the procedures used for inspection, testing, and corrective and preventative maintenance of tank gauging for hazardous liquid leak detection (CT44_4-17).
   a. Identify the prerequisites, competencies, and abnormal operating conditions (AOCs) associated with inspection, testing, and corrective and preventative maintenance of tank gauging.
   b. Describe the procedures used for inspection, testing, and corrective and preventative maintenance of tank gauging for hazardous liquid leak detection.

Performance Tasks

Under the supervision of the instructor, you should be able to do the following:

1. Inspect, test, and perform corrective and preventative maintenance of tank gauging for hazardous liquid leak detection (CT44_4-17).
   - Identify potential abnormal operating conditions (AOCs) that may occur during performance of this CT, and know the appropriate actions to take in response to them.
   - Utilize the appropriate personal protective equipment (PPE) according to relevant company procedures.
   - Determine the device's number and nameplate data and confirm the calibration point values for the device.
   - Notify control center and/or affected personnel before work begins.
   - Visually inspect the device and any associated equipment.
   - Confirm that anything that could disturb the tank is shut off, that inlets and outlets are closed, and that the liquid inside the tank is still.
   - Confirm the test equipment's current certification status.
   - Gauge the tank with a certified device (handline, etc.) and compare its result to the tank gauge output.
   - Calibrate the tank gauge following the manufacturer's procedure.
   - Perform all steps required to resume normal operation. Notify control center and/or affected personnel that the unit has been returned to normal operation.
   - Complete appropriate documentation as required by operator's procedures.

Trade Terms

Abnormal operating conditions (AOCs)
API gravity
Custody transfer
Fieldbus
Gauge
Roll
Still pipe

Industry Recognized Credentials

If you are training through an NCCER-accredited sponsor, you may be eligible for credentials from NCCER's Registry. The ID number for this module is CT44_4-17. Note that this module may have been used in other NCCER curricula and may apply to other level completions. Contact NCCER's Registry at 1.888.622.3720 or go to **www.nccer.org** for more information.

Contents

1.0.0 Custody Transfer and Tank Gauging 1
1.1.0 The Transfer Process 1
1.1.1 Metered Custody Receipts 1
1.1.2 Gauged Receipts 2
1.1.3 Metered Custody Deliveries 2
1.1.4 Gauged Custody Deliveries 3
1.1.5 Preparing Tickets 3
1.2.0 Tank Gauging Devices 3
1.2.1 Hand Gauging 3
1.2.2 Mechanical Tank Gauge 4
1.2.3 Servo Tank Gauge 4
1.2.4 Radar Tank Gauge 4
1.2.5 Sonar Tank Gauge 5
1.2.6 Associated Technologies 6
1.3.0 Servicing Tank Gauging Devices 6
1.3.1 Test Equipment 6
2.0.0 Working With Tank Gauging Devices 8
2.1.0 Prerequisites, Competencies, and Abnormal Operating Conditions 8
2.1.1 Prerequisites and Competencies 8
2.1.2 Abnormal Operating Conditions (AOCs) 9
2.2.0 Procedure for Inspecting, Testing, and Maintaining Tank Gauging Devices 9

Figures

Figure 1 Handlines used for tank gauging 3
Figure 2 A mechanical tank gauge system 4
Figure 3 A servo tank gauge 5
Figure 4 A radar tank gauge 5
Figure 5 Radar tank gauge antennas 5
Figure 6 A handheld communicator 6

SECTION ONE

1.0.0 CUSTODY TRANSFER AND TANK GAUGING

Objective

Identify and describe custody transfer and the various types of tank gauging devices associated with it as well as how they are serviced.

a. Describe custody transfer and its relationship to tank gauging.
b. Identify and describe various types of tank gauging devices.
c. Describe how tank gauging devices are serviced.

Trade Terms

API gravity: A density measurement system used by the American Petroleum Institute. API gravity numbers are the inverse of specific gravity and scaled so a value of 10° equals a specific gravity of 1.

Custody transfer: The act of passing the responsibility for the possession, care, supervision, control and/or ownership of a volume of crude oil or other product to another entity (individual or company).

Fieldbus: A simple digital communications network and protocol set designed to connect instruments, controllers, and other intelligent devices together. There are several different fieldbus standards.

Gauge: The process of measuring the height of the liquid product in a storage tank. This information is the basis of other calculated measurements, such as volume.

Roll: The tendency for the contents of a tank to swirl for some time after transfer. Roll can cause measurement inaccuracies.

Still pipe: A perforated pipe that runs from the top to the bottom of a tank. Still pipes are designed to provide a non-turbulent place from which to take measurements.

When a volume or batch of product or crude oil is shipped through the pipeline, the pipeline company is solely responsible for it and must maintain an exact volume of product. Only when the shipment has reached its destination and has been switched or transferred into the tanks or other facilities of the customer does control pass from the original holder. It is during this time that accuracy is vitally important. Measuring instruments are therefore critical to the transfer process.

1.1.0 The Transfer Process

Custody transfer is the term applied to the act of passing the responsibility for the possession, care, supervision, control, and/or ownership of a volume of crude oil or refined product to another entity (a customer or another pipeline company). It is the sole mode of exchange for pipeline operations because the pipeline business only transports crude oil and products for other companies.

1.1.1 Metered Custody Receipts

Operations personnel must **gauge** the tank before receipt. They must measure the tank, take the temperature at designated levels, and record the **API gravity** immediately before receipt begins. This procedure gives an accurate measurement of what was originally in the tank. The results can be recorded on the delivery log. Also, these results may be needed later to verify exactly what was delivered. If custody meter failure occurs during the shipment, it will be necessary to have this information for accounting purposes. Writing these measurements on the ticket is a useful practice employed by many operations personnel.

As soon as is practical after delivery is complete, tanks should be gauged, temperature and API gravity taken again, and the information recorded. Be aware, however, that under certain conditions, particularly with low flow rates, the contents of a tank can stratify, leading to varying readings at different locations within the same tank.

All other activities into the affected tank should be stopped during the process, if possible. If other activities are not stopped, the operator must gauge all the other tanks involved and record totalizer readings and other meters so that the number of barrels received can be accounted for separately. API gravity must be tested to verify the shipment quality, and the result must be sent to the flow computer so that correction factors may be added to the total calculation for volume received.

Finally, operations personnel are responsible for seeing that composite samples are collected. The automatic sampler settings should be verified, and the temperature and pressure readings should be read, recorded, and compared to those on the transmitted pumping orders. The concluding act requires operations personnel to close the pipeline valve, check for leaks, and make certain that the final meter reading is taken and recorded on the delivery ticket.

1.1.2 Gauged Receipts

Additional concerns for gauged receipts relate to tankage. Operations personnel must select tanks with enough volume available to hold the receipt safely. If switching is necessary, a time schedule must be calculated for the switch. A valve alignment must be worked out or selected from the charts. The valve lineup is often recorded on the ticket and receipt log for verification. For gauged receipts, the plan must include isolating the tanks being used from all nonessential connections so that exact measurement can be achieved. If the shipment is to arrive through a common receipt and delivery manifold, the operator must verify that the appropriate valve is open and all other valves on the delivery side are closed.

The same steps of sampling and testing must be followed for gauged receipts that were followed for metered receipts. Closing gauges are not normally taken for at least two hours after receipt due to a condition called roll. Product will often continue swirling around in the tank for several hours. This motion can affect the accuracy of a number of measurements. Waiting time is necessary to allow the product in the tanks to stop swirling. Since smaller tanks will remain turbulent for some time after filling at a high flow rate, some companies mandate four hours as a necessary settling time.

1.1.3 Metered Custody Deliveries

Metered custody deliveries make up large portions of custody transfer. Metered deliveries employ the same common procedures as receipts with a few exceptions. In most cases, personnel handle the transfer from a remote control center rather than being physically present at the terminal.

Operations personnel must determine from the schedule what product is to be delivered. The volume of the delivery must be calculated. The time frame of the operation (starting and stopping times) should be obtained from the scheduling department. Scheduling information is transmitted to the shipper as soon as possible to give the shipper time to prepare for the delivery. It is the responsibility of operations personnel to contact the shipper to notify or confirm the expected delivery. This is done at least one day or, preferably, two days in advance so that adequate preparation can be made.

Operations personnel must then plan the local operation. The pumping orders must be obtained from the scheduling department. Operations personnel must select the tanks from which to draw the shipment, prepare the valve alignment chart for the shipment, and prepare a set of backup plans to rely on in case problems occur in the yard.

The tanks from which the shipment will be moved must be gauged, their temperatures must be taken, and their API gravities must be tested immediately before the shipment begins in case a meter failure occurs. If there are other shipments being moved into or out of the tank, the operator must make certain that those meter readings are taken and recorded along with the opening gauge.

Before the shipment starts, operations personnel must select the appropriate valve alignment from the charts and make certain that the exact sequence is opened and closed with the exception of the main delivery valve to the pipeline. Once pumping begins, it is essential that the product be tested to confirm specifications. As with other shipments, all tests common to the particular product must be performed and recorded, including API gravity, temperature, line pressure, and meter operations. A composite sample is always taken, usually with an automatic sampler. Immediately preceding the beginning of the shipment, the operator should contact the shipper to make certain that the operation can proceed.

When pumping is finished, the pumping equipment is shut down. The meter readings are recorded on the ticket. Confirmation of receipt is the shipper's signature on the ticket. Operations personnel have the responsibility of verifying that all tickets issued are correct. If it is discovered that a ticket is incorrect, a corrected ticket can sometimes be issued.

1.1.4 Gauged Custody Deliveries

In some cases, gauged custody deliveries do occur, though they are not as common as metered shipments. The same basic steps are followed as those for metered custody deliveries. Opening and closing gauges and tests are the same as described for custody receipts.

1.1.5 Preparing Tickets

Metered custody deliveries make up large portions of custody transfer. Every product passing through the pipeline must have a ticket providing a record of product identification, volume, and disposition or delivery. Some tickets are printed automatically, while others are handwritten.

Immediately prior to the shipment or receipt, the tanks involved in the shipment must be gauged, their temperatures must be taken, and their API gravities must be tested in case a meter failure occurs. The same gauging procedures should apply both to the originating tank and the tank receiving the shipment. If there are other shipments being moved into or out of the tank, operations personnel must make certain that those meter readings are taken and recorded along with the opening gauge. A custody ticket may include the date of transaction, ticket number, opening and closing of gross and net barrels, correction factors, meter factors, type of product, and the shipper. Many operations personnel include the valve alignment on the ticket as a further verification of proper handling. If requested or required, the meter calibration information should also be recorded and a time frame should be determined for witnessing purposes. If the meter readings are not automatically stamped, they must be recorded on the ticket by hand.

Although manual gauging and measurement was previously the most common way of doing business on the pipeline, the majority of custody transfers are now handled by electronic equipment. The measurements of product, temperature, pressure, gravity, sediment and water (S&W), and other parameters are taken electronically. Due to the vast volumes and speed of handling, it would be nearly impossible to take these measurements by hand. Companies enter agreements for electronic measurement and handling of shipments so that all are in harmony in the business. Although it does involve a tremendous amount of electronic equipment, the level of accuracy is impressive, largely because the instruments and equipment are maintained in extremely good condition by pipeline E&I technicians.

1.2.0 Tank Gauging Devices

Gauging a tank is simply the process of determining the height of the stored liquid's surface relative to the bottom of the tank. This information, in turn, can be used to calculate other measurements, such as the actual volume of product in the tank. Tank gauging involves a number of instruments, some of them very simple and some rather complex.

1.2.1 Hand Gauging

The simplest tank gauging method uses a handline (*Figure 1*). This is nothing more than a weight attached to a measuring tape. The weight is lowered from a hatch on the top of the tank either

Figure 1 Handlines used for tank gauging.

until it touches the surface of the liquid (called outage gauging) or until it touches a special plate on the bottom of the tank (called innage gauging). If outage gauging is performed, the liquid height must be calculated using the known tank height, whereas innage gauging gives a direct answer. Hand gauging is quick and simple. It is often used to check more sophisticated automatic tank gauging (ATG) technologies.

1.2.2 Mechanical Tank Gauge

The mechanical tank gauge, shown in *Figure 2*, is essentially an automatic handline. It consists of a float attached to a tape. A mechanism at the top of the tape keeps it under tension as the float rises and sinks in response to the liquid level in the tank. A mechanical counter tracks the tape's up-and-down motion. This information, in turn, can drive a display or a transmitter that converts the information to one of the common electronic formats, such as 4–20 mA or a digital **fieldbus** protocol.

1.2.3 Servo Tank Gauge

A servo gauge, shown in *Figure 3*, is a variation on the mechanical tape gauge. It uses a weight attached to a line that is wound around a motor-controlled drum. The motor lowers the weight until a sensor detects that the weight is touching the liquid's surface. The gauge's controller uses the motor to adjust the weight's position continually so it's always just touching the liquid's surface. This position information is then translated into a liquid height value and transmitted in an appropriate format to a display or control system.

1.2.4 Radar Tank Gauge

A radar tank gauge, shown in *Figure 4*, is a much more sophisticated way to gauge a tank's contents. The radar tank gauge is used by bouncing microwave radio frequency energy off the liquid's surface and timing the reflections. This is the same principle used in radars of all kinds. The advantage to a radar tank gauge is that there are no moving parts to wear out or malfunction. It is also not affected by vapors in the tank, the liquid's thickness (viscosity), or the ambient temperature. Some units are even sophisticated enough to avoid being fooled by foams floating on top of the liquid's surface. Radar units are extremely accurate, reporting the surface position to within just a few millimeters.

Typical radar tank gauges usually have interchangeable antennas (*Figure 5*) that alter the shape of the microwave beam. Cone/horn and parabolic antennas are designed to be used in larger tanks with plenty of space around the antenna. In a **still pipe** where the space is much more limited, an array or rod antenna can be used. It is important to select and install the best antenna for the individual situation. Each manufacturer provides complete information to guide the selection process.

All radar gauges must go through a setup process designed to eliminate false readings due to echoes from tank seams and welds. Once this process has been completed, however, the radar gauge system requires relatively little maintenance or adjustment. Setup often involves using manufacturer-supplied software and an attached notebook computer.

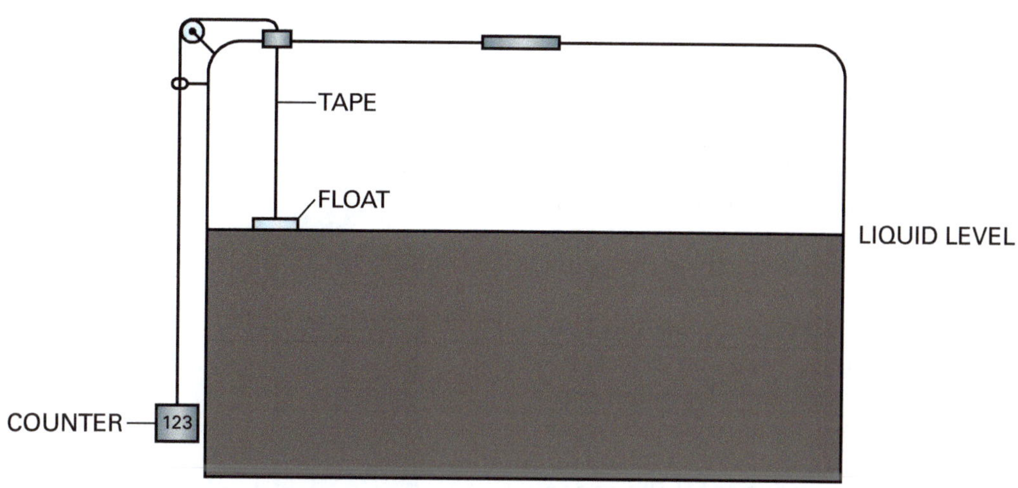

Figure 2 A mechanical tank gauge system.

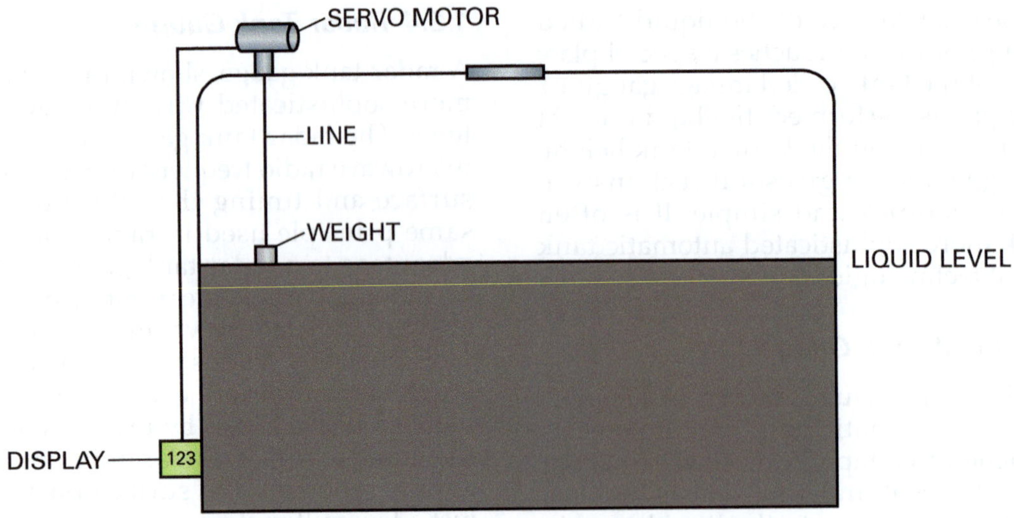

Figure 3 A servo tank gauge.

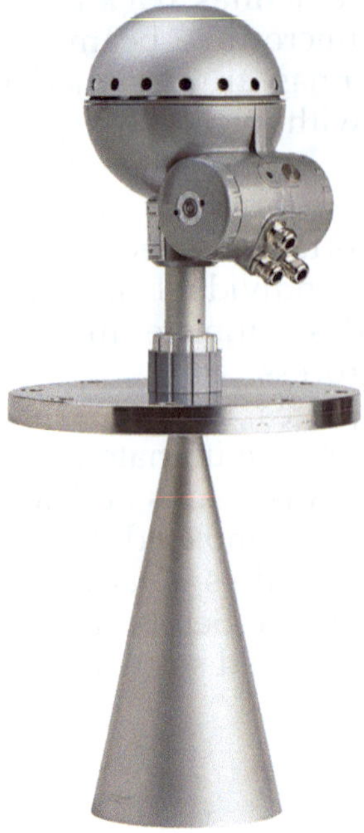

Figure 4 A radar tank gauge.

1.2.5 Sonar Tank Gauge

The sonar tank gauge works in much the same way as the radar gauge except that it uses high-frequency sound waves instead of radio waves. Like the radar gauge, it has no moving parts and relies on measuring echoes to determine the position of the liquid's surface. Unlike the radar gauge, however, sonar gauge accuracy can be affected by temperature and vapors since both can affect the speed of sound. These factors must be taken into account in order to obtain accurate results. Some sonar gauges include temperature sensors to help compensate for this component. Like radar gauges, sonar gauges must go through a setup process to eliminate the effects of tank construction on the echoes.

Figure 5 Radar tank gauge antennas.

1.2.6 Associated Technologies

Tank gauging and custody transfer usually require other information to be gathered as well. Product temperature, API gravity, and water content are common examples. Various kinds of sensors are available to perform these tasks. In fact, many of the companies that make tank gauges also offer sensors to make these related measurements. These are often integrated by proprietary software into a complete tank management system.

1.3.0 Servicing Tank Gauging Devices

Tank gauges may or may not require servicing, adjustment, and calibration. Much depends on the underlying technology. Units that are largely mechanical will require more attention than purely electronic units. In each case, the pipeline E&I technician should consult the manufacturer's manual for recommendations and guidelines. The following material provides a general overview of some of the tasks associated with tank gauging devices.

1.3.1 Test Equipment

Calibrating tank gauges requires detailed preparation to ensure successful results. Mechanical units will have a much different procedure than analog electronic ones. Similarly, smart devices have a procedure distinctly different from either mechanical or analog electronic ones. In some cases, the manufacturer may provide special software (usually run on a regular notebook computer) that must be used for setup and calibration if the unit is to provide reliable results.

One of the most important steps in this process is the selection of the correct test equipment for the device to be tested or calibrated. The selection is based on the type of instrument, type of test signal, and the range of test parameters. The manufacturer's instruction manual and recommendations for the instrument provide a procedure as well as data sheets that will identify values used in performing service and calibration tasks. Using this information, the correct test equipment for the device can be selected.

The following are examples of the uses of the various types of test equipment:

- Electronic simulators are used to calibrate electromechanical level transmitters.
- Handheld communicators (*Figure 6*) are used to connect to smart devices.
- Pneumatic power supplies and simulators provide variable test pressure for calibration of pressure-type level devices.
- Standing pipe level rigs are used as a reference for comparing the level instrument being tested against a known level.
- Digital multimeters provide a means of accurately measuring electrical values. Obtaining these values and comparing them to information from data sheets may be one of the calibration procedure steps.
- Multifunction calibrators can be used to read the output of transmitters and other similar devices. Some can also provide precision reference signals, eliminating the need for other calibration tools.

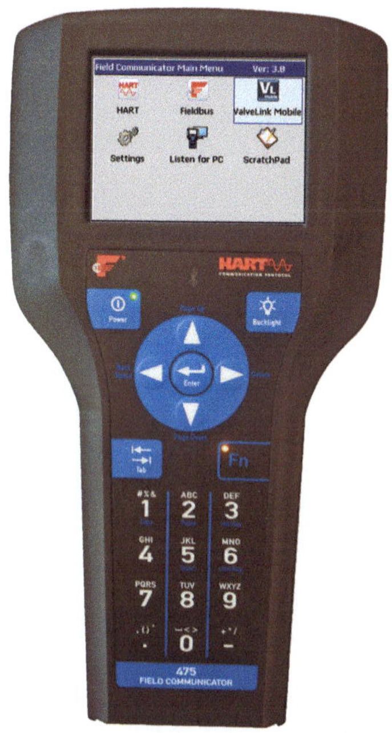

Figure 6 A handheld communicator.

Additional Resources

It is assumed that you will learn to perform these procedures under the direct supervision of your crew leader, foreman, or instructor. For independent study, refer to the recommended Additional Resources at the back of this module.

1.0.0 Section Review

1. Motion in a tank that can occur for several hours after a transfer is called _____.
   a. swirl
   b. churn
   c. roll
   d. surge

2. Which of the following tank gauges relies on timing reflections to determine the height of the stored liquid?
   a. Radar
   b. Mechanical
   c. Servo
   d. Handline

3. A smart transmitter will probably require the technician to use a notebook computer or a _____.
   a. wattmeter
   b. pneumatic dip meter
   c. IP converter
   d. handheld communicator

Section Two

2.0.0 Working With Tank Gauging Devices

Objective

Describe the procedures used for inspection, testing, and corrective and preventative maintenance of tank gauging for hazardous liquid leak detection (CT44_4-17).

a. Identify the prerequisites, competencies, and abnormal operating conditions (AOCs) associated with inspection, testing, and corrective and preventative maintenance of tank gauging.
b. Describe the procedures used for inspection, testing, and corrective and preventative maintenance of tank gauging for hazardous liquid leak detection.

Performance Tasks

Inspect, test, and perform corrective and preventative maintenance of tank gauging for hazardous liquid leak detection (CT44_4-17).

- Identify potential abnormal operating conditions (AOCs) that may occur during performance of this CT, and know the appropriate actions to take in response to them.
- Utilize the appropriate personal protective equipment (PPE) according to relevant company procedures.
- Determine the device's number and nameplate data and confirm the calibration point values for the device.
- Notify control center and/or affected personnel before work begins.
- Visually inspect the device and any associated equipment.
- Confirm that anything that could disturb the tank is shut off, that inlets and outlets are closed, and that the liquid inside the tank is still.
- Confirm the test equipment's current certification status.
- Gauge the tank with a certified device (handline, etc.) and compare its result to the tank gauge output.
- Calibrate the tank gauge following the manufacturer's procedure.
- Perform all steps required to resume normal operation. Notify control center and/or affected personnel that the unit has been returned to normal operation.
- Complete appropriate documentation as required by operator's procedures.

Trade Terms

Abnormal operating conditions (AOCs): Conditions within a pipeline system, identified by the operator, that may indicate (1) a malfunction of a component or (2) a deviation from normal operations that may indicate a condition exceeding design limits or result in a hazard to persons, property, or the environment.

Tank gauging is an important task associated with custody transfer. Consequently, it's vital that tank gauges continue to operate reliably and yield accurate data. Depending on their underlying technology, gauges may or may not require regular service. This section summarizes some of the issues and procedures associated with tank gauge maintenance.

2.1.0 Prerequisites, Competencies, and Abnormal Operating Conditions

Personnel must possess a variety of skills in order to successfully inspect, test, and maintain tank gauges. The various operations require a series of steps, which a pipeline E&I technician must be competent to perform. Personnel must also be aware of the particular abnormal operating conditions (AOCs) that may occur during the procedure and know the proper responses to them.

2.1.1 Prerequisites and Competencies

Personnel must possess the following basic knowledge and skills prior to inspecting, testing, and maintaining tank gauges:

- Familiarity with the tank gauging system in question and knowledge of its specific function (custody transfer, leak detection, overfill protection, etc.)
- Familiarity with the technologies behind different types of tank gauging systems
- Skill and safe handling of electrical systems

- Familiarity with the control system and its instrumentation, including alarm, control, indication, and recording functions
- Knowledge of tank types, fill/drain operations, and product transfer methods
- Familiarity with control center communications protocols
- Knowledge of operator gauge testing skills and knowledge of alarms and their significance
- Ability to use test equipment properly as well as to verify its certification status
- Knowledge of company documentation standards and methods
- Familiarity with the end device that the tank gauge ultimately drives

Personnel must also understand the company's safety policies and procedures, including the proper use of personal protective equipment (PPE).

2.1.2 Abnormal Operating Conditions (AOCs)

Personnel inspecting, testing, and maintaining tank gauges must be able to identify potential abnormal operating conditions (AOCs) and know the proper response to each, should they occur. The following are some potential AOCs, and the appropriate action personnel should take in response to them:

- *A tank liquid level that is too high or low occurs* – Personnel should alert the control center or other appropriate personnel to take the requisite corrective action.
- *A non-functioning tank gauging system occurs* – Personnel should alert appropriate personnel to take the requisite corrective action.
- *Damage to the tank roof or its components occurs* – Personnel should alert appropriate personnel to take the requisite corrective action.
- *Inappropriate material such as water, product, or other debris in the floating tank roof occurs* – Personnel should alert appropriate personnel to take the requisite corrective action.
- *Change to a static tank's liquid level with no obvious explanation occurs* – Personnel should alert appropriate personnel to take the requisite corrective action.

2.2.0 Procedure for Inspecting, Testing, and Maintaining Tank Gauging Devices

The following steps outline the general procedure for inspecting, testing, and maintaining tank gauging devices for hazardous liquid leak detection:

Step 1 Identify potential abnormal operating conditions (AOCs) that may occur, and know the appropriate actions to take in response to them.

Step 2 Utilize the appropriate personal protective equipment (PPE) according to relevant company procedures.

Step 3 Determine the device's number and nameplate data in order to verify that it is the correct unit to calibrate/maintain. If necessary, consult company documentation or drawings in order to be sure.

Step 4 Using company documentation, confirm the calibration point values for the device.

Step 5 Before beginning the process, notify the control center, local personnel, and any other personnel who might be affected by the calibration/maintenance procedure. The control center may have to provide feedback about tank gauge values or alarms during the process. It may also have to override automatic shutdowns or safety systems to protect the process from interruption. These events are often known as Critical Interlock and Trips (CIATs). They usually require higher authorization to override. Pipeline E&I technicians should be aware of these rules and know what they're allowed to do.

Step 6 Visually inspect the device and any associated equipment. Check for mechanical condition, corrosion, leakage, electrical connection problems, or problems with associated electronics. If the gauge is mechanical, confirm that all parts move freely. Repair or replace anything that isn't performing to specification.

Step 7 The tank's contents must not be moving during the testing process. Confirm that anything that could disturb the tank is shut off, that inlets and outlets are closed, and that the liquid inside the tank is still.

Step 8 Confirm the test equipment's current certification status. Only equipment that is within its certification date may be used.

WARNING!

Before performing the next step, determine if additional or special PPE is required with the tank in question. If the tank contains sour crude oil (oil containing hydrogen sulfide) or if highly aromatic hydrocarbons are present, additional protective gear will be necessary.

Step 9 Gauge the tank with a certified device (handline, etc.) and compare its result to the tank gauge output. Be sure to read both the gauge's and the end device's display. This task will determine if the tank gauge requires calibration.

Step 10 Document the results of the test along with all data. Follow company procedure and documentation standards.

Step 11 Calibrate the tank gauge following the manufacturer's procedure. Confirm correct values both at the gauge itself and at the end device.

Step 12 Document the results of the calibration procedure along with all data. Follow company procedure and documentation standards.

Step 13 Perform all steps required to resume normal operation. Follow company procedure.

Step 14 Notify the control center, local personnel, and any affected personnel that the unit has been returned to normal operation.

Step 15 Complete appropriate documentation as required by operator's procedures.

Additional Resources

It is assumed that you will learn to perform these procedures under the direct supervision of your crew leader, foreman, or instructor. For independent study, refer to the recommended Additional Resources at the back of this module.

2.0.0 Section Review

1. Besides being familiar with the underlying technology, what other aspect of a tank gauging system must the pipeline E&I technician understand?
   a. Its function
   b. Its malfunction history
   c. Its capacity
   d. Its age

2. What might the control center have to do during tank gauge calibration?
   a. Override an automatic shutdown.
   b. Disable an alarm.
   c. Set SCADA to ignore mode.
   d. Insert a calibrator into the circuit.

SUMMARY

Accurate tank gauging is at the heart of custody transfer. Since accuracy is a major necessity in the pipeline industry, both for economic as well as reputation reasons, well maintained tank gauges are similarly necessary. Tank gauges come in a number of types, some of which require more maintenance than others. This module summarized the various maintenance issues and provided a basic overview of the procedure for inspecting, testing, and maintaining these devices.

Review Questions

1. What must operations personnel gauge before receipt of product?
   a. The prover
   b. The pipeline
   c. The tank
   d. The control system

2. For how long do some companies mandate a small tank be allowed to stop swirling before a measurement occurs?
   a. Half an hour
   b. One hour
   c. Three hours
   d. Four hours

3. A tank gauging device that uses a float, a tape, and a counter, is called a _____.
   a. handline gauge
   b. mechanical tank gauge
   c. servo tank gauge
   d. sonar tank gauge

4. To help calibrate smart electronic tank gauges the manufacturer will often provide _____.
   a. special software
   b. a handheld communicator
   c. an electronic simulator
   d. a precision resistor

5. Besides knowing how to use test equipment, the pipeline E&I technician must be able to verify its _____.
   a. age
   b. certification status
   c. last use
   d. underlying technology

6. What is it considered if a supposedly static tank shows an unexplained level change?
   a. Drain oscillation
   b. Excessive roll
   c. An AOC
   d. Supply/demand precession

7. During testing and calibration, a tank's contents should be _____.
   a. vapor locked
   b. pressurized
   c. still
   d. rolling

Trade Terms Quiz

Fill in the blank with the correct term that you learned from your study of this module.

1. ________ is the tendency for the contents of a tank to swirl for some time after transfer. It can cause measurement inaccuracies.
2. The act of passing the responsibility for the possession, care, supervision, control and/or ownership of a volume of crude oil or other product to another entity (individual or company) is called ________.
3. ________ is the process of measuring the height of the liquid product in a storage tank. This information is the basis of other calculated measurements, such as volume.
4. Conditions within a pipeline system that may indicate (1) a malfunction of a component or (2) a condition exceeding design limits or resulting in a hazard to persons, property, or the environment are called ________.
5. A(n) ________ is a perforated pipe that runs from the top to the bottom of a tank. It is designed to provide a non-turbulent place from which to take measurements.
6. ________ is a simple digital communications network and protocol set designed to connect instruments, controllers, and other intelligent devices together. It has several different standards.
7. ________ is a density measurement system used by the American Petroleum Institute. This system's numbers are the inverse of specific gravity and scaled so a value of 10° equals a specific gravity of 1.

Trade Terms

Abnormal operating conditions (AOCs)
API gravity
Custody transfer
Fieldbus
Gauge
Roll
Still pipe

Trade Terms Introduced in This Module

Abnormal operating conditions (AOCs): Conditions within a pipeline system, identified by the operator, that may indicate (1) a malfunction of a component or (2) a deviation from normal operations that may indicate a condition exceeding design limits or result in a hazard to persons, property, or the environment.

API gravity: A density measurement system used by the American Petroleum Institute. API gravity numbers are the inverse of specific gravity and scaled so a value of 10° equals a specific gravity of 1.

Custody transfer: The act of passing the responsibility for the possession, care, supervision, control and/or ownership of a volume of crude oil or other product to another entity (individual or company).

Fieldbus: A simple digital communications network and protocol set designed to connect instruments, controllers, and other intelligent devices together. There are several different fieldbus standards.

Gauge: The process of measuring the height of the liquid product in a storage tank. This information is the basis of other calculated measurements, such as volume.

Roll: The tendency for the contents of a tank to swirl for some time after transfer. Roll can cause measurement inaccuracies.

Still pipe: A perforated pipe that runs from the top to the bottom of a tank. Still pipes are designed to provide a non-turbulent place from which to take measurements.

Additional Resources

The following recommended resources can provide additional helpful information related to the requirements for this covered task:

API Recommended Practice 1161, Third Edition, © 2014 American Petroleum Institute. API Publishing Services, 1220 L Street, NW, Washington, DC 20005. Distributed by Thompson Reuters (Scientific) LLC, **www.techstreet.com**.

The Condensed Handbook of Measurement and Control, N. E. Battikha. Third Edition. 2006. Research Triangle Park, NC: The International Society of Automation.

NCCER Module AOCCC-17, *Abnormal Operating Conditions Control Center*

NCCER Module AOCFG-17, *Abnormal Operating Conditions Field and Gas*

NCCER Module 60102-02, *Pipeline E&I Safety*

NCCER Module 60108-02, *Field Measurement*

NCCER Module 63_1-17, *Start-up of a Liquid Pipeline (Field)*

NCCER Module 63_2-17, *Shutdown of a Liquid Pipeline (Field)*

NCCER Module 63_4-17, *Locally Operate Valves on a Liquid Pipeline System*

49 *CFR* 192, **www.ecfr.gov**.

49 *CFR* 195, **www.ecfr.gov**.

The following websites offer resources for products and training:

The International Society of Automation, **www.isa.org**

Figure Credits

© Totalpics/Dreamstime.com, Module Opener

U.S. Tape Company, Inc., Figure 1

Emerson Process Management, Figures 4–6

Section Review Answer Key

Answer	Section Reference	Objective
Section One		
1. c	1.1.2	1a
2. a	1.2.4	1b
3. d	1.3.1	1c
Section Two		
1. a	2.1.1	2a
2. a	2.2.0	2b

NCCER CURRICULA — USER UPDATE

NCCER makes every effort to keep its textbooks up-to-date and free of technical errors. We appreciate your help in this process. If you find an error, a typographical mistake, or an inaccuracy in NCCER's curricula, please fill out this form (or a photocopy), or complete the online form at **www.nccer.org/olf**. Be sure to include the exact module ID number, page number, a detailed description, and your recommended correction. Your input will be brought to the attention of the Authoring Team. Thank you for your assistance.

Instructors – If you have an idea for improving this textbook, or have found that additional materials were necessary to teach this module effectively, please let us know so that we may present your suggestions to the Authoring Team.

NCCER Product Development and Revision

13614 Progress Blvd., Alachua, FL 32615

Email: curriculum@nccer.org
Online: www.nccer.org/olf

❑ Trainee Guide ❑ Lesson Plans ❑ Exam ❑ PowerPoints Other ____________

Craft / Level: Copyright Date:

Module ID Number / Title:

Section Number(s):

Description:

Recommended Correction:

Your Name:

Address:

Email: Phone:

CT44_5-16

Prove Flow Meters for Hazardous Liquid Leak Detection

Overview

Flow meters measure the product moving through a pipeline. As with all instruments, flow meters must be checked regularly to ensure that their readings match the actual amount of product passing through the pipeline. The checking process is called *proving*, and it's accomplished using one of several tools. This module describes the technology used for meter proving, and it explains the basic process for proving a flow meter.

Trainees with successful module completions may be eligible for credentialing through the NCCER Registry. To learn more, go to **www.nccer.org** or contact us at 1.888.622.3720. Our website has information on the latest product releases and training, as well as online versions of our *Cornerstone* magazine and Pearson's product catalog.

Your feedback is welcome. You may email your comments to **curriculum@nccer.org**, send general comments and inquiries to **info@nccer.org**, or fill in the User Update form at the back of this module.

This information is general in nature and intended for training purposes only. Actual performance of activities described in this manual requires compliance with all applicable operating, service, maintenance, and safety procedures under the direction of qualified personnel. References in this manual to patented or proprietary devices do not constitute a recommendation of their use.

CT44_5-17

Prove Flow Meters for Hazardous Liquid Leak Detection

Objectives

When you have completed this module, you will be able to do the following:

1. Identify various types of provers and how they are applied and serviced.
   a. Identify and describe how pipe provers are used.
   b. Identify and describe how a master meter prover is used.
   c. Describe how provers are serviced and calibrated.
2. Describe the procedures used to prove flow meters for hazardous liquid leak detection (CT44_5-17).
   a. Identify the prerequisites, competencies, and abnormal operating conditions (AOCs) associated with proving flow meters.
   b. Describe the procedures used to prove flow meters for hazardous liquid leak detection.

Performance Tasks

Under the supervision of the instructor, you should be able to do the following:

1. Prove flow meters for hazardous liquid leak detection (CT44_5-17).
   - Identify potential abnormal operating conditions (AOCs) that may occur during performance of this CT, and know the appropriate actions to take in response to them.
   - Utilize the appropriate personal protective equipment (PPE) according to relevant company procedures.
   - Determine the flow meter's number and nameplate data, and confirm the correct flow rate range for the meter in question.
   - Notify control center and/or affected personnel before work begins.
   - Confirm the correct flow rate range for the meter in question by consulting the manufacturer's specifications.
   - If using a portable prover, confirm that the pipeline pressure is within the prover's operating range and make the appropriate connections to the meter.
   - Line up the prover valves to fill and pressurize the prover and purge out all air. Check the equipment integrity to confirm that everything is working as intended.
   - Line up the prover valves for a proving run, check the block and bleed valves for seal integrity, and verify a stable temperature, pressure, and flow rate at the prover and meter.
   - Measure or otherwise ascertain the product's density or API gravity value and stability.
   - Start the proving run.
   - Confirm accurate input data and generate a meter factor through appropriate calculations or by entering the information into the flow computer.
   - Confirm that the new meter factor is appropriate and implement it as required.
   - If using a portable prover, set valves to appropriate positions to release pressure and empty the unit. Disconnect the unit and terminate all connections as necessary.
   - Return all valves to the correct positions for normal operation.
   - Notify the control center, local personnel, and any other affected personnel that the process is complete and normal operations are restored.
   - Complete appropriate documentation as required by operator's procedures.

Trade Terms

Abnormal operating conditions (AOCs)
Flow computer
K factor
Meter factor (MF)
Unity
Waterdraw method

Industry Recognized Credentials

If you are training through an NCCER-accredited sponsor, you may be eligible for credentials from NCCER's Registry. The ID number for this module is CT44_5-17. Note that this module may have been used in other NCCER curricula and may apply to other level completions. Contact NCCER's Registry at 1.888.622.3720 or go to **www.nccer.org** for more information.

Contents

1.0.0 Provers 1
1.1.0 Pipe Provers 1
1.1.1 Bidirectional Pipe Provers 1
1.1.2 Unidirectional Pipe Provers 1
1.1.3 Compact Pipe Provers 2
1.2.0 Master Meter Provers 3
1.2.1 Possible Inaccuracies 3
1.3.0 Prover Components and Calibration 4
1.3.1 Prover Pipe 5
1.3.2 Valves 5
1.3.3 Displacer and Sensors 6
1.3.4 Pulse Generator and Counting Circuits 6
1.3.5 Timing Units 6
1.3.6 Miscellaneous Components 6
1.3.7 Calibration 7
2.0.0 Proving Flow Meters 9
2.1.0 Prerequisites, Competencies, and Abnormal Operating Conditions 9
2.1.1 Prerequisites and Competencies 10
2.1.2 Abnormal Operating Conditions 10
2.2.0 Procedure 10

Figures

Figure 1 Permanently installed bidirectional pipe prover 2
Figure 2 Layout of a traditional bidirectional pipe prover 2
Figure 3 Aboveground traditional bidirectional prover loop 3
Figure 4 Unidirectional pipe prover 4
Figure 5 Compact pipe prover 5
Figure 6 Flow-reversal valve in a traditional bidirectional prover loop 5
Figure 7 Waterdraw calibration setup for a piston-type prover 7

SECTION ONE

1.0.0 PROVERS

Objective

Identify various types of provers and how they are applied and serviced.

a. Identify and describe how pipe provers are used.
b. Identify and describe how a master meter prover is used.
c. Describe how provers are serviced and calibrated.

Trade Terms

K factor: The number of pulses per unit volume generated by a meter.

Meter factor (MF): A factor determined during meter proving that is based on the prover's volume and the meter's measured volume. The meter factor is used to convert a meter's reading into an actual product quantity reading.

Unity: A condition in which a measurement is scaled so it has a value of 1.0000. Unity condition allows readings to be converted to other measurement units easily.

Waterdraw method: A method of calibrating provers by using water as a test medium. Water is drawn from the prover and measured with a calibrated container.

A prover is used to calibrate a meter by passing a known volume of fluid through the meter at a known rate and comparing the measured volume to a standard. There are several different types of provers. One very common type is the pipe, or continuous flow, prover. In the pipeline industry, pipe provers are usually bidirectional, unidirectional, or compact. A second category of pipeline prover is the master meter. The master meter is sometimes used to calibrate other provers.

1.1.0 Pipe Provers

There are many versions and variations of pipe provers on the market, but most use the same basic principle—a displacer forces product through a precisely calibrated pipe called the measuring section. Pipe provers can be classified into categories in a number of ways. One such category is the traditional pipe prover. Traditional pipe provers can be further classified by the possible direction of the product flow—they are either bidirectional or unidirectional. The difference between these two types is that bidirectional provers can measure volume in both directions, while the unidirectional pipe prover repeats the cycle in one direction only.

Another category of pipe provers is the compact or small volume prover. The compact prover also has a displacer passing through the pipe, but the type of displacer and method of meter volume calculation are different from those of traditional pipe provers. Most pipe provers are permanently installed as shown in *Figure 1.*

1.1.1 Bidirectional Pipe Provers

A traditional bidirectional pipe prover system involves a pipe, a displacer, a pair of detector switches, and a four-way valve for reversing flow and forcing the displacer to move in both directions. After the displacer has moved from one end of the prover pipe to the other, the valve is switched, and the displacer is forced to move back to the other end of the pipe. The displacer is usually an inflated ball or balloon device, called a sphere, which can be enlarged to slightly larger than the pipe for a tight seal to prevent a possible release. *Figure 2* shows the layout of a traditional bidirectional pipe prover.

As the displacer travels through the prover, it engages the two detecting sensors, one at the beginning of the run and the other at the end of the run. The first sensor engages the counting circuit, counting the pulses sent by the meter or pulse generator between the sensors. The second sensor disengages the counter. The signals counted between sensors give an exact measurement of the metered flow. By comparing this volume with the known volume between the prover sensors, the **meter factor (MF)** can be calculated.

Many prover loops are buried in the ground. *Figure 3* shows a traditional bidirectional prover loop that is aboveground.

1.1.2 Unidirectional Pipe Provers

The unidirectional pipe prover, shown in *Figure 4*, is simply a variation of the traditional bidirectional pipe prover. The basic difference is that in the unidirectional prover, the displacer or sphere is only permitted to measure volume in one direction. With the aid of a sphere-handling valve,

Figure 1 Permanently installed bidirectional pipe prover.

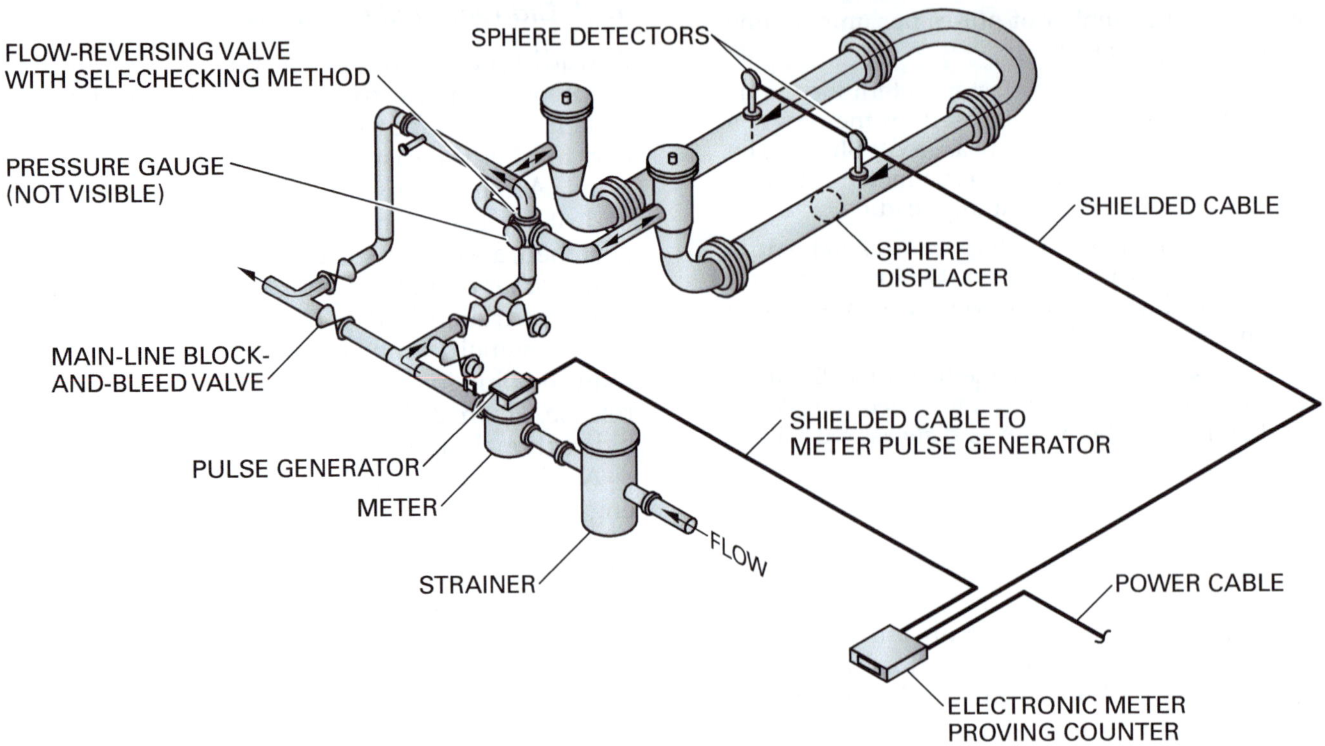

Figure 2 Layout of a traditional bidirectional pipe prover.

which replaces the four-way reversing valve, the displacer is recirculated around the prover loop in one direction only.

As seen in the illustrations, the product in the prover is separated from that in the pipeline and meter by a four-way valve or block and bleed valves (for the unidirectional prover). Care must be taken to recreate the exact circumstances, including temperature and pressure, for both areas. For the pipeline meter and prover, temperature and pressure are adjusted to 60°F (16°C) and zero pressure. If a problem develops with the pipeline flow during proving, the procedure should be discontinued. A new procedure should be initiated when operations have returned to steady state.

1.1.3 Compact Pipe Provers

Compact (small volume) pipe provers employ the same principles used by other provers but use a piston-type displacer. They are sometimes called *ballistic provers* to distinguish them from pipe provers that use a sphere as the displacer. However, small volume pipe provers extract the measurement of the meter's volume by dividing

Figure 3 Aboveground traditional bidirectional prover loop.

the usual meter signal into very small segments. The meter pulse and the displacer travel are both timed. Dividing the number of meter pulses during the proving flow by the time between the opening and closing sensor signals gives a pulse-per-unit rate, also known as the **K factor**.

Timing units for small volume pipe provers calculate to the nearest microsecond. Because the signals divide measurements into such precise units, the volume of the prover can be greatly reduced without affecting the precision of the meter factor. Although small volume pipe provers are much smaller and faster than traditional provers, care must still be taken to make certain that the product flowing through the unit and the pipeline meter remains in steady state. Pipeline variations have much more pronounced effects on a small prover, and therefore, more proving cycles should be performed to guarantee repeatability. An example of a compact pipe prover is shown in *Figure 5*.

1.2.0 Master Meter Provers

A master meter prover is used in situations in which proving with a pipe prover is not possible. The master meter is selected, maintained, and operated to serve as the reference for the proving of other meters. The master meter prover must be placed in line with the meter to be proved and must be close enough to minimize corrections for volume during the proving period. Before proving begins, the meters must be operated at the desired flow rate until the system is purged and a steady temperature and pressure have been achieved. A comparison of the two meter outputs is the basis of the master meter proving method.

Any one of a fixed battery of meters that has suitable manifolding can be assigned to be a master meter. A meter selected to be a master meter prover must be a meter that will have little meter factor change over a wide range of flow. It is kept in extremely good condition and used only to prove other meters. Alternatively, a selected portable meter or a meter at a test station can be assigned to be a master meter with the requirement that it must have demonstrated reliability and consistency. A portable meter must be adequately protected from damage during transportation, installation, and handling.

Master meters are involved in two types of proving procedures: direct and indirect. In the direct proving procedure, the master meter is proved against a pipe prover. In the indirect proving procedure, the master meter is used to prove another meter. The master meter itself has been proved through the direct method.

Two types of master meter provers are used within the petroleum industry: turbine meters and positive displacement meters. The only requirement for either choice is that it must be possible to express the value of the resulting meter factor to four decimal places, such as 1.0026 or 0.9989. A discrimination of 1 in 10,000 may be achieved by using a meter whose output represents 0.1 barrel and then using a minimum proof-run of 1,000 barrels. The same discrimination level may be achieved by installing devices that generate high-resolution output to obtain the required discrimination. Turbine meters are often used because they are normally equipped with high-speed counters. As with all metering activities, master meters must be set up in compliance with *API Manual of petroleum measurement standard*, Chapter 4. Several precautions must be observed when operating a turbine master meter, including:

- A turbine master meter must be a complete assembly comprised of flow-straightening devices and the meter itself. The assembly must remain intact during transportation and operation.
- When both meters are turbines, they should both have flow-straightening devices so that one meter does not adversely affect the other.
- Both counters must be started and stopped by the same signal. Electronic gating circuits are recommended.

1.2.1 Possible Inaccuracies

One problem associated with using a master meter is that it moves the meter being proved an extra step away from the original standard. The master meter must also be proved before it can prove an-

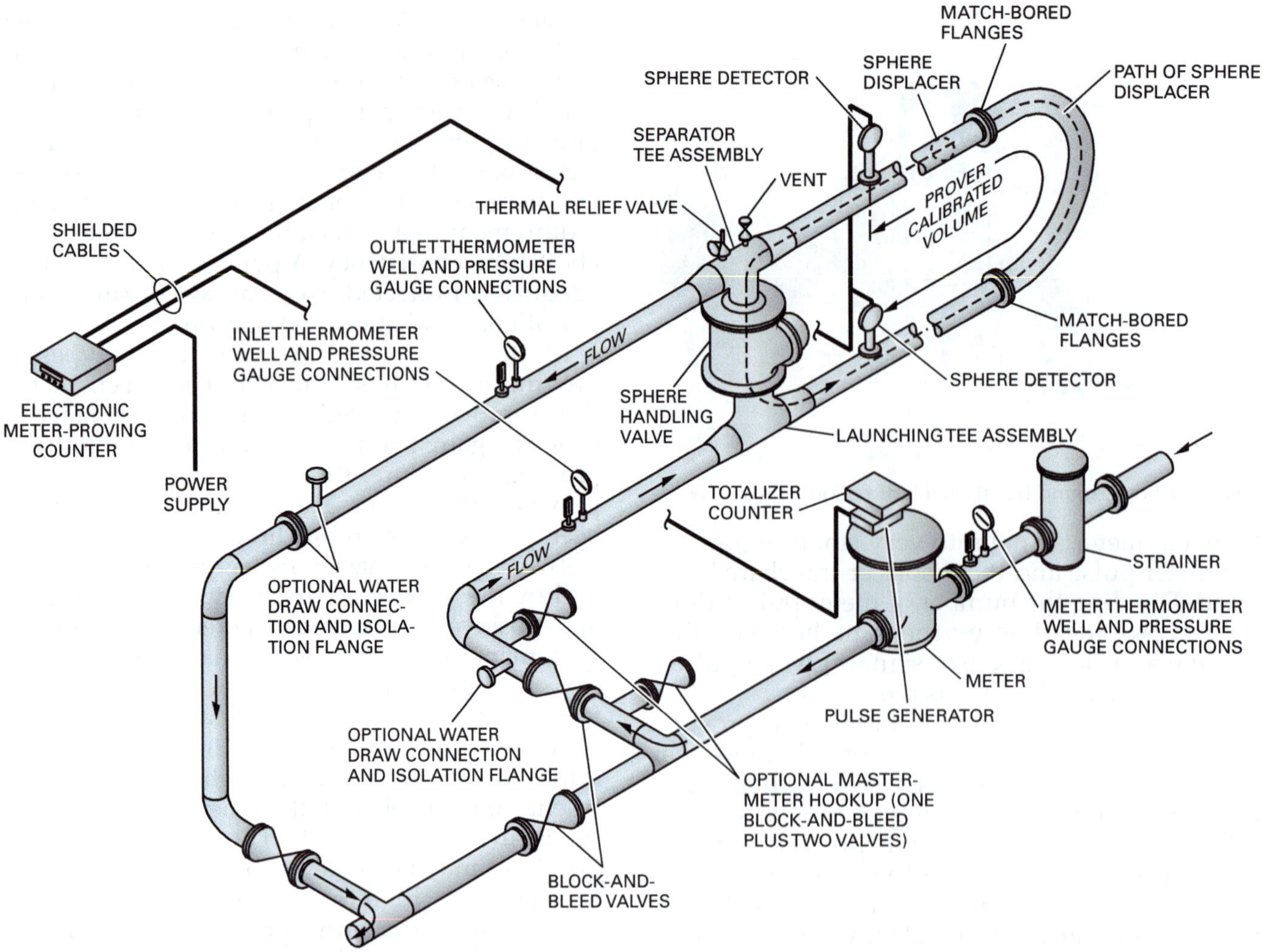

Figure 4 Unidirectional pipe prover.

other one. A proven meter proving another meter doubles the inaccuracies normally associated with the process. Great care must be exercised when setting up this proving procedure so that conditions under which the meter was proved can be duplicated exactly. If changes in temperature and pressure are not closely monitored, they will cause the original meter factor to change. Any change in the master meter's factor will pass a built-in error along to the second meter.

The type of meter selected to serve as the master meter may vary. However, any problems that might arise with it will likely be the same as those common to other meters of the same type. Pipeline E&I technicians should refer to the procedures and manufacturer's recommendations for that particular type and model of meter.

1.3.0 Prover Components and Calibration

Pipeline E&I technicians are not responsible for mechanical repairs to the prover valves and other equipment unless company procedures include them in the job description. However, prover electronics and instrumentation need the attention of a pipeline E&I technician. Technicians should be thoroughly familiar with the way a prover operates in order to interpret the data acquired through testing. They should also understand the various calibration procedures.

> **NOTE**
>
> All operations involving a prover must be documented according to company policy and legal standards. Records should be filled out completely.

1.3.1 Prover Pipe

The prover pipe, sometimes called the measuring pipe, is the part of the prover between the sensors that is calibrated so that the exact volume is known. When the prover is operated, this calibrated volume is measured through the meter. The metered volume is then compared to the calibrated volume of the prover to determine the meter factor.

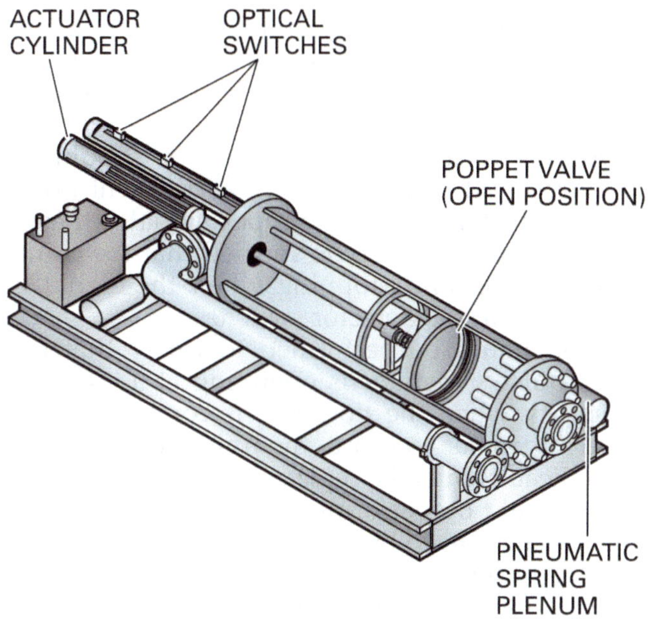

Figure 5 Compact pipe prover.

1.3.2 Valves

The flow-reversal valve, shown in *Figure 6*, is a four-way valve for reversing flow and forcing the displacer to move in both directions. After the displacer has moved from one end of the prover pipe to the other, the valve is switched, and the displacer is forced to move back to the other end of the pipe. Almost every type of prover has a valve that accomplishes this action. It is important that this valve be inspected closely for leaks because it separates the product in the prover from that in the pipeline. After visual inspection, it should be confirmed that there is no pressure shown on the gauge when this valve is seated and flow is directed through the valve.

The meter flow control valves control the flow from the pipeline into the prover. They are the prover inlet and outlet valves and the meter bypass valve. These must be inspected for leaks, using a block and bleed test, because all flow through the meter must be directed through the inlet valve. The inlet valve can be used to adjust the flow to the prover. This practice is not typical, however, and the need to do so often indicates a problem. Normally the inlet valve is fully open or fully closed. If it is used as a throttle, the meter factor will be accurate only at that particular flow, limiting its usefulness.

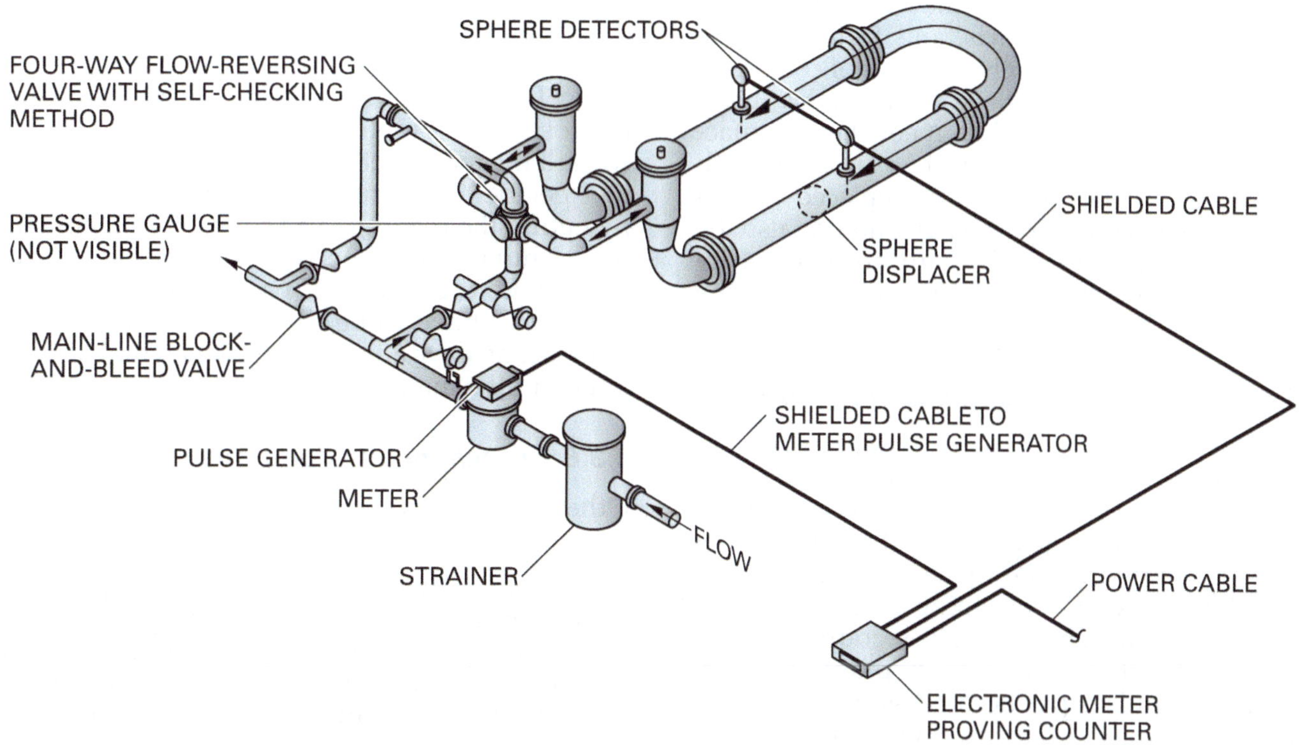

Figure 6 Flow-reversal valve in a traditional bidirectional prover loop.

1.3.3 Displacer and Sensors

The displacer is either an inflatable ball slightly larger (by about 2 percent) than the prover pipe or a piston fitted for a tight seal to prevent a possible product release. As the displacer travels through the prover, it engages the two sphere detectors, or sensors, one at the beginning of the run and the other at the end of the run. The first sensor engages the counting circuit, causing it to start counting the pulses sent by the meter or pulse generator while the sphere travels between the sensors. The second sensor disengages the counter. The signals counted between sensor triggers give an exact measurement of the metered flow. The meter factor can be calculated by comparing this value with the known volume between the prover sensors (refer to *Figure 6*).

> **NOTE**
>
> The pressure on both sides of the displacer should be checked to verify equality. If there is a difference in pressure, the displacer should be examined. A difference in the inlet and outlet pressure indicates the existence of some type of problem with the displacer that must be solved before proving proceeds.

The purpose of the detecting sensors is to precisely mark the volume measured in the prover's calibrated measuring pipe. These detectors are usually microswitches and may include several electrical connections, parts, and fittings. These should be examined closely for signs of corrosion, physical damage, and loose connections. Their output should be verified to meet the manufacturer's specifications. If there is a problem with one of the switches, it must be replaced with a perfectly matched duplicate. It should also be confirmed that the switches are not being triggered by interference from other equipment. The prover must then be recalibrated.

1.3.4 Pulse Generator and Counting Circuits

The pulse generator is the electrical device installed on the meter to signal the number of revolutions that the meter makes. Knowing the volume of the prover pipe and the number of pulses generated by the meter provides the information needed to determine an exact volume based on the number of pulses counted. Consider a meter that is sending 20 pulses per second (72,000 pulses per hour). Assume that each pulse represents 0.1 barrels. Calculations would show that the pipeline flow is proceeding at 7,200 barrels per hour. If that were the flow entering a tank, the calculation would only need to be multiplied by the number of hours for the delivery in order to find the total delivery volume.

The counting circuit counts the number of pulses that are sent from the meter during the time that the displacer travels between the starting switch and the ending switch. This number is used to calculate the meter factor from the volume of the prover. Generally, the K factor is adjusted to yield a meter factor of 1.0000, a condition known as **unity**. While adjusting for unity is not mandatory, this condition is often preferred as it looks normal when one is inspecting many meters and looking to spot anomalies.

Outputs from the pulse generator should be measured with a meter at several different flows to make certain that the proper signal is being received. It should also be verified that the counter is starting and stopping when the signal is received from the detector switches.

Tests should be performed to verify that the meter pulse generator is operating at a constant frequency when the flow rate is constant. If the pulse rate is *not* constant, the unit and gear train (if applicable) must be replaced. Tests for constant pulse should be performed again. Erratic test results can be caused by electrical interference from equipment nearby. The source of the interference must be found and eliminated before the meter can be trusted. If the source cannot be eliminated, arrangements to move the meter should be made.

1.3.5 Timing Units

Typical timing units can calculate to the nearest microsecond. Pipeline E&I technicians should consult the manufacturer's instruction manual for the specific timer unit to secure its specifications. By taking the readings and comparing them with the manufacturer's specifications, the technician can isolate the source of problems. Some of the units may have to be discarded and replaced. Others may have adjustments, which will be mentioned in the manual.

1.3.6 Miscellaneous Components

Other components of provers include the following:

- *Grounding system* – Provers are required to have an appropriate grounding system intact, whether permanent or portable. Technicians should examine it and its connections while on site to make certain that it is in good working order.

- *Leveling devices* – Leveling devices are installed on portable provers to allow adjustment of the unit to perfect level before proving is done. The operation of the system and the level should be examined and repaired as needed.

1.3.7 Calibration

All provers must be calibrated when new. Pipe provers are to be calibrated by the most recent **waterdraw method** (*Figure 7*) as certified by the National Institute of Standards and Technology (NIST). Permanent provers must be calibrated one year after installation and then at least once every five years thereafter. Portable provers must be calibrated at least once every three years. All provers must be calibrated after any work or maintenance that affects the calibrated section of pipe or the prover detectors. If the provers are used often, they should be calibrated more often. They should also be calibrated any time changes in volume are discovered, including the discovery of coating damage or corrosion. It is often during such calibration that problems with the electrical components of a prover are discovered.

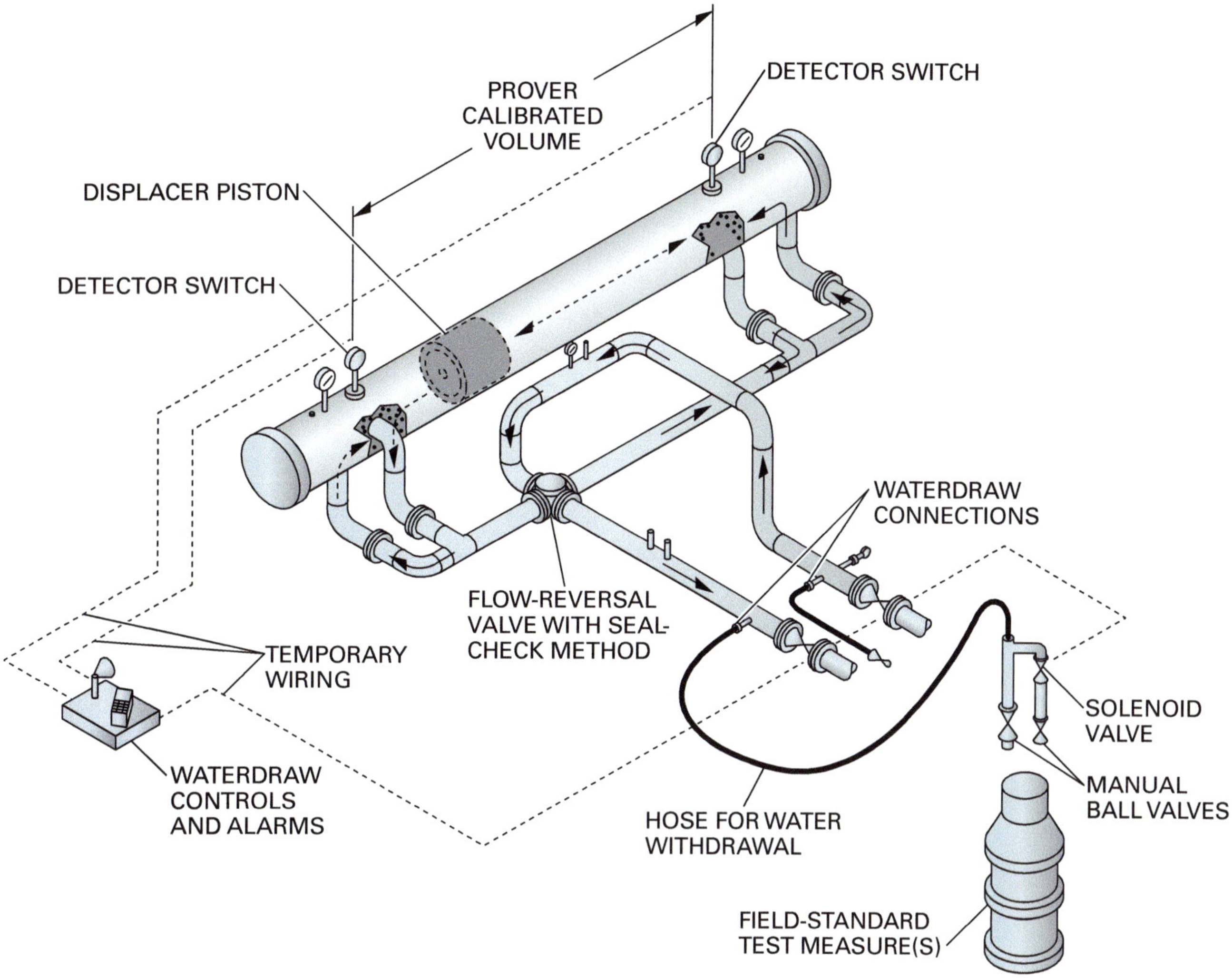

Figure 7 Waterdraw calibration setup for a piston-type prover.

Additional Resources

It is assumed that you will learn to perform these procedures under the direct supervision of your crew leader, foreman, or instructor. For independent study, refer to the recommended Additional Resources at the back of this module.

1.0.0 Section Review

1. The detectors in a bidirectional pipe prover engage and disengage a _____.
   a. counter
   b. displacer
   c. sphere detector
   d. pulse generator

2. Proving with a master meter doubles _____.
   a. accuracy
   b. inaccuracy
   c. resolution
   d. range

3. Portable provers must be calibrated at least once every _____.
   a. year
   b. two years
   c. three years
   d. five years

2.0.0 PROVING FLOW METERS

Objective

Describe the procedures used to prove flow meters for hazardous liquid leak detection (CT44_5-17).

a. Identify the prerequisites, competencies, and abnormal operating conditions (AOCs) associated with proving flow meters.
b. Describe the procedures used to prove flow meters for hazardous liquid leak detection.

Performance Tasks

Prove flow meters for hazardous liquid leak detection (CT44_5-17).

- Identify potential abnormal operating conditions (AOCs) that may occur during performance of this CT, and know the appropriate actions to take in response to them.
- Utilize the appropriate personal protective equipment (PPE) according to relevant company procedures.
- Determine the flow meter's number and nameplate data, and confirm the correct flow rate range for the meter in question.
- Notify control center and/or affected personnel before work begins.
- Confirm the correct flow rate range for the meter in question by consulting the manufacturer's specifications.
- If using a portable prover, confirm that the pipeline pressure is within the prover's operating range and make the appropriate connections to the meter.
- Line up the prover valves to fill and pressurize the prover and purge out all air. Check the equipment integrity to confirm that everything is working as intended.
- Line up the prover valves for a proving run, check the block and bleed valves for seal integrity, and verify a stable temperature, pressure, and flow rate at the prover and meter.
- Measure or otherwise ascertain the product's density or API gravity value and stability.
- Start the proving run.
- Confirm accurate input data and generate a meter factor through appropriate calculations or by entering the information into the flow computer.
- Confirm that the new meter factor is appropriate and implement it as required.
- If using a portable prover, set valves to appropriate positions to release pressure and empty the unit. Disconnect the unit and terminate all connections as necessary.
- Return all valves to the correct positions for normal operation.
- Notify the control center, local personnel, and any other affected personnel that the process is complete and normal operations are restored.
- Complete appropriate documentation as required by operator's procedures.

Trade Terms

Abnormal operating conditions (AOCs): Conditions within a pipeline system, identified by the operator, that may indicate (1) a malfunction of a component or (2) a deviation from normal operations that may indicate a condition exceeding design limits or result in a hazard to persons, property, or the environment.

Flow computer: A compact industrial computer with dedicated programming that combines data received from pipeline sensors to calculate the volume of product flowing through the pipeline in a given time.

The purpose of proving a flow meter is to obtain an up-to-date meter factor. Since meter factors are used to calculate actual product flow, they are crucial to accurate custody transfer and product management. The following material discusses the proving process.

2.1.0 Prerequisites, Competencies, and Abnormal Operating Conditions

Personnel must possess a variety of skills in order to successfully prove flow meters. The various operations require a number of steps, which a pipeline E&I technician must be competent to perform. Personnel must also be aware of the particular abnormal operating conditions (AOCs) that may occur during the procedure and know the proper responses to them.

2.1.1 Prerequisites and Competencies

Personnel must possess the following basic knowledge and skills prior to proving flow meters:

- Prover types and their correct operations
- Calculating meter factors and obtaining the various pieces of data required to do so including:
  - Product tables and material selection
  - Prover volume
  - K factor
  - Temperature and pressure coefficients

Personnel must also understand the company's safety policies and procedures, including the proper use of personal protective equipment (PPE).

2.1.2 Abnormal Operating Conditions

Personnel proving flow meters must be able to identify abnormal operating conditions (AOCs) and know the proper response to each, should they occur. The following are some potential AOCs, and the appropriate action personnel should take in response to them:

- *Introduction of air to the pipeline system by a prover occurs– Personnel should* alert the control center or other appropriate personnel to take the appropriate action.
- *An inoperable meter or a meter not communicating with the end device occurs– Personnel should* alert the control center or other appropriate personnel to take the appropriate action.
- *Unexpected release of hazardous liquid or carbon dioxide occurs* – Personnel trained to do so should try to identify and isolate the source of the leak. Proper notifications should be made as soon as possible, including calling 911 if necessary. If there is evidence of a leak, stop task activities, move to a safe distance, and notify appropriate pipeline personnel.

2.2.0 Procedure

Proving a flow meter involves connecting it to the prover after performing all preliminary steps to notify appropriate personnel and ready the meter for isolation. The pipeline E&I technician then performs a number of proving runs, ultimately obtaining a new meter factor, if successful.

The following steps outline the general procedure for proving a flow meter:

Step 1 Identify potential abnormal operating conditions (AOCs) that may occur, and know the appropriate actions to take in response to them.

Step 2 Utilize the appropriate personal protective equipment (PPE) according to relevant company procedures.

Step 3 Determine the flow meter's number and nameplate data in order to verify that it is the correct meter to be proved.

Step 4 Before beginning the proving process, notify the control center, local personnel, and anyone else who might be affected by the operation.

Step 5 Confirm the correct flow rate range for the meter in question by consulting the manufacturer's specifications. Proving must occur within this range in order to obtain an accurate meter factor.

Step 6 If you are using a portable prover, confirm that the pipeline pressure is within the prover's operating range. The prover must meet or exceed the pipeline pressure.

Step 7 If you are using a portable prover, make the appropriate connections to the meter (piping, pulse wiring, power, and grounding), as appropriate.

Step 8 Line up the prover valves to fill and pressurize the prover and purge out all air.

Step 9 Check the equipment integrity to confirm that everything is working as intended.

Step 10 Line up the prover valves for a proving run. This step diverts meter flow.

Step 11 Check the block and bleed valves for seal integrity.

Step 12 Verify a stable temperature, pressure, and flow rate at the prover and meter.

Step 13 Measure or otherwise ascertain the product's density or API gravity value and stability. Stable conditions are essential if the proving run is to be accurate.

Step 14 Start the proving run. Repeat as many times as is required to obtain valid and sufficient information.

Step 15 Confirm accurate input data. This includes information that is used to manually calculate the meter factor or that is provided to the flow computer so it can calculate the meter factor automatically.

Step 16 Using the information obtained, generate a meter factor through appropriate calculations or by entering the information into the flow computer. Also create a proving report.

Step 17 Confirm that the new meter factor is appropriate and implement it as required. This may be done by programming it into the flow computer or by applying a manual meter ticket.

Step 18 If using a portable prover, set valves to appropriate positions to release pressure and empty the unit.

Step 19 If using a portable prover, disconnect the unit, essentially reversing the tasks performed in Step 7.

Step 20 Return all valves to the correct positions for normal operation.

Step 21 Notify the control center, local personnel, and any other affected personnel that the process is complete and normal operations are restored.

Step 22 Complete appropriate documentation as required by operator's procedures.

Additional Resources

It is assumed that you will learn to perform these procedures under the direct supervision of your crew leader, foreman, or instructor. For independent study, refer to the recommended Additional Resources at the back of this module.

2.0.0 Section Review

1. Which piece of information is important to know when proving a meter in order to obtain a new meter factor?
   a. The prover's compensatory coefficient
   b. The weather during the proving run
   c. The meter's K factor
   d. The last time the meter was proved

2. A portable prover's range must meet or exceed the pipeline's _____.
   a. volume
   b. pressure
   c. temperature
   d. cycle time

SUMMARY

Proving flow meters is a necessary and routine task that provides up-to-date meter factors. These are critical for determining how much product is actually flowing through the pipeline. This module describes types of provers used to perform this task. The process of proving a meter, along with possible AOCs associated with this task was presented.

Review Questions

1. A bidirectional pipe prover uses a _____.
   a. recycle valve
   b. sphere-handling valve
   c. four-way valve
   d. circulator valve

2. To displace the fluid contained in the pipe a compact pipe prover uses a _____.
   a. sphere
   b. piston
   c. pig
   d. balloon

3. An essential quality of a master meter prover is that it must _____.
   a. have little meter factor change
   b. be a turbine meter
   c. be calibrated weekly
   d. always be proved with another master meter prover

4. When proving using a master meter prover, to obtain accurate results, what must be true?
   a. The master meter must have been proved within six months
   b. The master meter must be in parallel with the meter being proved
   c. The block and bleed valves must be open
   d. The original proving conditions must be duplicated

5. The part of a prover that must be calibrated so that its exact volume is known is called the _____.
   a. sphere detectors
   b. prover pipe
   c. prover valves
   d. SCADA interface

6. To calculate a meter factor, the pipeline E&I technician must know or be able to determine the _____.
   a. prover volume
   b. meter's repair history
   c. SCADA system's propagation delay
   d. sphere detector lag time

7. During the proving process, what must be checked to obtain an accurate meter factor?
   a. The counter circuit's pulse offset
   b. The flow computer's offset
   c. The meter's correct flow rate range
   d. The meter's last calibration

Trade Terms Quiz

Fill in the blank with the correct term that you learned from your study of this module.

1. The number of pulses per unit volume generated by a meter is called the _________.
2. _________ is a condition in which a measurement is scaled so it has a value of 1.0000. This condition allows readings to be converted to other measurement units easily.
3. Conditions within a pipeline system that may indicate (1) a malfunction of a component or (2) a condition exceeding design limits or resulting in a hazard to persons, property, or the environment are called _________.
4. A factor determined during proving that is based on the prover's volume and the meter's measured volume is the _________. It is used to convert a meter's reading into an actual product quantity reading.
5. The _________ is a method of calibrating provers by using water as a test medium. Water is drawn from the prover and measured with a calibrated container.
6. A compact industrial computer with dedicated programming that combines data received from pipeline sensors to calculate the volume of product flowing through the pipeline in a given time is called a(n) _________.

Trade Terms

Abnormal operating conditions (AOCs)
Flow computer
K factor
Meter factor (MF)
Unity
Waterdraw method

Trade Terms Introduced in This Module

Abnormal operating conditions (AOCs): Conditions within a pipeline system, identified by the operator, that may indicate (1) a malfunction of a component or (2) a deviation from normal operations that may indicate a condition exceeding design limits or result in a hazard to persons, property, or the environment.

Flow computer: A compact industrial computer with dedicated programming that combines data received from pipeline sensors to calculate the volume of product flowing through the pipeline in a given time.

K factor: The number of pulses per unit volume generated by a meter.

Meter factor (MF): A factor determined during proving that is based on the prover's volume and the meter's measured volume. The meter factor is used to convert a meter's reading into an actual product quantity reading.

Unity: A condition in which a measurement is scaled so it has a value of 1.0000. Unity condition allows readings to be converted to other measurement units easily.

Waterdraw method: A method of calibrating provers by using water as a test medium. Water is drawn from the prover and measured with a calibrated container.

Additional Resources

The following recommended resources can provide additional helpful information related to the requirements for this covered task:

API Recommended Practice 1161, Third Edition, © 2014 American Petroleum Institute. API Publishing Services, 1220 L Street, NW, Washington, DC 20005. Distributed by Thompson Reuters (Scientific) LLC, **www.techstreet.com**.

The Condensed Handbook of Measurement and Control, N. E. Battikha. Third Edition. 2006. Research Triangle Park, NC: The International Society of Automation.

49 *CFR* 192, **www.ecfr.gov**.

49 *CFR* 195, **www.ecfr.gov**.

NCCER Module AOCCC-17, *Abnormal Operating Conditions Control Center.*

NCCER Module AOCFG-17, *Abnormal Operating Conditions Field and Gas.*

NCCER Module 60102-02, *Pipeline E&I Safety.*

NCCER Module CT25_2-17, *Inspect, Test and Calibrate Pressure Transmitters.*

NCCER Module CT44_3-17, *Inspect, Test, and Maintain Flow Computer for Hazardous Liquid Leak Detection.*

NCCER Module CT44_8-17, *Inspect, Test and Maintain Temperature Transmitters for Hazardous Liquid Leak Detection.*

NCCER Module CT63_1-17, *Start-up of a Liquid Pipeline (Field).*

NCCER Module CT63_2-17, *Shutdown of a Liquid Pipeline (Field).*

NCCER Module CT63_3-17, *Monitor Pressures, Flows, Communications, and Line Integrity and Maintain Them Within Allowable Limits on a Liquid Pipeline System (Field).*

NCCER Module CT63_4-17, *Locally Operate Valves on a Liquid Pipeline System.*

The following websites offer resources for products and training:

The International Society of Automation, **www.isa.org.**

Figure Credits

© Curraheeshutter/Dreamstime.com, Module Opener

WEAMCO, Figures 1 and 3

Section Review Answer Key

Answer	Section Reference	Objective
Section One		
1. a	1.1.1	1a
2. b	1.2.1	1b
3. c	1.3.7	1c
Section Two		
1. c	2.1.1	2a
2. b	2.2.0	2b

NCCER CURRICULA — USER UPDATE

NCCER makes every effort to keep its textbooks up-to-date and free of technical errors. We appreciate your help in this process. If you find an error, a typographical mistake, or an inaccuracy in NCCER's curricula, please fill out this form (or a photocopy), or complete the online form at **www.nccer.org/olf**. Be sure to include the exact module ID number, page number, a detailed description, and your recommended correction. Your input will be brought to the attention of the Authoring Team. Thank you for your assistance.

Instructors – If you have an idea for improving this textbook, or have found that additional materials were necessary to teach this module effectively, please let us know so that we may present your suggestions to the Authoring Team.

NCCER Product Development and Revision

13614 Progress Blvd., Alachua, FL 32615

Email: curriculum@nccer.org
Online: www.nccer.org/olf

❑ Trainee Guide ❑ Lesson Plans ❑ Exam ❑ PowerPoints Other ____________________

Craft / Level: Copyright Date:

Module ID Number / Title:

Section Number(s):

Description:

Recommended Correction:

Your Name:

Address:

Email: Phone:

CT44_6-17

Maintain Flow Meters for Hazardous Liquid Leak Detection

Overview

Measuring product flow is a critical task in the pipeline industry. An accurate understanding of flow is crucial for custody transfer and other product management tasks. Pipeline engineers and technicians have a large number of flow-measuring technologies at their disposal. Each has its strengths and weaknesses. Like all instrumentation, flow meters require regular maintenance in order to keep them working properly and producing accurate results. This module describes types of flow meters and the procedures used to maintain them.

Trainees with successful module completions may be eligible for credentialing through the NCCER Registry. To learn more, go to **www.nccer.org** or contact us at 1.888.622.3720. Our website has information on the latest product releases and training, as well as online versions of our *Cornerstone* magazine and Pearson's product catalog.

Your feedback is welcome. You may email your comments to **curriculum@nccer.org**, send general comments and inquiries to **info@nccer.org**, or fill in the User Update form at the back of this module.

This information is general in nature and intended for training purposes only. Actual performance of activities described in this manual requires compliance with all applicable operating, service, maintenance, and safety procedures under the direction of qualified personnel. References in this manual to patented or proprietary devices do not constitute a recommendation of their use.

CT44_6-17

Maintain Flow Meters for Hazardous Liquid Leak Detection

Objectives

When you have completed this module, you will be able to do the following:

1. Describe flow measuring technologies.
   a. Describe positive displacement flow meters.
   b. Describe turbine flow meters.
   c. Describe magnetic flow meters.
   d. Describe ultrasonic flow meters.
   e. Describe Coriolis flow meters.
   f. Describe vortex flow meters.
   g. Describe orifice flow meters.
2. Describe the procedures used to maintain flow meters for hazardous liquid leak detection (CT44_6-17).
   a. Identify the prerequisites, competencies, and abnormal operating conditions (AOCs) associated with maintaining flow meters.
   b. Describe the procedures used to maintain flow meters for hazardous liquid leak detection.

Performance Tasks

Under the supervision of the instructor, you should be able to do the following:

1. Maintain flow meters for hazardous liquid leak detection (CT44_6-17).
   - Identify potential abnormal operating conditions (AOCs) that may occur during performance of this CT, and know the appropriate actions to take in response to them.
   - Utilize the appropriate personal protective equipment (PPE) according to relevant company procedures.
   - Determine the flow meter's number and nameplate data in order to verify that it is the correct meter to be maintained.
   - Notify control center and/or affected personnel before work begins.
   - Visually examine the flow meter and assess its physical and mechanical condition. Identify and perform any preventative maintenance as required.
   - Identify any diagnostic steps specified by the manufacturer's documentation and perform them as appropriate.
   - If the meter needs to be disconnected for maintenance, isolate the meter, release pressure, empty the meter and lines, and disconnect the meter.
   - Complete the necessary repairs to the meter.
   - If the meter has been disconnected for the repair, reconnect it correctly.
   - Notify the control center that the meter is being brought online and that commissioning and startup activities have commenced.
   - Bring the meter online by purging it and filling the pipeline.
   - Check the meter and its associated hardware and connections.
   - Notify the control center that the meter is ready for startup and that it should be proved.
   - Complete appropriate documentation as required by operator's procedures.

Trade Terms

Abnormal operating conditions (AOCs)
Cavitation
Coriolis force
K factor
Meter factor (MF)
Slippage
Von Karman vortex principle

Industry Recognized Credentials

If you are training through an NCCER-accredited sponsor, you may be eligible for credentials from NCCER's Registry. The ID number for this module is CT44_6-17. Note that this module may have been used in other NCCER curricula and may apply to other level completions. Contact NCCER's Registry at 1.888.622.3720 or go to **www.nccer.org** for more information.

Contents

1.0.0 Flow Meters 1
1.1.0 Positive Displacement (PD) Flow Meters 1
1.1.1 Key Principles 1
1.2.0 Turbine Flow Meters 2
1.2.1 Key Principles 2
1.3.0 Magnetic Flow Meters 6
1.3.1 Key Principles 6
1.4.0 Ultrasonic Flow Meters 6
1.4.1 Key Principles 6
1.5.0 Coriolis Flow Meters 8
1.5.1 Key Principles 8
1.6.0 Vortex Flow Meter 8
1.6.1 Key Principles 8
1.7.0 Orifice Flow Meters 9
1.7.1 Key Principles 10
2.0.0 Maintaining Flow Meters 13
2.1.0 Prerequisites, Competencies, and Abnormal Operating Conditions 13
2.1.1 Prerequisites and Competencies 13
2.1.2 Abnormal Operating Conditions 14
2.2.0 Procedure 14

Figures

Figure 1 Positive displacement meter manifold 1
Figure 2 Principle of meter slippage 2
Figure 3 Schematic of a typical PD meter and drive train 3
Figure 4 Typical meter stack of a PD meter 4
Figure 5 Cut-away view of a turbine meter 4
Figure 6 Turbine meter signal detection systems 5
Figure 7 Typical meter run layout 5
Figure 8 Magnetic flow meter components 6
Figure 9 Ultrasonic flow meter 7
Figure 10 Coriolis flow meter 8
Figure 11 Layout of a Coriolis flow and density meter with meter prover 9
Figure 12 The von Karman vortex principle 9
Figure 13 Vortex flow meter 9
Figure 14 Orifice plates come with different hole shapes 10
Figure 15 Cut-away view of an orifice plate meter installation 10
Figure 16 Common orifice plate holders 11

SECTION ONE

1.0.0 FLOW METERS

Objective

Describe flow measuring technologies.

a. Describe positive displacement flow meters.
b. Describe turbine flow meters.
c. Describe magnetic flow meters.
d. Describe ultrasonic flow meters.
e. Describe Coriolis flow meters.
f. Describe vortex flow meters.
g. Describe orifice flow meters.

Trade Terms

Cavitation: Vapor pockets that form in a fluid flow as a result of rapid pressure changes. They cause shockwave-like pulsations that can lead to inaccuracies and instrument damage.

Coriolis force: A deflective effect created in certain kinds of rotating systems. Named for the French scientist Gaspard-Gustave Coriolis.

K factor: The number of pulses per unit volume generated by a meter.

Meter factor (MF): A factor determined during meter proving that is based on the prover's volume and the meter's measured volume. The meter factor is used to convert a meter's reading into an actual product quantity reading.

Slippage: A cause of inaccuracy in a positive displacement meter. It results from fluid leaking around the edge of the meter rotor.

Von Karman vortex principle: The tendency for a fluid flow to form vortices (swirling sections of fluid) after passing across an obstruction in the flow path. Named for the Hungarian-American physicist Theodore von Kármán.

Flow meters come in many different forms and rely on different physical principles to measure a product flowing through a pipeline. Each meter type has a setting in which it works best, although multiple types may be suitable options in a given situation. This module describes the major types of flow meters used in the pipeline industry.

1.1.0 Positive Displacement (PD) Flow Meters

Positive displacement flow meters are mechanical devices that rotate as fluid passes through them. Their actions can be compared to those of an old-fashioned waterwheel, so they rotate in proportion to the amount of liquid passing through.

1.1.1 Key Principles

Measured amounts of liquid are caught in distinct segments as the fluid passes through a PD flow meter. Segment sizes vary depending on the meter size, and many different meter sizes are available. While one portion of liquid is entering the meter another portion is being measured, and a third portion is exiting the meter back into the line. The shaft attached to the rotor transmits the motion to a geared unit that, in turn, operates the meter. Each time the unit rotates a sensor is activated, incrementing a counter, somewhat like the odometer in an automobile. Electronic transmitters or signaling devices can also be connected to the meter to relay information to computers, recorders, and other equipment that will perform recordkeeping. A PD meter manifold is shown in *Figure 1*.

One of the problems with PD meters relates to design. In most models, the blade or vane that causes the meter to turn does not actually touch the casing, and a minute amount of fluid, called **slippage**, is allowed to pass by it (*Figure 2*). Slippage ranges from almost zero for high viscosity fluids to a measureable percentage (0.25 percent) for liquids like gasoline. Temperature and viscosity seriously affect the accuracy of a PD meter, but due to the fact that refined products

Figure 1 Positive displacement meter manifold.

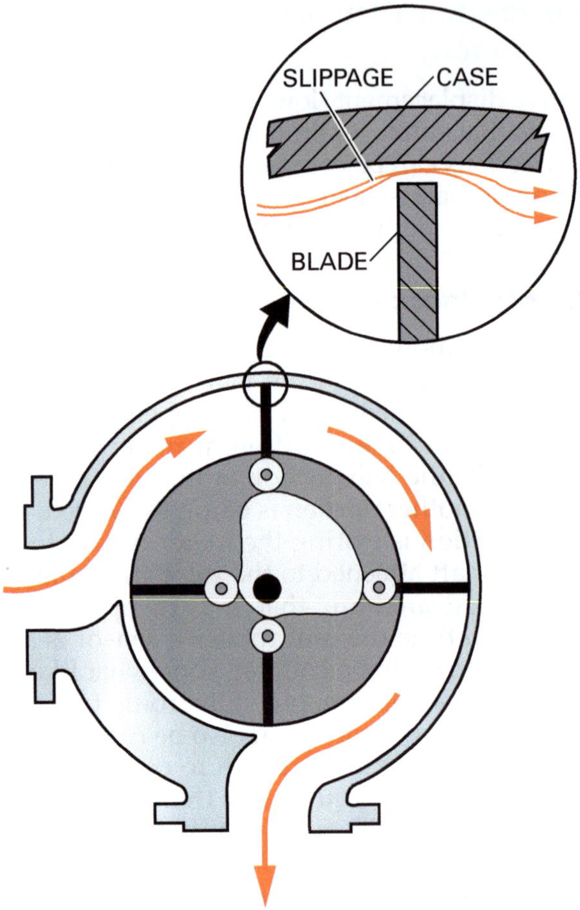

Figure 2 Principle of meter slippage.

are consistent, the effect is reduced and the **meter factor (MF)** can be calculated accurately.

The PD meter is relatively high maintenance when compared to some of the other flow meter technologies. It has a significant number of moving parts that require regular servicing. The meter may incur considerable damage in the case of a surge. A PD meter can become jammed and even block the line.

Some of the problems experienced by a PD meter relate to the gear train that drives the meter stack, including the counter and signaling devices. Gears must be selected and installed to accommodate the type of measurement or counting that is desired from the meter. Meters can be set up to give readings in any standard unit of measurement (gallons, liters, or barrels) by changing the gear ratio. Many PD meters are equipped with complicated double-clutch mechanisms called calibrators that adjust to provide accurate meter readings. A modified version with no adjustment is installed when other types of temperature compensation are utilized. A schematic of a typical PD meter and drive train is shown in *Figure 3*.

The meter stack creates drag due to the power required for turning the devices located above the calibrator, limiting the range of small meters. The devices turned include the meter counters, registers, right-angle drives, and compensators. Right-angle drives are connected to a high-resolution pulse generator on a temporary basis to provide the signal for proving operations when transmission equipment is not used. Transmission equipment provides this pulse generation, eliminating the need for the temporary unit. Any accessory that might be added to the meter must be carefully assessed because extra drag may affect the meter adversely. A typical meter stack is shown in *Figure 4*.

Regular maintenance of PD meters and meter stacks is essential. For older units, the counter drive train and the shaft seal must be lubricated regularly to keep drag as low as possible and to prevent release of product around the drive shaft. Newer units rely on materials like PTFE (Teflon®) and therefore do not require lubrication in these locations. Since much of a PD meter involves mechanical components, many repairs involve bearings, gears, and standard mechanical procedures. Technicians must follow the manufacturer's recommendations for all such repairs.

1.2.0 Turbine Flow Meters

Turbine flow meters were invented in the nineteenth century, but widespread use did not begin until the second half of the twentieth century. Their popularity is due to their mechanical simplicity and the fact that they require little maintenance. However, they become very undependable with heavier crude products.

1.2.1 Key Principles

Turbine meters do not measure the volume of fluid directly. Instead, they measure its velocity, from which volume may be calculated. A turbine meter, shown in a cut-away view in *Figure 5*, consists of a multi-bladed rotor on a free-running bearing suspended in the fluid stream. The axis of rotation of the rotor is perpendicular to the flow direction, and the rotor blades sweep out virtually the full bore of the meter. The fluid impinging on the rotor blades causes the rotor to revolve. Within the linear flow range of the meter, the angular speed of rotation is directly proportional to the volumetric flow rate.

The speed of rotation is monitored by an electromagnetic pickup coil, which is fitted to the outside of the meter housing. Two types of pickup

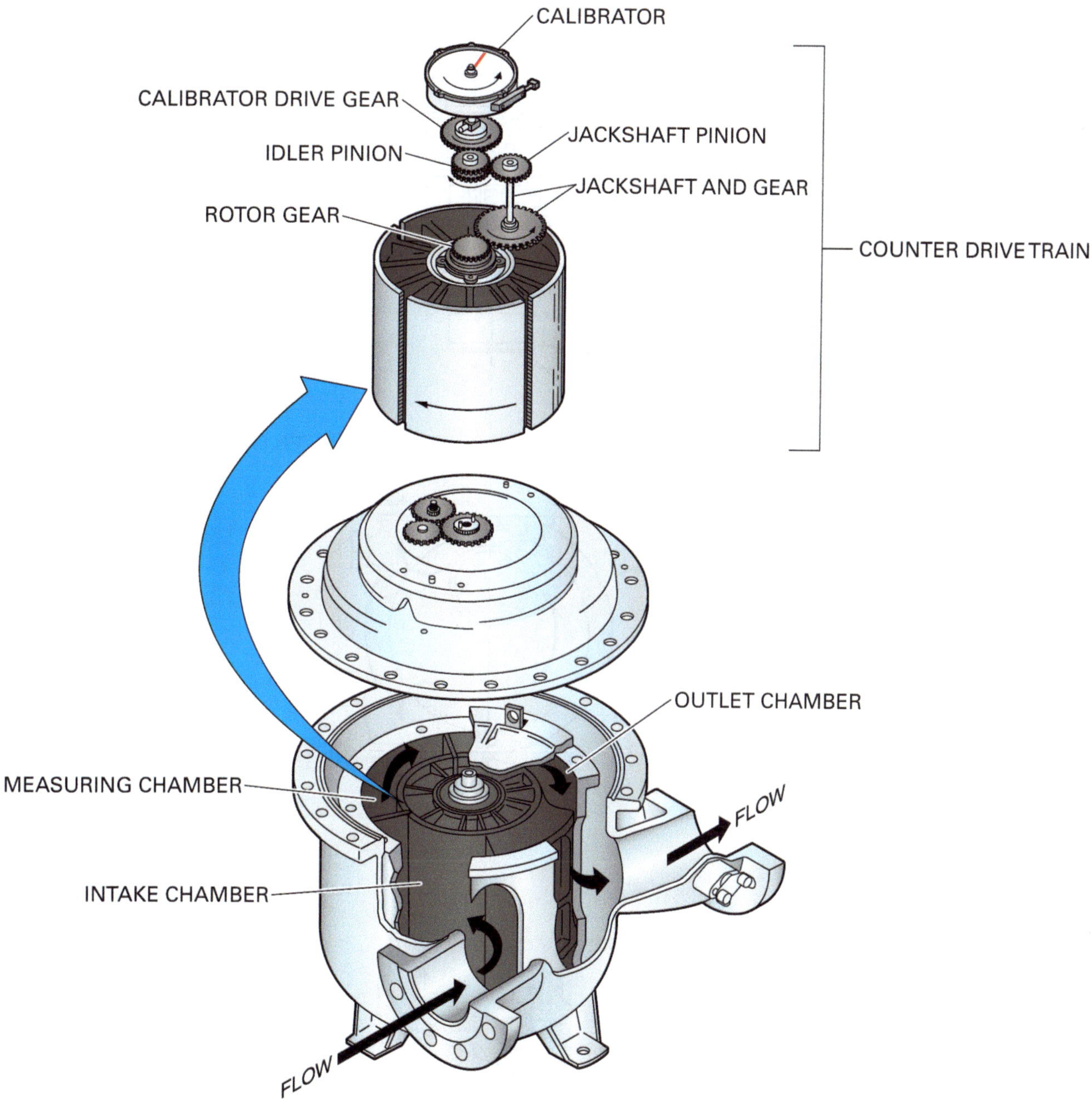

Figure 3 Schematic of a typical PD meter and drive train.

coils are primarily used: reluctance and inductance (*Figure 6*). Both operate on the principle of a magnetic field moving through a coil.

In the reluctance pickup coil system [*Figure 6 (A)*], a permanent magnet is part of the coil assembly. The field produced is concentrated to a small point by the cone. The turbine rotor blades are made of a paramagnetic material. As a blade approaches the cone point, its magnetic properties deflect the magnetic field. This deflection causes a voltage to be induced in the coil. As the blade passes under the cone point, the voltage decays, only to be built back up in the opposite polarity as the departing blade deflects the magnetic field in the opposite direction. Thus, each blade produces a separate and distinct voltage pulse as it passes the cone.

In the inductance pickup coil system [*Figure 6 (B)*], the rotor itself is a permanent magnet. As it rotates, its magnetic field cuts through the coil, inducing a varying voltage. The output of this kind of pickup is a sine wave, with each cycle representing one complete rotation of the turbine.

Only two basic parameters must be constant for the turbine meter to operate in a dependable manner. First, the velocity of the stream must be in exact proportion to the volumetric flow of the pipeline. Second, the velocity of the stream at the rotor must be equal to the velocity of the entire stream in the pipeline.

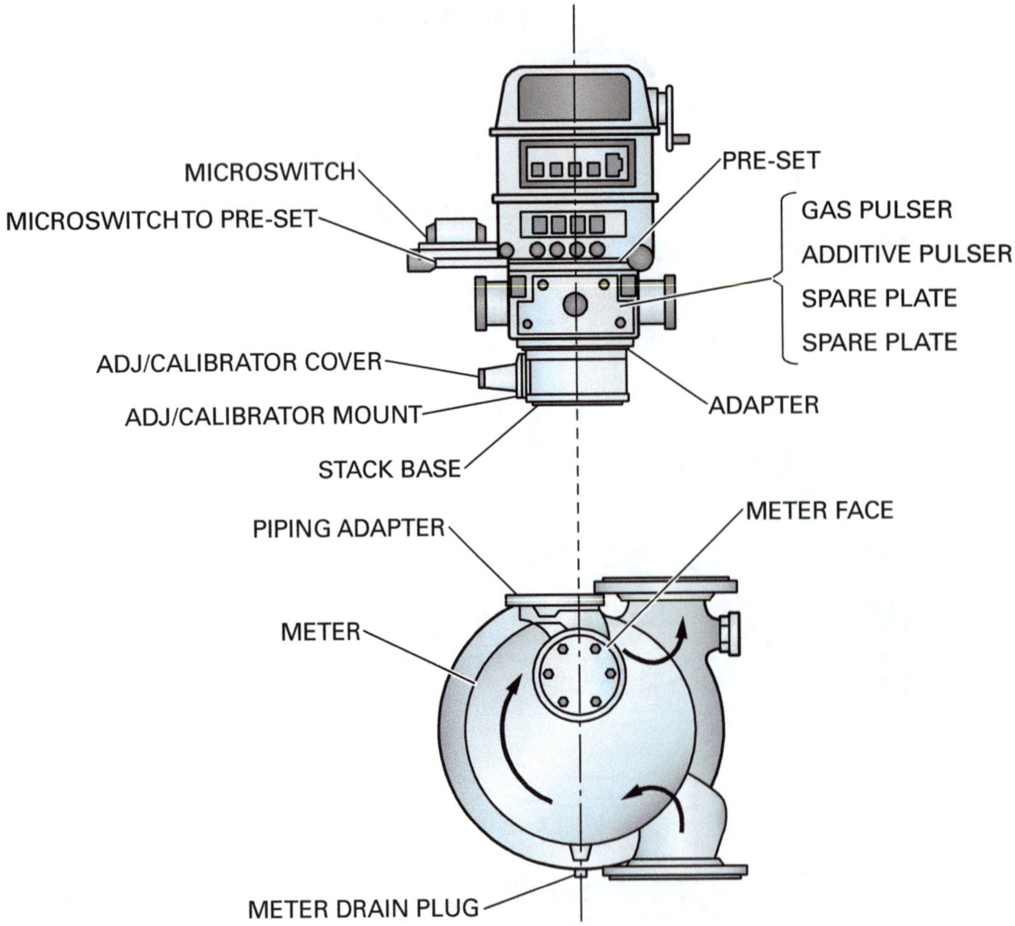

Figure 4 Typical meter stack of a PD meter.

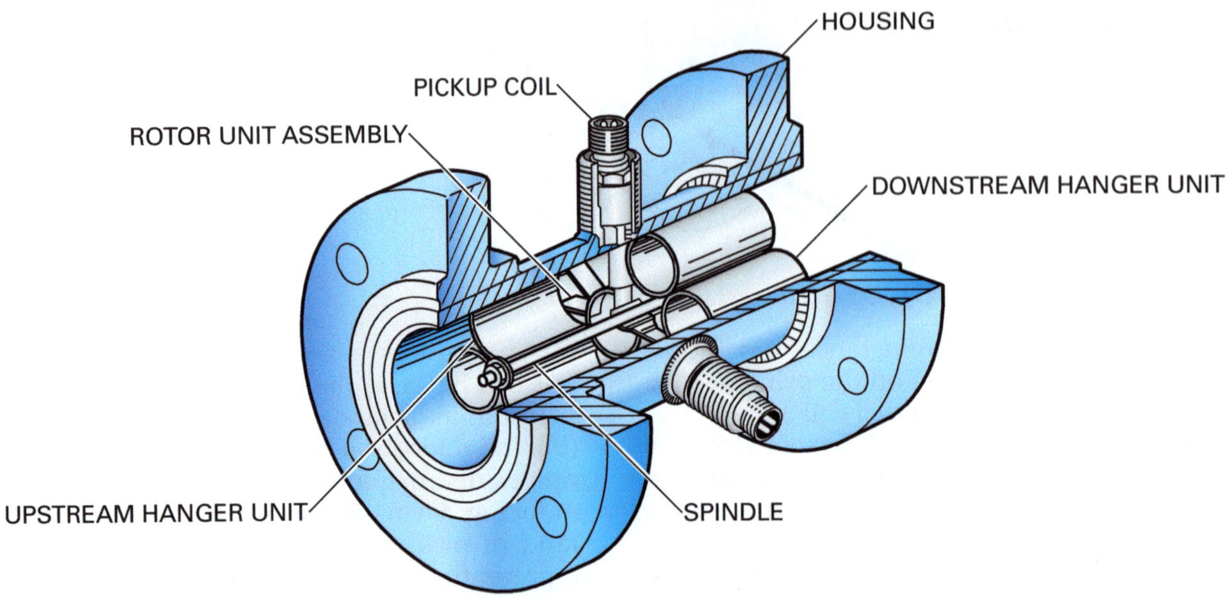

Figure 5 Cut-away view of a turbine meter.

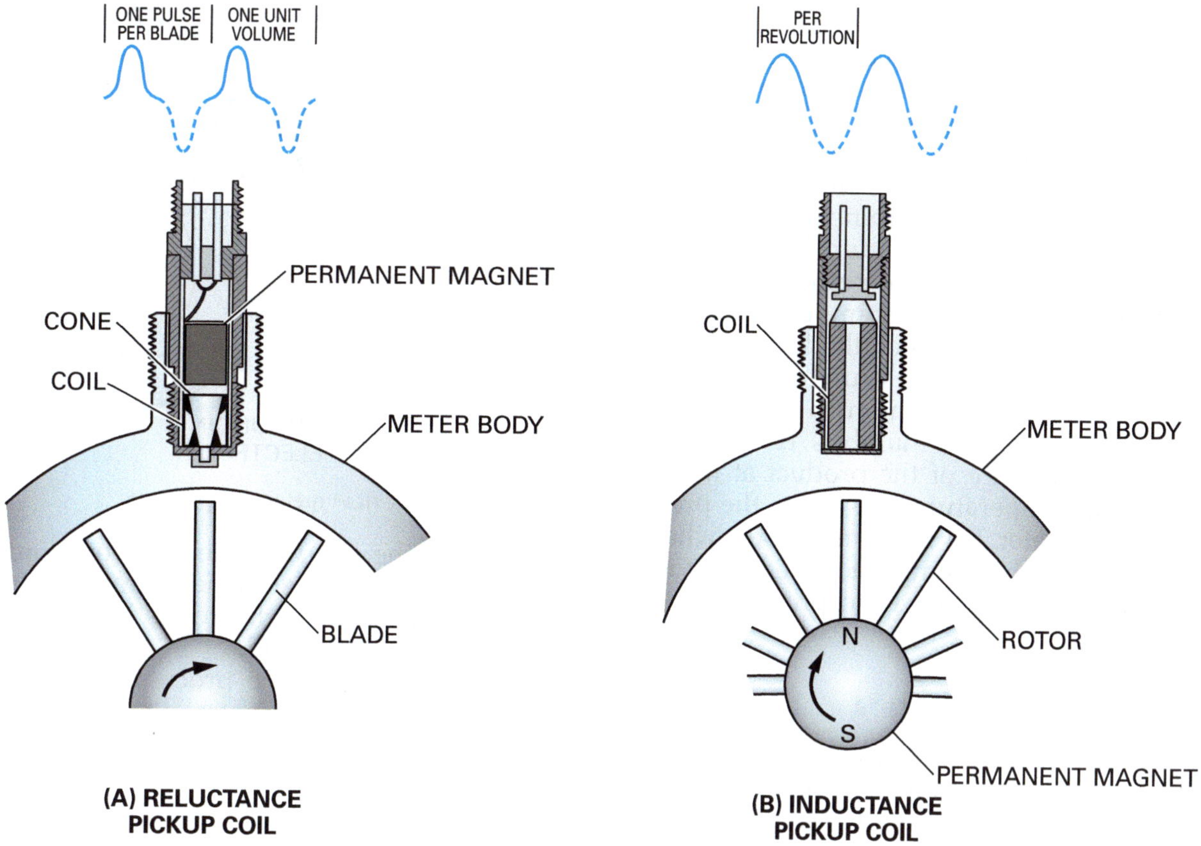

Figure 6 Turbine meter signal detection systems.

API standards require that the meter be installed in a prescribed way to ensure straightened flow. The proper installation includes straight pipe of the same dimensions as the meter reaching at least ten pipe diameters upstream and five pipe diameters downstream. Newer types of pipes are fitted with straightening vanes. A typical meter run layout is shown in *Figure 7*.

Knowing the area of the pipe and the elapsed time of flow, volume can easily be calculated from fluid velocity. The turbine meter measures velocity, but the readout registers in volume. Rotors are fitted with signaling devices, and each time the unit turns one revolution, a signal is sent to the electronic counter. The sum of the pulses is divided by the pulse frequency per unit of volume.

A meter is rated by a **K factor**. The K factor is the number of times the meter turns for each unit of volume. For example, if the K factor for a particular meter were 150 pulses for each barrel of liquid, then the volume of liquid measured would be equal to the number of pulses divided by the K factor. For example, a popular meter model generates 49 pulses per gallon of fluid. The unit then divides the total number of pulses by 49 to obtain total gallons. This value can be the basis for further calculations as required. Since many modern systems include a PLC, calculations of this type

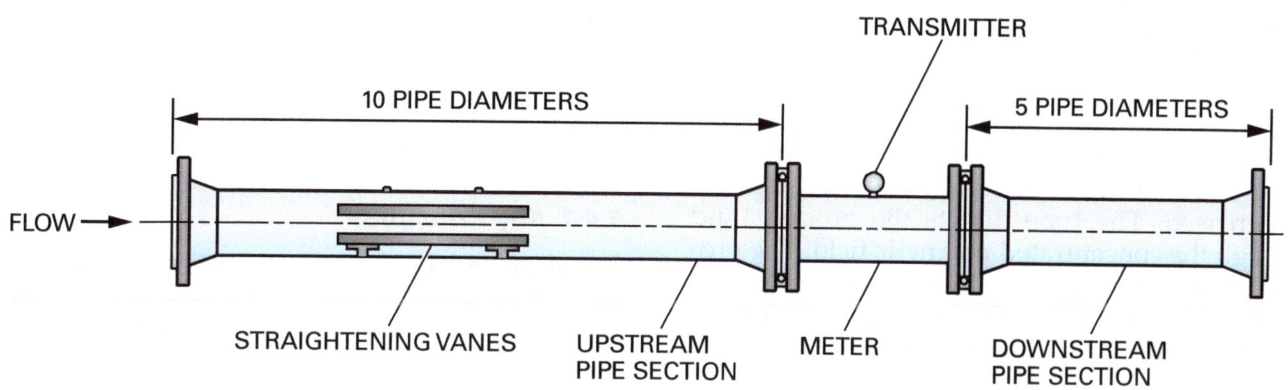

Figure 7 Typical meter run layout.

are handled by the program, which stores the meter factor in memory. Changes are simple, involving nothing more than a modification of this stored value. In many cases, the PLC converts the meter value to a number in the range of 0 to 1, which makes conversion to other units simple.

The biggest enemy of the turbine meter is cavitation. If the pressure of the product is allowed to drop below the vapor pressure point and vaporization occurs, the meter will spin out of control. In some cases, a meter will literally destroy itself. Maintaining enough backpressure will prevent such occurrences. The minimum safe backpressure required as a rule of thumb is 125% of the absolute vapor pressure of the product at maximum operating temperature plus twice the pressure drop at the meter. For the same reason, when purging air out of the piping after meter servicing, perform the action slowly to prevent meter damage.

1.3.0 Magnetic Flow Meters

The magnetic flow meter was developed to measure the volumetric flow rate of electrically conductive fluids. It is particularly useful for measuring the flow rate of fluids that present very difficult handling problems, such as corrosive acids, sewage, slurries, and paper pulp stock.

1.3.1 Key Principles

The magnetic flow meter operates on the principle of Faraday's law of induction. This law states that any time a conductor is moved through a magnetic field at right angles, an electrical potential (a voltage) is developed across it. The magnitude of the induced voltage is proportional to the magnetic flux density, the length of the conductor in the magnetic field, and the velocity of the moving conductor.

A typical magnetic flow meter consists of a tube, a set of coils, a laminated iron core (armature), and an electrode assembly. These components and others are shown in *Figure 8*.

Current flow through the coil produces a magnetic field. The core around which the coil is wound concentrates the magnetic lines of flux of this field around the tube. By concentrating these lines of flux, the magnetic flux density is increased. The tube directs the process fluid through the concentrated magnetic field. The process fluid is the moving electrical conductor. As the fluid passes through the magnetic field, a voltage is induced that is sensed by the electrode

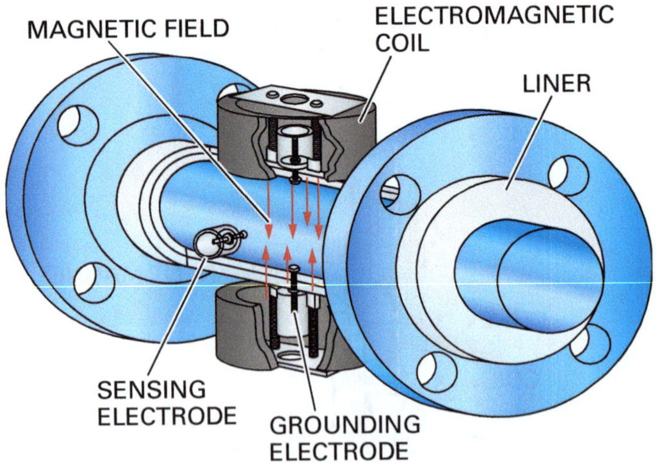

Figure 8 Magnetic flow meter components.

assembly. If the magnetic field strength and the distance between the electrodes are constant, the voltage sensed by the electrodes is proportional to the velocity of the fluid.

The tube carrying the process fluid must be made of a nonmagnetic material to allow the field to penetrate to the liquid. In *Figure 8*, the flow meter has an insulating pipe liner to prevent voltage induced in the fluid from being short-circuited by the pipe. Typical materials used in the construction of the flow meter tube are fiberglass reinforced plastic or nonmagnetic stainless steel. When stainless steel is used, the interior of the tube is lined with a nonconducting material, such as PTFE (Teflon®), polyurethane, or glass.

The electrodes are flush with the tube interior. They actually come in contact with the process fluid. The electrodes must be made of good conducting material. They are usually made of type 316 stainless steel, but platinum electrodes are often used with highly corrosive fluids. It is essential that these electrodes remain free of dirt. Dirt acts as an electrical insulator and reduces the accuracy of measurement.

1.4.0 Ultrasonic Flow Meters

Ultrasonic flow meters employ high-frequency sound waves and are used for gas flow measurement in situations involving a wide range of flow rates and/or pressures. They are a newer technology and are considered highly accurate.

1.4.1 Key Principles

An ultrasonic flow meter, shown in *Figure 9*, employs transducers that emit and receive sound waves. As the product flows through the pipeline,

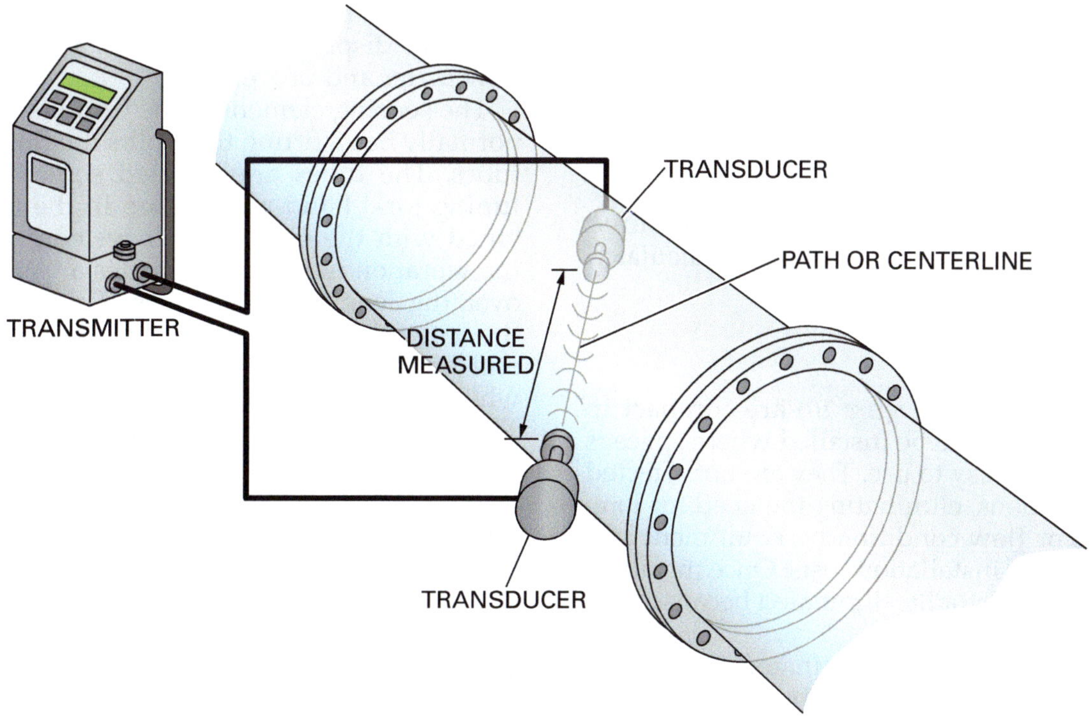

Figure 9 Ultrasonic flow meter.

it causes changes to these waves, which the meter circuitry detects and uses to determine the flow rate. The speed of the sound waves is directly affected by the product through which they are traveling. Speed is also affected by the temperature and pressure of the product. Travel rate is calculated by dividing the distance by the time and is expressed in feet-per-second or meters-per-second. The time that is required for sound to travel from the emitting transducer to the receiving transducer is called the transit time. To calculate the flow rate of the gas, the differences in the travel rates are measured, and a comparison with the speed of sound for known gas properties is made.

The centerline of the sound waves as they travel between the transducers is called the *path*. Multi-path meters can be set up to initiate signals down the center of the unit but require a specially designed and calculated shape to accommodate the process. Units can also be set up to employ parallel or straight-across signals. Parallel units do not require a specially designed and calculated shape, because the profile of the meter is divided into sections by the signals. Under certain controlled conditions, multi-path meters can be used for custody transfer.

Ultrasonic meters come in several varieties including spooled meter bodies, hot-tapped meter bodies, and clamp-ons. The spooled types are constructed of thick cast metal or fabricated of pipe-and-sleeve construction and installed in the pipe as any other component might be. They may be of a single-path or multi-path configuration. Single-path meters are used for check meters and guidance tools for equipment control units, system balance controls, leak detection, and custody transfer control.

Hot-tapped ultrasonic flow meters are tapped and welded directly into an actively operating pipeline. Correct installation is crucial for this variety, and their accuracy can be harmed if it's done improperly. During the welding procedure, the meter must be perfectly aligned along the X, Y, and Z axes. For every one-half degree of error in the angle of installation, there will be a resultant measurement error of as much as one percent during operation. It must be noted that the pipe into which a hot-tapped meter is installed is thin-walled compared to the meter's construction. Pressure and temperature will have a marked effect due to the expansion or contraction of the pipe, affecting the length and angle of the sound signal pathway. Most of this error can be corrected by the meter's circuitry.

A clamp-on ultrasonic flow meter is attached to the outside of the pipe using stainless steel clamps or straps. These units can be portable or permanently mounted on the pipe with the external transponders providing data to a flow computer.

1.5.0 Coriolis Flow Meters

Coriolis flow meters are becoming more prominent in liquid and gas applications as their reputation for accuracy and dependability becomes more widely known. They are accurate over a wide range of flow rates. One major advantage they have is that the mass flow rate is measured directly from the flow instead of being calculated using other sensor readings.

1.5.1 Key Principles

Coriolis flow meters (*Figure 10*) are compact in size, allowing them to be installed where space is limited, and are easy to use. They are not affected by inlet conditions, eliminating the need for long meter runs or flow conditioning equipment and greatly reducing installation costs. Once installed, maintenance is virtually eliminated because they have no moving parts.

Meter factor change is eliminated. All meters are proved as a matter of good business practice; however, tests have shown that Coriolis flow meters experience no detectable meter factor change over several years of use. Meters are calibrated and the meter factor established at the factory. Only liquids that cause corrosion or erosion of the meter's tubes cause a meter factor change. Since natural gas operations are clean, this problem would never occur in gas applications.

Coriolis meters include only two basic components. The first is an in-line sensing element that detects flow. The second is a transmitter that receives the data from the sensor and converts it into a useable form. Coriolis meters can be set up to output digital, analog, and pulse-type outputs according to the application and its requirements. *Figure 11* displays the layout of a Coriolis mass flow meter and prover.

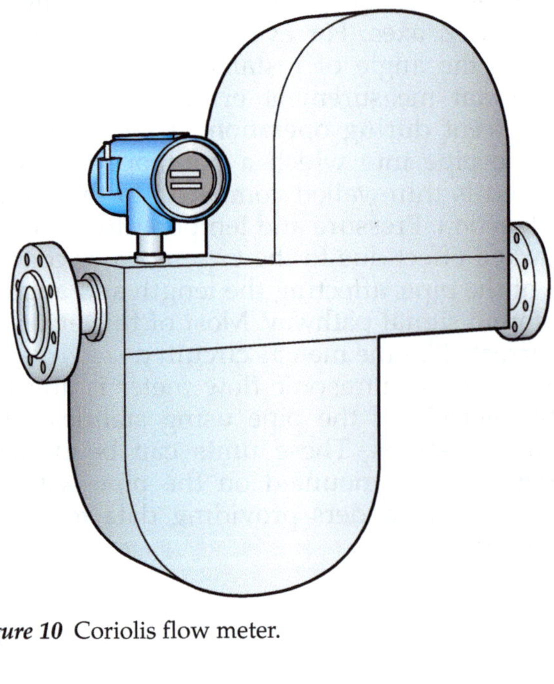

Figure 10 Coriolis flow meter.

The sensing element of the Coriolis flow meter normally has a fitting that splits the flow into two tubes. The tubes are vibrated somewhat like a tuning fork. Fluid momentum in the tubes combined with the oscillatory movement created by the vibration in the tubes creates a **Coriolis force** over the length of the tubes. This force causes them to twist as they vibrate. Sensors at each end of the tubes produce a pair of sine waves whose phase relationship depends on the flow. The greater the mass flow, the greater the phase difference becomes. Coriolis flow meters can also provide density information about the fluid flowing through them since density is proportional to the tubes' oscillation frequency.

1.6.0 Vortex Flow Meter

A vortex flow meter is used to measure flow rate in both liquids and gases. These meters are rugged instruments that are accurate, consistent, and reliable. A vortex meter has no moving parts, making it a low-maintenance unit. This type of meter consists of a meter body containing a specially machined shedder bar and an amplifier assembly.

1.6.1 Key Principles

Vortex meters employ the **von Karman vortex principle** to measure velocity. This principle (*Figure 12*) is illustrated by the back-and-forth waving of a flag in the wind caused by air passing across the flagpole. The vortices of air shed off the flagpole and cause the flag to take on the waving, rippling motion of the vortices. As wind velocity increases, the flag will go from a soft rippling or waving motion to a more harsh fluttering motion because the frequency of the wind vortices increases. The velocity of the wind can then be calculated by measuring the frequency of the flag's fluttering motion.

The shedder bar of the vortex meter causes this same vortex effect in the product passing by the bar. As each vortex begins to peel away from the shedder bar, a temporary low-pressure area is created on the bar, putting a very slight side stress on it. The bar, embedded with pressure-sensitive piezoelectric crystals, creates a voltage spike. The meter detects and counts the spikes, using their frequency to calculate the fluid velocity. After measuring the velocity of the stream, the meter (*Figure 13*) then computes the flow by

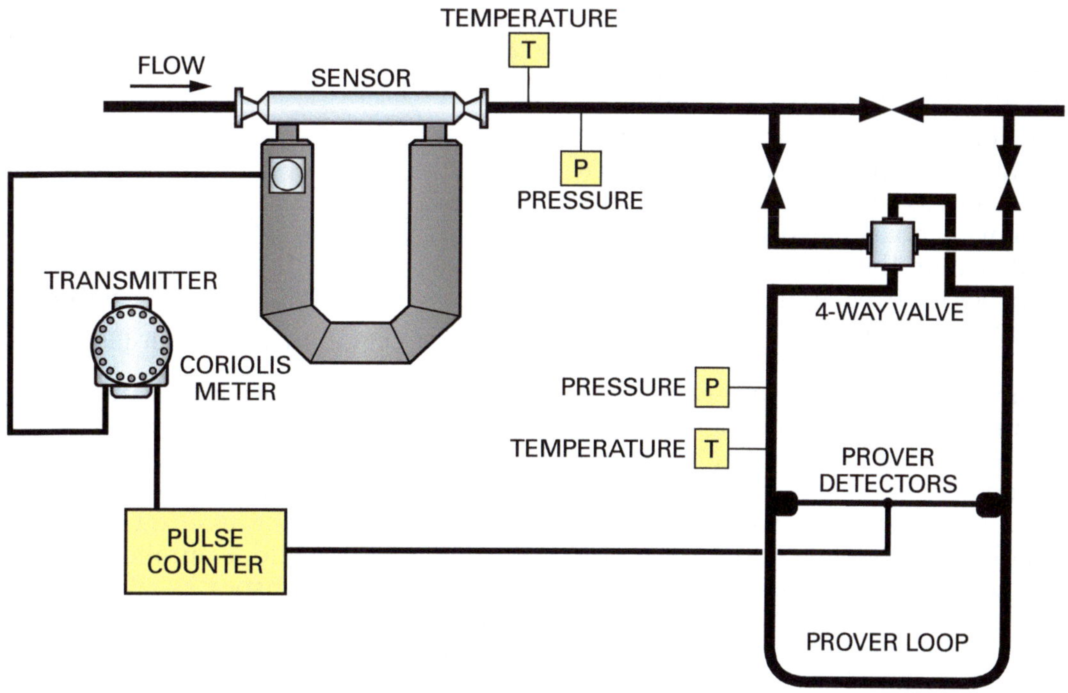

Figure 11 Layout of a Coriolis flow and density meter with meter prover.

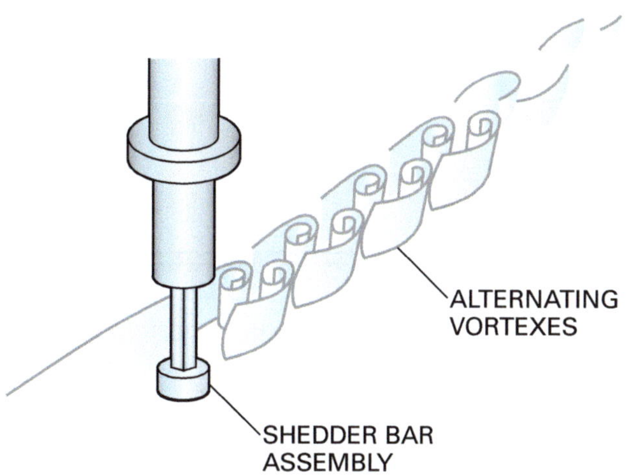

Figure 12 The von Karman vortex principle.

multiplying the cross-sectional area of the meter by the velocity of the fluid or gas.

The relationship between the velocity, the flow rate, and the vortex frequency is determined at the factory, establishing a meter factor and K factor for each meter. The K factor is given in the number of pulses per gallon but can be set up to give readings in any standard unit of measurement. The transmitter converts the measurement to the required signal type and transmits the data to the control center. In the case of a compressible flow, such as steam, the units can be set up to correct for temperature and pressure.

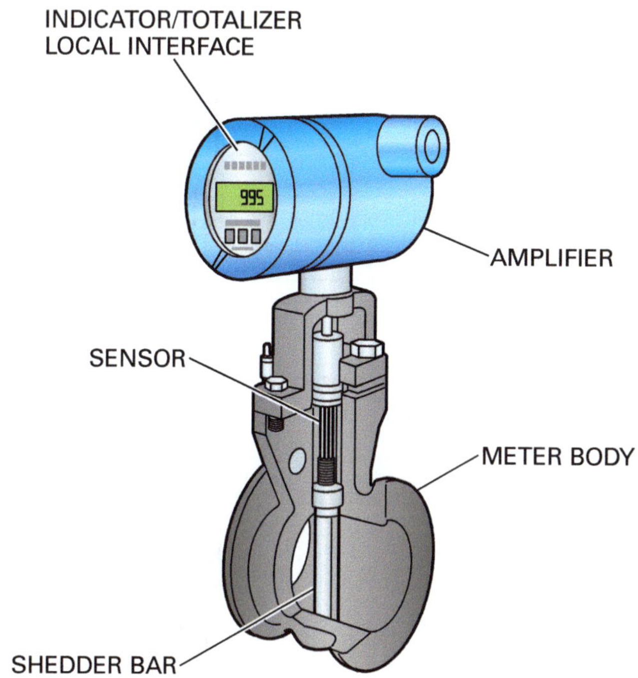

Figure 13 Vortex flow meter.

1.7.0 Orifice Flow Meters

Orifice flow meters rely on a principle of physics in which a flowing fluid's pressure drops as it passes through a constriction in the pipeline. Orifice meters consist of two key components to produce and measure this effect: an orifice plate and a differential pressure-sensing system.

1.7.1 Key Principles

The orifice plate is the vital component in the orifice meter. It is the part that restricts the flow to develop the difference in pressure that is measured to calculate the actual flow. Orifice plates are flat, plate-like devices that are installed in the pipeline. They have a hole of a specific size that's somewhat smaller than the pipe diameter. Plates vary from 2.5 inches to 30 inches in diameter (6–76 cm). There are many different styles with different hole shapes and positions (*Figure 14*). Expert installation and maintenance are necessary to ensure the continued accuracy of the meter. *Figure 15* shows a cut-away view of an orifice plate in a meter installation.

Orifice plates are secured between halves or flanges of machined carriers or in the body of a special orifice meter housing constructed so that the plate can be removed for inspection. Some meters allow the plate to be removed without stopping the flow. *Figure 16* shows some common orifice plate holders designed for easy removal. The orifice meter is made up of the carrier housing or flanges, the orifice plate, gaskets and O-rings, devices for removal, and pressure fittings. Pressure taps are drilled at specific locations on either side of the orifice plate.

A pressure-sensing device connects to the taps and measures the pressure difference across the orifice plate. While a number of different pressure-detecting sensors can be used, the differential pressure (DP) cell is quite common. A transmitter sends the data from the cell to the associated display or measuring device. The pressure value is used, in turn, to calculate the flow rate.

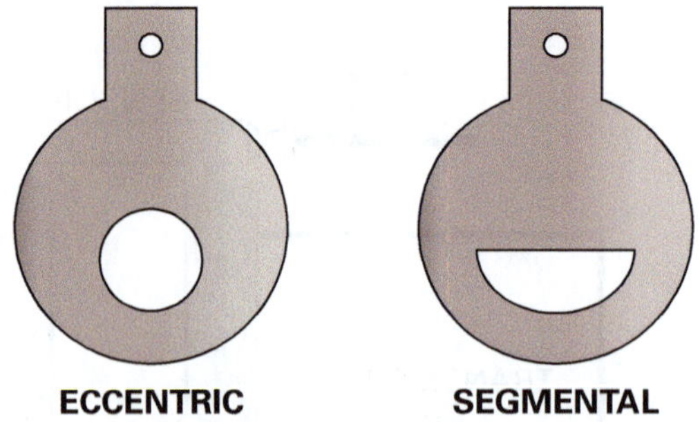

Figure 14 Orifice plates come with different hole shapes.

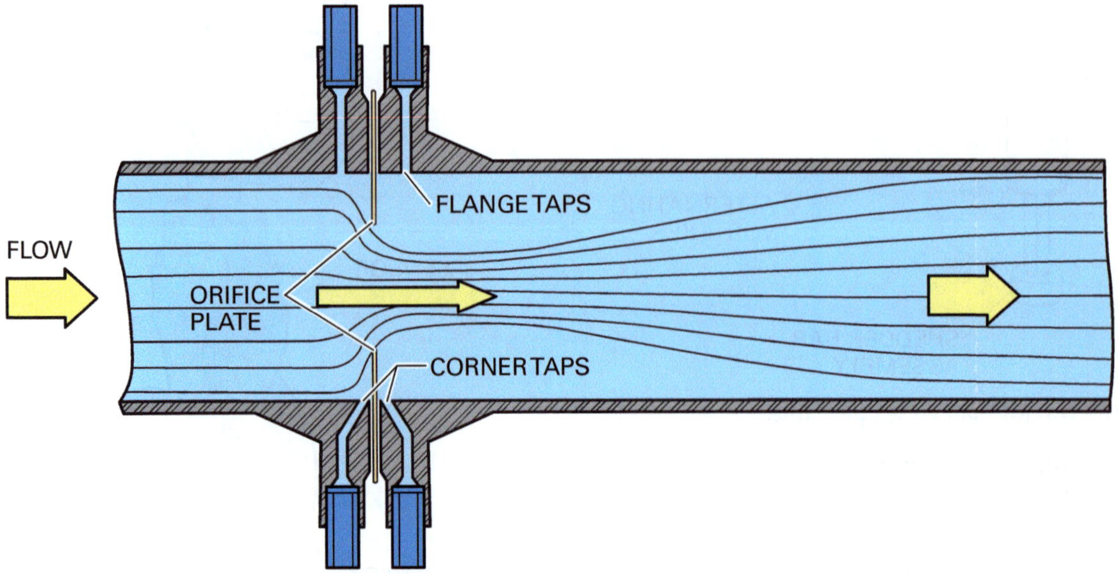

Figure 15 Cut-away view of an orifice plate meter installation.

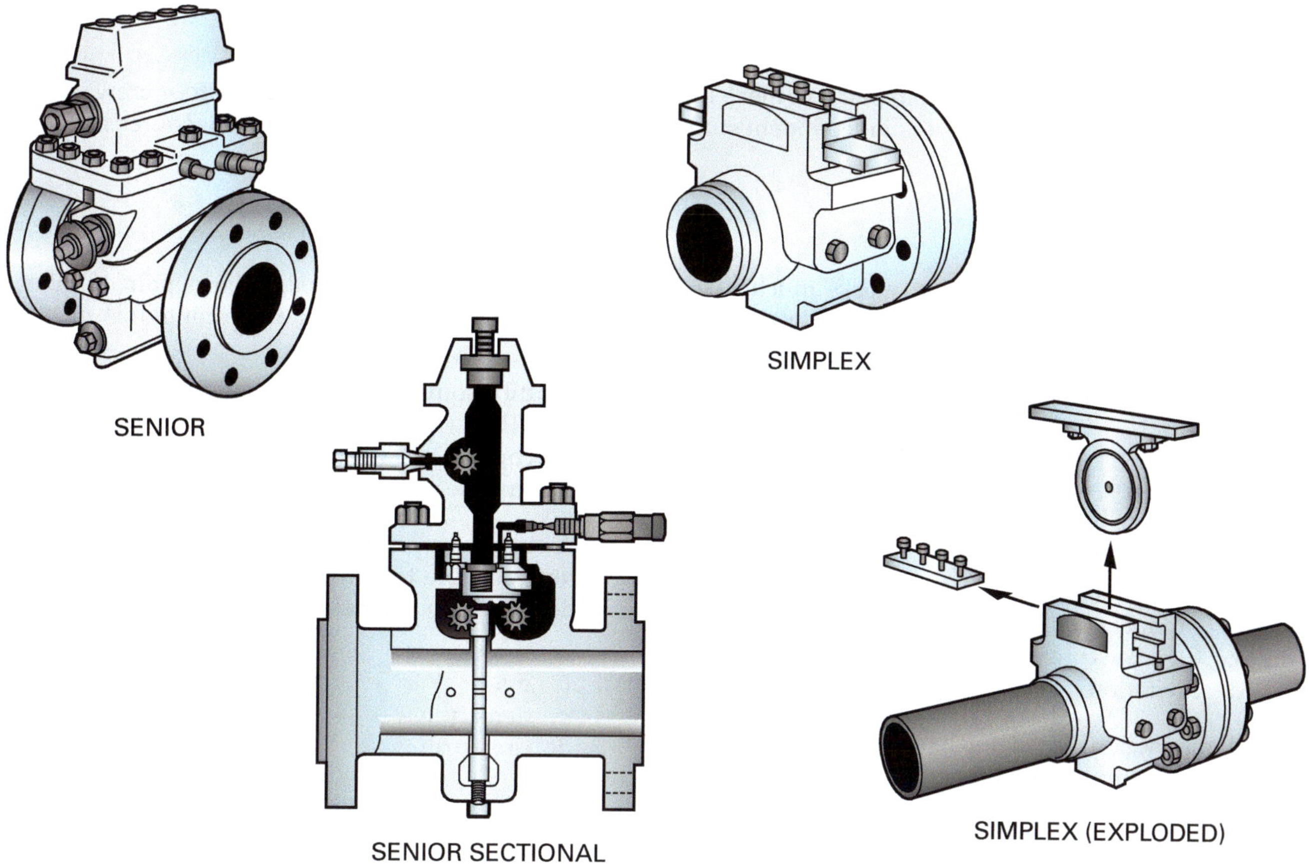

Figure 16 Common orifice plate holders.

Additional Resources

It is assumed that you will learn to perform these procedures under the direct supervision of your crew leader, foreman, or instructor. For independent study, refer to the recommended Additional Resources at the back of this module.

1.0.0 Section Review

1. The type of flow meter that works much like the odometer in a car is the _____.
   a. orifice
   b. positive displacement
   c. coriolis
   d. vortex

2. A turbine flow meter can be destroyed by _____.
   a. magnetism
   b. product condensation
   c. high backpressure
   d. cavitation

3. A magnetic flow meter requires that the fluid flowing through the pipeline _____.
   a. be electrically conductive
   b. be free from particles
   c. have low viscosity
   d. have high viscosity

4. The kind of ultrasonic flow meter that will be inaccurate if its X, Y, and Z alignment is off is the _____.
   a. spooled
   b. hot-tapped
   c. clamp-on
   d. tri-beam

5. The flow meter that relies on a pair of tubes vibrating somewhat like a tuning fork is the _____.
   a. ultrasonic
   b. PD
   c. coriolis
   d. turbine

6. A vortex flow meter uses the von Karman vortex principle to generate deflections in a shedder bar proportional to the product's _____.
   a. velocity
   b. mass
   c. density
   d. pressure

7. An orifice flow meter measures the effect created by the orifice plate's presence in the flow by using a device called a _____.
   a. temperature sensor
   b. vortex detector
   c. Coriolis force sensor
   d. differential pressure cell

SECTION TWO

2.0.0 MAINTAINING FLOW METERS

Objective

Describe the procedures used to maintain flow meters for hazardous liquid leak detection (CT44_6-17).

a. Identify the prerequisites, competencies, and abnormal operating conditions (AOCs) associated with maintaining flow meters.
b. Describe the procedures used to maintain flow meters for hazardous liquid leak detection.

Performance Tasks

Maintain flow meters for hazardous liquid leak detection (CT44_6-17).

- Identify potential abnormal operating conditions (AOCs) that may occur during performance of this CT, and know the appropriate actions to take in response to them.
- Utilize the appropriate personal protective equipment (PPE) according to relevant company procedures.
- Determine the flow meter's number and nameplate data in order to verify that it is the correct meter to be maintained.
- Notify control center and/or affected personnel before work begins.
- Visually examine the flow meter and assess its physical and mechanical condition. Identify and perform any preventative maintenance as required.
- Identify any diagnostic steps specified by the manufacturer's documentation and perform them as appropriate.
- If the meter needs to be disconnected for maintenance, isolate the meter, release pressure, empty the meter and lines, and disconnect the meter.
- Complete the necessary repairs to the meter.
- If the meter has been disconnected for the repair, reconnect it correctly.
- Notify the control center that the meter is being brought online and that commissioning and startup activities have commenced.
- Bring the meter online by purging it and filling the pipeline.
- Check the meter and its associated hardware and connections.
- Notify the control center that the meter is ready for startup and that it should be proved.
- Complete appropriate documentation as required by operator's procedures.

Trade Term

Abnormal operating conditions (AOCs): Conditions within a pipeline system, identified by the operator, that may indicate (1) a malfunction of a component or (2) a deviation from normal operations that may indicate a condition exceeding design limits or result in a hazard to persons, property, or the environment.

Most flow meters require a certain amount of maintenance to keep them working correctly and returning accurate flow data. Some meter technologies are fairly complex, with many moving parts. These will require more maintenance than simpler designs with minimal moving parts. In each case, the pipeline E&I technician must be familiar with the meter technology in question and know what's appropriate for it. Having access to the manufacturer's service manuals is essential to doing the job correctly.

2.1.0 Prerequisites, Competencies, and Abnormal Operating Conditions

Personnel must possess a variety of skills in order to successfully maintain flow meters. The various operations require a number of steps, which a pipeline E&I technician must be competent to perform. Personnel must also be aware of the particular abnormal operating conditions (AOCs) that may occur during the procedure and know the proper responses to them.

2.1.1 Prerequisites and Competencies

Personnel must possess the following basic knowledge and skills prior to maintaining flow meters:

- Familiarity with the purpose of a meter and different flow meter types, including all of the ones described in this module.
- Possess electrical systems knowledge and safety.

- Familiarity with the system in question, including instruments, alarms, controllers, indicators, and recorders.
- Understand equipment associated with flow meters such as flow conditioners/straighteners and filters/strainers.
- Familiarity with control center communications.
- Knowledge of different alarms, particularly those indicating that an operating parameter has been exceeded.
- Competent with test/diagnostic equipment and familiar with their calibration procedures (where applicable).
- Familiarity with end devices, such as PLCs, SCADA, and HMIs, that receive the flow meter's data.
- Able to differentiate between prover and meter problems.

Personnel must also understand the company's safety policies and procedures, including lockout/tagout procedures and proper use of personal protective equipment (PPE).

2.1.2 Abnormal Operating Conditions

Personnel maintaining flow meters must be able to identify abnormal operating conditions (AOCs) and know the proper response to each, should they occur. The following are some potential AOCs, and the appropriate action personnel should take in response to them.

- *An inoperable meter or meter not communicating with the end device is discovered* – Personnel should alert the control center or other appropriate personnel to take the requisite corrective action.
- *Unexpected release of hazardous liquid or carbon dioxide occurs* – Personnel should try to identify and isolate the source of the leak. Proper notifications should be made as soon as possible, including calling 911 if necessary. If there is evidence of a leak, stop task activities, move to a safe distance, and notify appropriate pipeline personnel.
- *A fire or explosion occurs* – Personnel should evacuate the area immediately and call 911 and the company emergency center.

2.2.0 Procedure

When maintaining a flow meter, it's crucial that the pipeline E&I technician obtain appropriate manufacturer's manuals before beginning any maintenance procedure. Each flow meter has its own unique requirements, which cannot be fully addressed by the general procedure outlined here. The following steps outline the general procedure for maintaining a flow meter:

Step 1 Identify potential abnormal operating conditions (AOCs) that may occur, and know the appropriate actions to take in response to them.

Step 2 Utilize the appropriate personal protective equipment (PPE) according to relevant company procedures.

Step 3 Determine the flow meter's number and nameplate data in order to verify that it is the correct meter to be maintained. Confirm that the manufacturer's manuals in-hand are correct for the meter in question.

Step 4 Before beginning the maintenance process, notify the control center, local personnel, and anyone else who might be affected by the operation.

Step 5 Identify any preventative maintenance steps specified by the manufacturer's documentation and perform them as appropriate.

Step 6 Visually examine the flow meter and assess its physical and mechanical condition. Check for corrosion and leakage. Check electrical connections and associated electronics. If the flow meter has any supporting equipment attached, inspect that as well.

Step 7 Identify any diagnostic steps specified by the manufacturer's documentation and perform them as appropriate. These may include meter factor trend analysis, pulse output checks, vibration/noise checks, and electrical/electronic connections checks. The manufacturer may provide software diagnostics that should be run as well.

CAUTION

Due to the wide variety of test equipment that may be encountered, always be sure to check the relevant specifications.

Step 8 The meter may have to be disconnected for maintenance. If you conclude that this task is necessary, proceed to the next step. Otherwise, skip to Step 11.

Step 9 Isolate the meter, release pressure, and empty the meter and lines, following the appropriate documentation and procedures.

Step 10 Disconnect the meter.

Step 11 Complete the necessary repairs to the meter. Consult the manufacturer's documentation and company procedure for specific guidelines.

Step 12 If the meter has been disconnected for the repair, reconnect it correctly, following all manufacturer and company guidelines.

Step 13 Notify the control center that the meter is being brought online and that commissioning and startup activities have commenced.

Step 14 Bring the meter online by purging it and filling the pipeline. Follow all manufacturer and company guidelines to protect the meter and the process itself.

Step 15 Check the meter and its associated hardware and connections to verify that there are no leaks and that everything is operating within the specification.

Step 16 Notify the control center that the meter is ready for startup and that it should be proved (if company or manufacturer guidelines so recommend).

Step 17 Complete appropriate documentation as required by operator's procedures.

Additional Resources

It is assumed that you will learn to perform these procedures under the direct supervision of your crew leader, foreman, or instructor. For independent study, refer to the recommended Additional Resources at the back of this module.

2.0.0 Section Review

1. What kind of end device might a flow meter be connected to?
   a. A ball valve
   b. A flow strainer
   c. A PLC
   d. A Coriolis screen

2. Before beginning any flow meter maintenance procedure, the pipeline E&I technician must _____.
   a. obtain the manufacturer's manual
   b. notify the control center that the meter is going offline
   c. isolate the meter from the process
   d. confirm proper SCADA communication

SUMMARY

Flow meters must be maintained in order for them to work correctly and provide reliable data. Pipeline E&I technicians must understand how the many different types of flow meters work before they attempt maintenance procedures. This module surveys the major meter types as well as outlines the general procedures recommended for maintaining meters. Above all, technicians must always follow manufacturer and company guidelines for all procedures. Having the manufacturer's manual for the meter in question is critical.

Review Questions

1. One problem specific to PD meters is _____.
   a. cavitation
   b. slippage
   c. SCADA pulse errors
   d. problematic K factors

2. Which kind of flow meter is undependable when it has to measure heavy crude products?
   a. PD flow meters
   b. Vortex flow meters
   c. Coriolis flow meters
   d. Turbine flow meters

3. A magnetic field is used to detect the rotating component in a(n) _____.
   a. ultrasonic flow meter
   b. PD flow meter
   c. turbine flow meter
   d. magnetic flow meter

4. Which kind of flow meter is very good at measuring slurries and other "problem" products?
   a. Magnetic flow meters
   b. PD flow meters
   c. Turbine flow meters
   d. Ultrasonic flow meters

5. In an ultrasonic flow meter, the centerline of the sound waves as they travel between transducers is called the _____.
   a. vertex
   b. channel
   c. path
   d. conduit

6. In addition to measuring flow, a Coriolis flow meter can also measure _____.
   a. density
   b. turbidity
   c. flammability
   d. vapor pressure

7. A vortex flow meter generates a series of vortices with a _____.
   a. piezoelectric crystal
   b. cyclic body
   c. von Karman ladder
   d. shedder bar

8. An orifice flow meter relies on the principle that a constriction in a pipeline causes a pressure _____.
   a. rise
   b. drop
   c. oscillation
   d. excursion

9. The flow meters that require the most maintenance are those containing _____.
   a. DP cells
   b. orifice plates
   c. moving parts
   d. Coriolis tubes

10. During flow meter maintenance, if an operating parameter is exceeded, there is likely to be a(n) _____.
    a. alarm
    b. PLC reset
    c. explosion
    d. SCADA halt

11. During a visual examination of a flow meter, always check for _____.
    a. faulty meter factors
    b. SCADA errors
    c. improper pressure
    d. corrosion

Trade Terms Quiz

Fill in the blank with the correct term that you learned from your study of this module.

1. Vapor pockets that form in a fluid flow as a result of rapid pressure changes that cause shockwave-like pulsations that can lead to inaccuracies and instrument damage are called ________.
2. ________ is a cause of inaccuracy in a positive displacement meter that results from fluid leaking around the edge of the meter rotor.
3. The number of pulses per unit volume generated by a meter is called the ________.
4. A deflective effect created in certain kinds of rotating systems is called the ________. Named for a French scientist.
5. The tendency for a fluid flow to form vortices (swirling sections of fluid) after passing across an obstruction in the flow path is called the ________. Named for a Hungarian-American physicist.
6. Conditions within a pipeline system that may indicate (1) a malfunction of a component or (2) a condition exceeding design limits or resulting in a hazard to persons, property, or the environment are called ________.
7. A factor determined during proving that is based on the prover's volume and the meter's measured volume is the ________. It is used to convert a meter's reading into an actual product quantity reading.

Trade Terms

Abnormal operating conditions (AOCs)
Cavitation
Coriolis force
K factor
Meter factor (MF)
Slippage
Von Karman vortex principle

Trade Terms Introduced in This Module

Abnormal operating conditions (AOCs): Conditions within a pipeline system, identified by the operator, that may indicate (1) a malfunction of a component or (2) a deviation from normal operations that may indicate a condition exceeding design limits or result in a hazard to persons, property, or the environment.

Cavitation: Vapor pockets that form in a fluid flow as a result of rapid pressure changes. They cause shockwave-like pulsations that can lead to inaccuracies and instrument damage.

Coriolis force: A deflective effect created in certain kinds of rotating systems. Named for the French scientist Gaspard-Gustave Coriolis.

K factor: The number of pulses per unit volume generated by a meter.

Meter factor (MF): A factor determined during meter proving that is based on the prover's volume and the meter's measured volume. The meter factor is used to convert a meter's reading into an actual product quantity reading.

Slippage: A cause of inaccuracy in a positive displacement meter. It results from fluid leaking around the edge of the meter rotor.

Von Karman vortex principle: The tendency for a fluid flow to form vortices (swirling sections of fluid) after passing across an obstruction in the flow path. Named for the Hungarian-American physicist Theodore von Kármán.

Additional Resources

The following recommended resources can provide additional helpful information related to the requirements for this covered task:

API Recommended Practice 1161, Third Edition, © 2014 American Petroleum Institute. API Publishing Services, 1220 L Street, NW, Washington, DC 20005. Distributed by Thompson Reuters (Scientific) LLC, **www.techstreet.com**.

49 *CFR* 192, **www.ecfr.gov**.

49 *CFR* 195, **www.ecfr.gov**.

The Condensed Handbook of Measurement and Control, N. E. Battikha. Third Edition. 2006. Research Triangle Park, NC: The International Society of Automation.

NCCER Module AOCCC-17, *Abnormal Operating Conditions Control Center*

NCCER Module AOCFG-17, *Abnormal Operating Conditions Field and Gas*

NCCER Module CT44_5-17, *Proving Flow Meters.*

NCCER Module 60102-02, *Pipeline E&I Safety.*

NCCER Module 63_1-17, *Start-up of a Liquid Pipeline (Field).*

NCCER Module 63_2-17, *Shutdown of a Liquid Pipeline (Field).*

NCCER Module 63_3-17, *Monitor Pressures, Flows, Communications, and Line Integrity and Maintain Them Within Allowable Limits on a Liquid Pipeline System (Field).*

NCCER Module 63_4-17, *Locally Operate Valves on a Liquid Pipeline System.*

The following websites provide resources for products and training:

The International Society of Automation, **www.isa.org**

Figure Credits

photostock77/Shutterstock.com, Module Opener

Satam, Figure 1

Section Review Answer Key

Answer	Section Reference	Objective
Section One		
1. b	1.1.1	1a
2. d	1.2.1	1b
3. a	1.3.0	1c
4. b	1.4.1	1d
5. c	1.5.1	1e
6. a	1.6.1	1f
7. d	1.7.1	1g
Section Two		
1. c	2.1.1	2a
2. a	2.2.0	2b

NCCER CURRICULA — USER UPDATE

NCCER makes every effort to keep its textbooks up-to-date and free of technical errors. We appreciate your help in this process. If you find an error, a typographical mistake, or an inaccuracy in NCCER's curricula, please fill out this form (or a photocopy), or complete the online form at **www.nccer.org/olf**. Be sure to include the exact module ID number, page number, a detailed description, and your recommended correction. Your input will be brought to the attention of the Authoring Team. Thank you for your assistance.

Instructors – If you have an idea for improving this textbook, or have found that additional materials were necessary to teach this module effectively, please let us know so that we may present your suggestions to the Authoring Team.

NCCER Product Development and Revision

13614 Progress Blvd., Alachua, FL 32615

Email: curriculum@nccer.org

Online: www.nccer.org/olf

❑ Trainee Guide ❑ Lesson Plans ❑ Exam ❑ PowerPoints Other ____________

Craft / Level: Copyright Date:

Module ID Number / Title:

Section Number(s):

Description:

Recommended Correction:

Your Name:

Address:

Email: Phone:

NCCER CURRICULA — USER UPDATE

NCCER makes every effort to keep its textbooks up-to-date and free of technical errors. We appreciate your help in this process. If you find an error, a typographical mistake, or an inaccuracy in NCCER's curricula, please fill out this form (or a photocopy), or complete the online form at www.nccer.org/olf. Be sure to include the exact module ID number, page number, a detailed description, and your recommended correction. Your input will be brought to the attention of the Authoring Team. Thank you for your assistance.

Instructors – If you have an idea for improving this textbook, or have found that additional materials were necessary to teach this module effectively, please let us know so that we may present your suggestions to the Authoring Team.

NCCER Product Development and Revision
13614 Progress Blvd., Alachua, FL 32615

Email: curriculum@nccer.org
Online: www.nccer.org/olf

❑ Trainee Guide ❑ Lesson Plans ❑ Exam ❑ PowerPoints Other ____________

Craft / Level: ____________ Copyright Date: ____________

Module ID Number / Title: ____________

Section Number(s): ____________

Description: ____________

Recommended Correction: ____________

Your Name: ____________

Address: ____________

Email: ____________ Phone: ____________

CT44_7-17

Inspect, Test, and Maintain Gravitometers/ Densitometers for Hazardous Liquid Leak Detection

Overview

Knowing a pipeline product's density is key to establishing its grade, and grade is associated with value. Therefore, being able to monitor density is an essential step in the pipeline process. Various instruments measure density using a number of different methods. This module describes these instruments and discusses the procedures used for inspecting, testing, and maintaining them.

Trainees with successful module completions may be eligible for credentialing through the NCCER Registry. To learn more, go to **www.nccer.org** or contact us at 1.888.622.3720. Our website has information on the latest product releases and training, as well as online versions of our *Cornerstone* magazine and Pearson's product catalog.

Your feedback is welcome. You may email your comments to **curriculum@nccer.org**, send general comments and inquiries to **info@nccer.org**, or fill in the User Update form at the back of this module.

This information is general in nature and intended for training purposes only. Actual performance of activities described in this manual requires compliance with all applicable operating, service, maintenance, and safety procedures under the direction of qualified personnel. References in this manual to patented or proprietary devices do not constitute a recommendation of their use.

CT44_7-17
Inspect, Test, and Maintain Gravitometers/ Densitometers for Hazardous Liquid Leak Detection

Objectives

When you have completed this module, you will be able to do the following:

1. Identify and describe various types of gravitometers and densitometers and how they are serviced.
   a. Identify and describe various types of gravitometers and densitometers.
   b. Describe how gravitometers and densitometers are serviced.
2. Describe the procedures used to inspect, test, and maintain gravitometers/densitometers for hazardous liquid leak detection (CT44_7-17).
   a. Identify the prerequisites, competencies, and abnormal operating conditions (AOCs) associated with inspecting, testing, and maintaining gravitometers/densitometers.
   b. Describe the procedures used to inspect, test, and maintain gravitometers/densitometers for hazardous liquid leak detection.

Performance Tasks

Under the supervision of the instructor, you should be able to do the following:

1. Inspect, test, and maintain gravitometers/densitometers for hazardous liquid leak detection (CT44_7-17).
   - Identify potential abnormal operating conditions (AOCs) that may occur during performance of this CT, and know the appropriate actions to take in response to them.
   - Utilize the appropriate personal protective equipment according to relevant company procedures.
   - Determine the device's number and nameplate data.
   - Verify certification that test equipment has been calibrated prior to performing calibrations.
   - Notify control center and/or affected personnel before work begins.
   - Visually inspect the device and any associated equipment.
   - Calibrate the device using a pycnometer or hydrometer.
   - Determine the density correction factor using a suitable method and produce an appropriate report.
   - Confirm that the density correction factor is acceptable and enter it into the flow computer.
   - If the unit needs to be removed from the pipeline for repair, determine the current live density value and enter an override to the leak detection system so it can continue operating normally during the maintenance process. Isolate the device, release pressure, empty the device and its piping, and disconnect the unit.
   - Inspect, clean, and repair the device as required.
   - Reconnect the device to the pipeline, taking care that it is correctly oriented for the flow direction (if applicable).
   - Place the device back in-line, purge, and refill the piping with product.
   - Check the instrument loop integrity to confirm that there are no leaks and that the unit is ready to return to service.
   - Notify the control center, local personnel, and any affected personnel that the unit is about to be commissioned and that startup is about to begin.
   - Clear the override density value so a live value will be available for calibration and repeat the calibration procedure.
   - Complete appropriate documentation as required by operator's procedures.

Trade Terms

Abnormal operating conditions (AOCs)
API gravity
Density correction factor (DCF)
Pycnometer

Industry Recognized Credentials

If you are training through an NCCER-accredited sponsor, you may be eligible for credentials from NCCER's Registry. The ID number for this module is CT44_7-17. Note that this module may have been used in other NCCER curricula and may apply to other level completions. Contact NCCER's Registry at 1.888.622.3720 or go to **www.nccer.org** for more information.

Contents

1.0.0 Gravitometers and Densitometers 1
1.1.0 Density Measurement 1
1.1.1 Specific Gravity Analyzer 1
1.1.2 Liquid Density Transducer 1
1.1.3 Nuclear Density Detector 1
1.1.4 Optical Density Detector 3
1.2.0 Servicing Gravitometers and Densitometers 3
1.2.1 Troubleshooting and Repairing Specific Gravity Analyzers 4
1.2.2 Troubleshooting and Repairing Liquid Density Transducers 4
2.0.0 Working With Gravitometers and Densitometers 6
2.1.0 Prerequisites, Competencies, and Abnormal Operating Conditions 7
2.1.1 Prerequisites and Competencies 7
2.1.2 Abnormal Operating Conditions 7
2.2.0 Procedure 7

Figures

Figure 1 Specific gravity analyzer. 2
Figure 2 Liquid density transducer. 2
Figure 3 Nuclear density detector. 3
Figure 4 A pycnometer. 3

SECTION ONE

1.0.0 GRAVITOMETERS AND DENSITOMETERS

Objective

Identify and describe various types of gravitometers and densitometers and how they are serviced.

a. Identify and describe various types of gravitometers and densitometers.
b. Describe how gravitometers and densitometers are serviced.

Trade Terms

API gravity: A density measurement system used by the American Petroleum Institute with values expressed in degrees (°). API gravity is essentially a modified form of the inverse of specific gravity.

Pycnometer: A container with a precisely known volume. It is used for determining a liquid's density by filling it with the liquid and weighing it.

One significant physical property assessed in product measurement is density. Density is the ratio of a material's mass/weight to its volume. Density units are a combination of mass and volume units, such as g/cm^3 or kg/m^3. Specific gravity is closely related to density. It is the ratio of a material's mass to the mass of an equal volume of water. Specific gravity numbers are essentially the same as density values, but they are unitless. A number of different instruments measure density and specific gravity.

1.1.0 Density Measurement

In the petroleum industry, knowing the density of a product is crucial. In many cases, however, this industry uses a special density measurement scale known as API gravity, which is essentially a modified form of the inverse of specific gravity. It has no true units, but is expressed in degrees. API gravity is also scaled so a value of 10° is equal to a specific gravity of 1. If a product has an API gravity of less than 10°, it is denser than water (has a specific gravity greater than 1). If it has an API gravity of greater than 10°, it is less dense than water (has a specific gravity less than 1). For example, crude oil, being less dense than water, might have an API gravity of 20°.

1.1.1 Specific Gravity Analyzer

API gravity is measured electronically by using a specific gravity analyzer (densitometer or gravitometer), shown in *Figure 1*. Specific gravity analyzers are installed on pipelines. Densitometers are designed to provide continuous on-line determination of liquid density and API gravity. Density is converted to specific gravity and degrees API by software and is displayed on the metering system's monitor. Laboratory samples from an automatic sampling unit confirm the exact specifications of a liquid. Operations personnel can monitor the API gravity and make comparisons with the published specifications of the product.

Accurate measurement of density and API gravity is accomplished by a vibrating element that is energized by an electric current. The densitometer measures the vibration frequency and converts it to a density value. Fluid passing over the surface of the element causes a decrease in the frequency emitted. Higher-density fluids cause the frequency to be lower. Calibration constants for the unit are determined by laboratory tests, and the constants for each unit are programmed into the computer. SCADA calculates, records, and displays all gravity information.

1.1.2 Liquid Density Transducer

Some pipelines in which flow rates are below 10 feet per second (3 m/s) use a liquid density transducer (*Figure 2*) that works on the principle of a tuning fork. The fork tines are mounted on a flange and inserted into the liquid. The flange also supports the electronic equipment for the transducer. One of the tines of the tuning fork is excited by a piezoelectric element. The frequency of the first tine is maintained at its natural resonant frequency and is sensed by the second tine. As product density varies, the change in product mass changes the resonant frequency. The transducer transmits the frequency to a computer, which calculates the density.

1.1.3 Nuclear Density Detector

A totally different way of determining product density involves sending gamma rays through the pipeline and measuring how they are absorbed. Denser products absorb more energy than less dense ones. This kind of detector is known

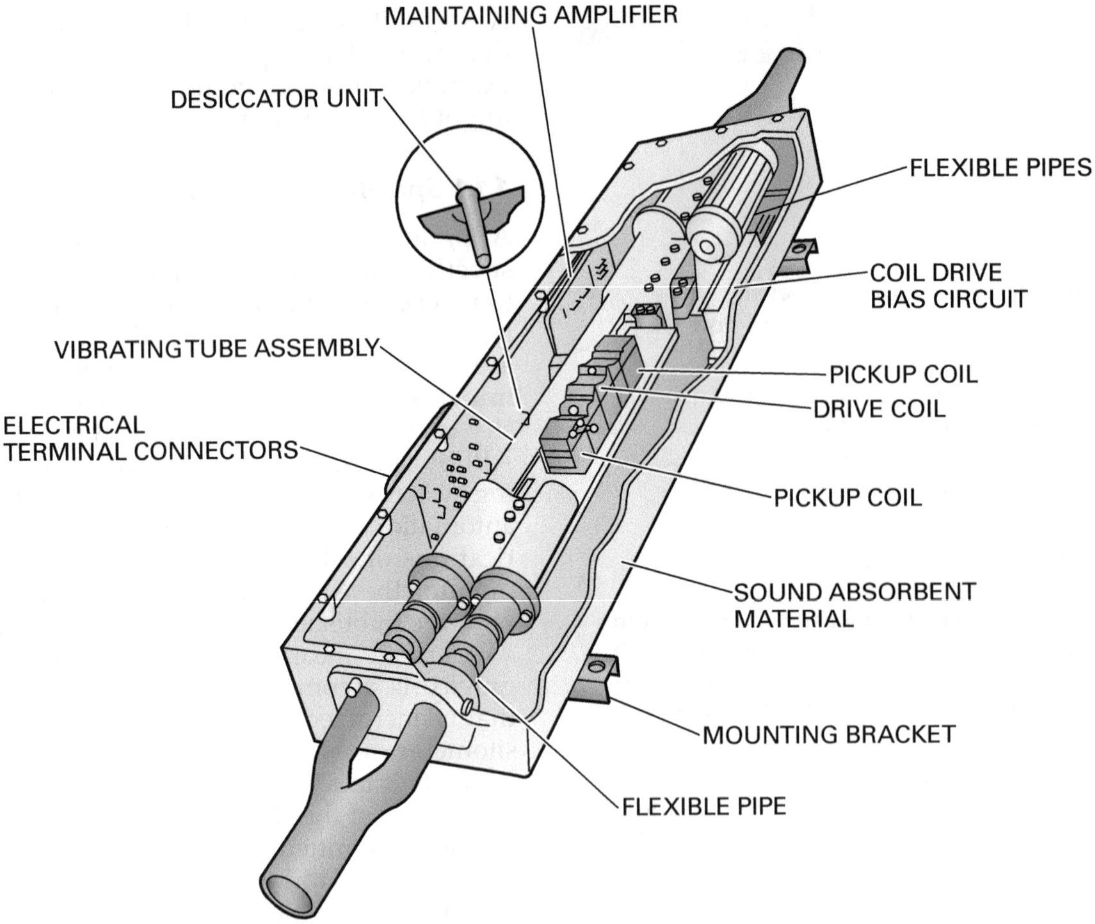

Figure 1 Specific gravity analyzer.

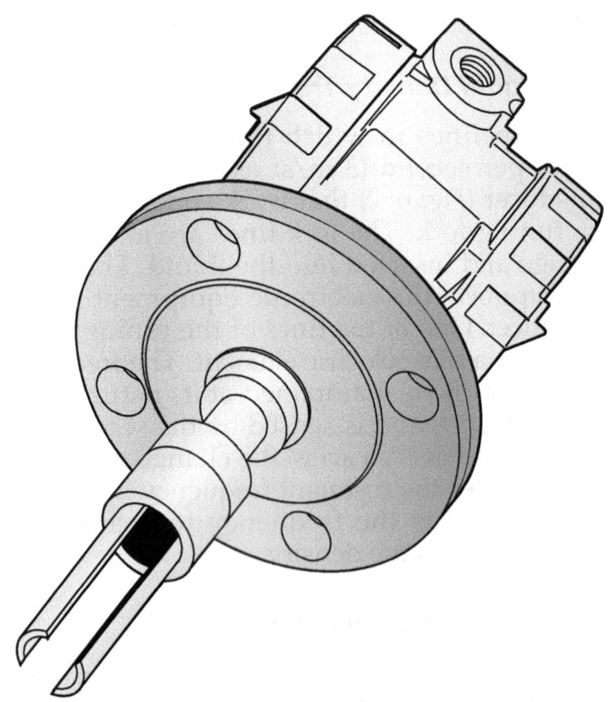

Figure 2 Liquid density transducer.

as a nuclear density detector. It consists of two parts, a nuclear source and a radiation detector. These are bolted onto opposite sides of the pipeline as shown in *Figure 3*. The source contains a radioactive material such as cesium-137 that emits gamma rays. The detector measures the energy that makes it through the pipeline and the product. Mathematical equations within the detector's controller calculate the product's density.

While nuclear density detectors are simple and convenient, they have the disadvantage of relying on a hazardous radioactive material with stringent disposal guidelines. Pipeline E&I technicians must never disassemble or otherwise service a nuclear density detector without proper training and certification. Cesium-137 is water soluble, making it particularly easy for it to get into the environment and living organisms, where it can cause harm.

> **WARNING!**
>
> Due to the hazards associated with nuclear devices, exercise special care around them. Be sure that you have the proper qualifications and certifications before you perform any kind of task involving one. If you are in doubt, consult your supervisor.

1.1.4 Optical Density Detector

Some density detectors rely on the fact that light is absorbed in varying degrees as a product's density changes. Optical density detectors pass a light beam through the product and measure it with a light detector. Based on the characteristics of the product and the light detector's output, the attached controller calculates the product density.

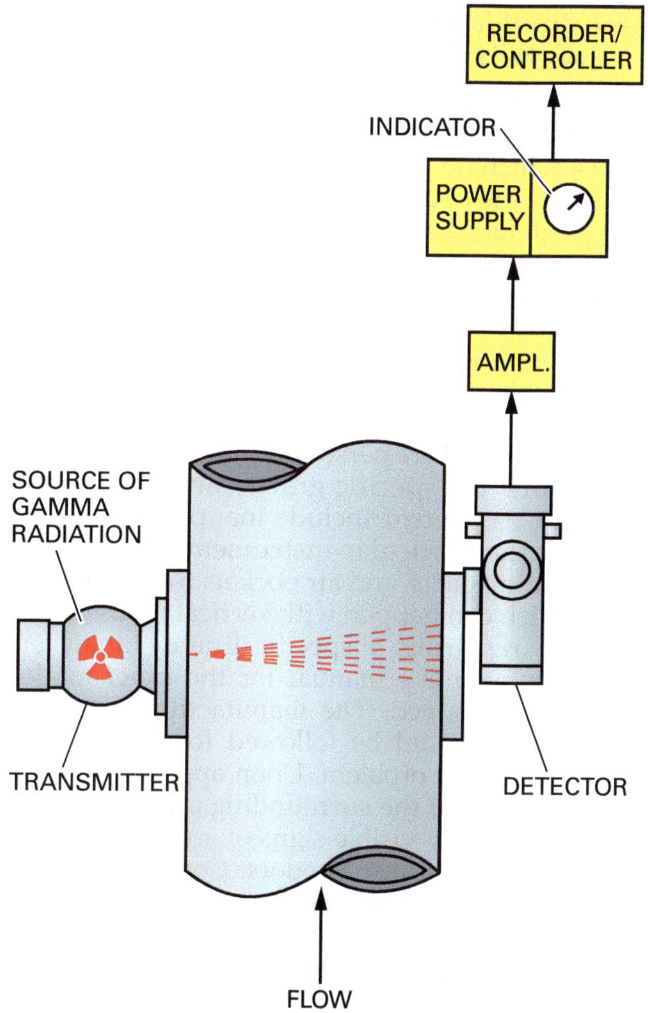

Figure 3 Nuclear density detector.

1.2.0 Servicing Gravitometers and Densitometers

Specific gravity analyzers and liquid density transducers must be tested to ascertain their accuracy. In order to do so, some companies use a **pycnometer**. A pycnometer (*Figure 4*) is a spherical or cylindrical double-walled container about 8 inches (20 cm) in diameter. It has an inlet and outlet valve that allows it to be connected in line with the product. Pycnometers measure density through weight comparison. After being properly cleaned, the unit is installed in the process line. A scale weighs a calibrated volume of two mediums: clean dry air or nitrogen, and the product in the pipeline. Then the density of these two mediums can be compared to the output of the specific gravity analyzer (densitometer/gravitometer) or the liquid density transducer.

The scale and any other equipment used for the testing process must be certified. Company policy and the manufacturer's recommendations must be followed for the successful use of the pycnometer.

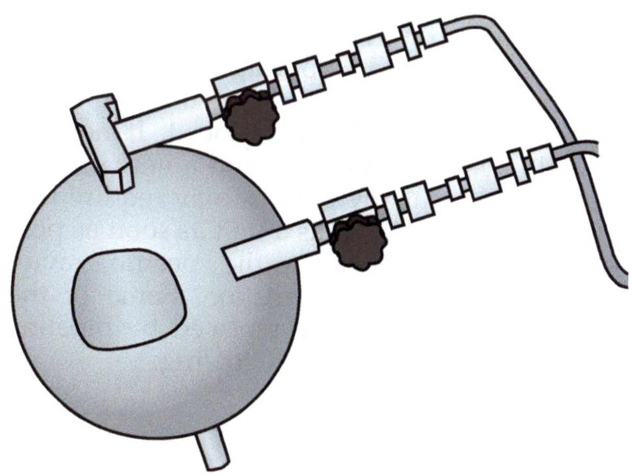

Figure 4 A pycnometer.

> **NOTE**
>
> Any leaks discovered void the pycnometer test. Generally, repairs to the pycnometer void its certification since they can change the weight. If a repair is necessary, the unit should be sent out for re-certification.

1.2.1 *Troubleshooting and Repairing Specific Gravity Analyzers*

Problems with specific gravity analyzers invariably reveal themselves through incorrect measurements, which do not correspond to the published gravity of the petroleum being shipped. It should first be determined if the variation is caused by equipment malfunction, incorrect information, or an unexplained change in the product. When it is confirmed that the problem is not due to inaccurate information and SCADA confirms that there is no problem with the transmission equipment, the problem is narrowed down to the field equipment. Pipeline E&I personnel can be dispatched to examine the specific gravity analyzer. Common causes of error include inappropriate flow rates for the particular instrument, deposits on the vibrating parts, and air pockets trapped in the unit (which can happen with vertical mountings).

When repairing a specific gravity analyzer, the manufacturer's manual for the exact model should be obtained. The manufacturer's recommendations should be followed for testing and diagnosis of the problem. Upon approaching the site, the unit and the surrounding area should be observed for any visible signs of a release, physical damage, or unusual odors. Company policy should be followed at all times.

Specific gravity analyzers can be tested with a pycnometer and application of the correct calibration medium. If the repeatability of the test cannot be established within company policy limits, it should be documented and reported. This test may be referred to experienced pipeline E&I technicians for accuracy.

The specific gravity analyzer's inputs and outputs should be tested according to the manufacturer's recommendations. In some units, a failing component may simply be replaced. If not, the unit will have to be removed and a new unit installed. Some company policies may require that the analyzer be replaced as a unit as soon as possible. Any work or repair to the specific gravity analyzer requires documentation. Some of the components that might be replaced by pipeline E&I technicians include the following:

- Transmitter (if applicable)
- Electric terminal connectors
- Electric wires
- Desiccator unit
- Amplifier card or unit
- Coil drive bias circuit
- Pickup coils
- Drive coil

In order to replace any component, standard company safety policy should be followed. All power should be turned off to the unit and the breakers locked out. The manufacturer's instruction manual should be followed. Circuit diagrams should be examined closely before proceeding.

1.2.2 *Troubleshooting and Repairing Liquid Density Transducers*

Liquid density transducers experience failure occasionally, but failure is limited because there are few moving parts. The parts that do experience failure are the tuning fork ends inserted in the liquid and the electronic units located in the top unit. Pipeline E&I technicians should secure the manufacturer's manual for the exact unit and test the output of each section carefully. Liquid density transducers may be calibrated in the field by alternating the medium from air to water.

Some units may detect both density and temperature. The individual test leads must be identified and isolated during testing to make certain that the appropriate readings are obtained. Company procedures must be followed and documented so that the calibration is acceptable. Some elements of the transducer that might cause it to fail include the following:

- Power supply
- Tuning fork
- Temperature sensor
- Piezoelectric pickup unit
- Piezoelectric drive unit
- Maintenance circuit
- Output/density signal unit

By comparing output readings to the manufacturer's manual, it can be determined which part of the transducer is the probable cause of the problem. If the forks have become inoperable, they can be replaced. However, if the transducer is inserted directly in the pipeline, the pipeline or at least that section will have to be shut down or bypassed and the pressure relieved so that the unit can be removed for tuning fork replacement. The repair should be scheduled through the control center, and company policy should be followed. Some transducers are installed parallel to the pipeline in a slipstream configuration and do not require a shutdown for servicing, but note that tuning forks must be oriented correctly for a valid reading. Flow must go through the fork tines. Problems with this kind of transducer are often traceable to improper installation.

Some of the repairs for problems discovered in the electronics unit can be performed without shutting down the pipeline. The manufacturer's technical manual should be followed when replacing any of the components. Exact replacement parts are required, and part numbers should be matched carefully. Some units cannot be repaired and are discarded and replaced when a problem develops.

Additional Resources

It is assumed that you will learn to perform these procedures under the direct supervision of your crew leader, foreman, or instructor. For independent study, refer to the recommended Additional Resources at the back of this module.

1.0.0 Section Review

1. How is API gravity related to specific gravity?
   a. It is scaled by a factor of 100.
   b. It is the inverse of specific gravity and scaled.
   c. It is the inverse of specific gravity.
   d. It is the same thing as specific gravity.

2. A common device used to test density measuring instruments is the _____.
   a. calibrated praximeter
   b. hygrometer
   c. pycnometer
   d. pipeline pragmatometer

Section Two

2.0.0 Working With Gravitometers and Densitometers

Objective

Describe the procedures used to inspect, test, and maintain gravitometers/densitometers for hazardous liquid leak detection (CT44_7-17).

a. Identify the prerequisites, competencies, and abnormal operating conditions (AOCs) associated with inspecting, testing, and maintaining gravitometers/densitometers.
b. Describe the procedures used to inspect, test, and maintain gravitometers/densitometers for hazardous liquid leak detection.

Performance Task

Inspect, test, and maintain gravitometers/densitometers for hazardous liquid leak detection (CT44_7-17).

- Identify potential abnormal operating conditions (AOCs) that may occur during performance of this CT, and know the appropriate actions to take in response to them.
- Utilize the appropriate personal protective equipment according to relevant company procedures.
- Determine the device's number and nameplate data.
- Verify certification that test equipment has been calibrated prior to performing calibrations.
- Notify control center and/or affected personnel before work begins.
- Visually inspect the device and any associated equipment.
- Calibrate the device using a pycnometer or hydrometer.
- Determine the density correction factor using a suitable method and produce an appropriate report.
- Confirm that the density correction factor is acceptable and enter it into the flow computer.
- If the unit needs to be removed from the pipeline for repair, determine the current live density value and enter an override to the leak detection system so it can continue operating normally during the maintenance process. Isolate the device, release pressure, empty the device and its piping, and disconnect the unit.
- Inspect, clean, and repair the device as required.
- Reconnect the device to the pipeline, taking care that it is correctly oriented for the flow direction (if applicable).
- Place the device back in-line, purge, and refill the piping with product.
- Check the instrument loop integrity to confirm that there are no leaks and that the unit is ready to return to service.
- Notify the control center, local personnel, and any affected personnel that the unit is about to be commissioned and that startup is about to begin.
- Clear the override density value so a live value will be available for calibration and repeat the calibration procedure.
- Complete appropriate documentation as required by operator's procedures.

Trade Terms

Abnormal operating conditions (AOCs): Conditions within a pipeline system, identified by the operator, that may indicate (1) a malfunction of a component or (2) a deviation from normal operations that may indicate a condition exceeding design limits or result in a hazard to persons, property, or the environment.

Density correction factor (DCF): A constant used to convert between density at a standard temperature and density at other temperatures.

Since knowing the density of a pipeline product is crucial to the process, it is equally essential to keep density measuring instruments working correctly. The following material discusses inspecting, testing, and maintaining these devices.

2.1.0 Prerequisites, Competencies, and Abnormal Operating Conditions

Personnel must possess a variety of skills in order to successfully inspect, test, and maintain gravitometers/densitometers. The various operations require a series of steps, which a pipeline E&I technician must be competent to perform. Personnel must also be aware of the particular abnormal operating conditions (AOCs) that may occur during the procedure and know the proper responses to them. Personnel must also understand the company's safety policies and procedures, including the proper use of personal protective equipment (PPE).

2.1.1 Prerequisites and Competencies

Personnel must possess the following basic knowledge and skills prior to inspecting, testing, and maintaining gravitometers/densitometers:

- Test equipment skills and knowing how to operate the following test equipment correctly:
  - Pycnometers
  - Multimeters
  - Certified hydrometers, thermometers, and scales
- Skill and safety with electrical systems
- Familiarity with the control system and its instrumentation, including alarm, control, indication, and recording functions
- Familiarity with density, pycnometers, using the appropriate density correction factor (DCF), and test equipment certification

2.1.2 Abnormal Operating Conditions

Personnel inspecting, testing, and maintaining gravitometers/densitometers must be able to identify potential abnormal operating conditions (AOCs) and know the proper response to each, should they occur. For example, if the device is not functioning or is not communicating with the end device, the pipeline E&I technician must alert the control center or other appropriate personnel to take the requisite corrective action.

2.2.0 Procedure

The following steps outline the general procedure for inspecting, testing, and maintaining gravitometers/densitometers:

Step 1 Identify potential abnormal operating conditions that may occur, and know the appropriate actions to take in response to them.

Step 2 Utilize the appropriate personal protective equipment according to relevant company procedures.

Step 3 Determine the device's number and nameplate data in order to verify that it is the correct unit to calibrate/maintain.

Step 4 Before beginning the calibration process, be sure that the test equipment has been certified, calibrated, and verified.

Step 5 Also before beginning the process, notify the control center, local personnel, and any other personnel who might be affected by the calibration/maintenance procedure.

Step 6 Visually inspect the device and any associated equipment. Check for mechanical condition, corrosion, leakage, electrical connection problems, or problems with associated electronics.

Step 7 Calibrate the device using a pycnometer or hydrometer, whichever is appropriate for the product and meets company procedure guidelines.

Step 8 Determine the density correction factor using a suitable method and produce an appropriate report. Follow company and industry guidelines when performing this task.

Step 9 Confirm that the density correction factor is acceptable and enter it into the flow computer. Alternatively, apply a manual meter ticket. Follow all applicable standards and guidelines.

Step 10 If the density correction factor is *not* acceptable, employ applicable tests and diagnostics to determine the proper corrective action.

Step 11 The unit may have to be removed from the pipeline for repair. If disconnection is *not* required, skip to Step 15.

Step 12 Determine the current live density value and enter an override to the leak detection system so it can continue operating normally during the maintenance process. Follow company guidelines and procedures.

> **NOTE**
>
> On multi-product pipelines, an override or manually entered (forced) factor should be valid for only a limited period of time, since a change of product in the pipeline renders it invalid. Be alert to this possibility and be careful to clear the factor as soon as it is not required. Always be alert to the possibility of product changes and be prepared to take appropriate action.

Step 13 Isolate the device, release pressure, and empty the device and its piping. Follow company guidelines and procedures.

Step 14 Disconnect the unit, following company guidelines and procedures.

Step 15 Inspect, clean, and repair the device as required. Follow the manufacturer's guidelines and company procedures.

Step 16 Reconnect the device to the pipeline, taking care that it is correctly oriented for the flow direction (if applicable).

Step 17 Place the device back in-line, purge, and refill the piping with product.

Step 18 Check the instrument loop integrity to confirm that there are no leaks and that the unit is ready to return to service.

Step 19 Notify the control center, local personnel, and any affected personnel that the unit is about to be commissioned and that startup is about to begin.

Step 20 Clear the override density value so a live value will be available for calibration.

Step 21 Repeat the calibration procedure as specified in Steps 7–10.

Step 22 Complete appropriate documentation as required by operator's procedures.

Additional Resources

It is assumed that you will learn to perform these procedures under the direct supervision of your crew leader, foreman, or instructor. For independent study, refer to the recommended Additional Resources at the back of this module.

2.0.0 Section Review

1. Which concept is important to understand as a prerequisite for calibrating a density measuring instrument?
   a. The density correction factor
   b. The meter's K factor
   c. The density inversion factor
   d. The meter's compensation factor

2. What must be done before taking a density measuring instrument offline to keep the leak detection system working correctly?
   a. Enter an override value.
   b. Disable the alarm.
   c. Set SCADA to ignore mode.
   d. Insert a calibrator into the circuit.

SUMMARY

Keeping density measuring instruments working reliably is key to having accurate information about pipeline product grade and value. A number of different instruments measure pipeline product density. This module described these instruments and explained some of their common maintenance issues. A basic overview of the procedures for inspecting, testing, and maintaining these devices was also presented.

Review Questions

1. A product with an API gravity value of 25° is _____.
   a. less dense than water
   b. more dense than water
   c. equal in density to water
   d. impossible to compare to water

2. Which kind of density measuring instrument works well in pipelines whose flow is less than 10 feet per second (3 m/s)?
   a. A specific gravity analyzer
   b. A sphere analyzer
   c. A depression densitometer
   d. A liquid density transducer

3. When performing a calibration with a pycnometer, what must be true?
   a. The pycnometer must be flushed with water.
   b. All instruments must be certified.
   c. The pycnometer must be calibrated within the last month.
   d. The product must be chilled to 45°F (7°C).

4. Which of the following is a test instrument that the pipeline E&I technician should be familiar with when maintaining density measuring instruments?
   a. A certified hydrometer
   b. A pneumatic calibrator
   c. A pressure source
   d. A USB-to-serial adapter

5. Before beginning any calibration activities on density measuring instruments, what should the pipeline E&I technician do?
   a. Isolate the instrument preparatory to depressurizing and draining it
   b. Check for electrical problems as well as check the associated electronics
   c. Confirm that the test equipment has been certified, calibrated, and verified
   d. Confirm that the density measuring device is free from leakage and corrosion

6. In addition to a pycnometer, what other instrument is sometimes used to calibrate a density measuring device?
   a. An API gravity gauge
   b. A density scale
   c. A hydrometer
   d. A pneumatic calibrator

Trade Terms Quiz

Fill in the blank with the correct term that you learned from your study of this module.

1. The ________ is a constant used to convert between density at a standard temperature and density at other temperatures.
2. Conditions within a pipeline system that may indicate (1) a malfunction of a component or (2) a condition exceeding design limits or resulting in a hazard to persons, property, or the environment are called ________.
3. A(n) ________ is a container with a precisely known volume. It is used for determining a liquid's density by filling it with the liquid and weighing it.
4. A density measurement system used by the American Petroleum Institute with values expressed in degrees (°) is called ________. It is essentially a modified form of the inverse of specific gravity.

Trade Terms

Abnormal operating conditions (AOCs)
API gravity
Density correction factor (DCF)
Pycnometer

Trade Terms Introduced in This Module

Abnormal operating conditions (AOCs): Conditions within a pipeline system, identified by the operator, that may indicate (1) a malfunction of a component or (2) a deviation from normal operations that may indicate a condition exceeding design limits or result in a hazard to persons, property, or the environment.

API gravity: A density measurement system used by the American Petroleum Institute with values expressed in degrees (°). API gravity is essentially a modified form of the inverse of specific gravity.

Density correction factor (DCF): A constant used to convert between density at a standard temperature and density at other temperatures.

Pycnometer: A container with a precisely known volume. It is used for determining a liquid's density by filling it with the liquid and weighing it.

Additional Resources

The following recommended resources can provide additional helpful information related to the requirements for this covered task:

API Recommended Practice 1161, Third Edition, © 2014 American Petroleum Institute. API Publishing Services, 1220 L Street, NW, Washington, DC 20005. Distributed by Thompson Reuters (Scientific) LLC, **www.techstreet.com**.

49 *CFR* 192, **www.ecfr.gov**.

49 *CFR* 195, **www.ecfr.gov**.

The Condensed Handbook of Measurement and Control, N. E. Battikha. Third Edition. 2006. Research Triangle Park, NC: The International Society of Automation.

NCCER Module 60102-02, *Pipeline E&I Safety*.

NCCER Module AOCCC-17, *Abnormal Operating Conditions Control Center.*

NCCER Module AOCFG-17, *Abnormal Operating Conditions Field and Gas.*

NCCER Module CT63_1-17, *Start-up of a Liquid Pipeline (Field).*

NCCER Module CT63_2-17, *Shutdown of a Liquid Pipeline (Field).*

NCCER Module CT63_3-17, *Monitor Pressures, Flows, Communications, and Line Integrity and Maintain Them Within Allowable Limits on a Liquid Pipeline System (Field).*

NCCER Module CT63_4-17, *Locally Operate Valves on a Liquid Pipeline System.*

The following websites offer resources for products and training:

The International Society of Automation, **www.isa.org.**

Figure Credits

© Dan Bannister/Dreamstime.com, Module Opener

Section Review Answer Key

Answer	Section Reference	Objective
Section One		
1. b	1.1.0	1a
2. c	1.2.0	1b
Section Two		
1. a	2.1.1	2a
2. a	2.2.0	2b

NCCER CURRICULA — USER UPDATE

NCCER makes every effort to keep its textbooks up-to-date and free of technical errors. We appreciate your help in this process. If you find an error, a typographical mistake, or an inaccuracy in NCCER's curricula, please fill out this form (or a photocopy), or complete the online form at **www.nccer.org/olf**. Be sure to include the exact module ID number, page number, a detailed description, and your recommended correction. Your input will be brought to the attention of the Authoring Team. Thank you for your assistance.

Instructors – If you have an idea for improving this textbook, or have found that additional materials were necessary to teach this module effectively, please let us know so that we may present your suggestions to the Authoring Team.

NCCER Product Development and Revision

13614 Progress Blvd., Alachua, FL 32615

Email: curriculum@nccer.org
Online: www.nccer.org/olf

❑ Trainee Guide ❑ Lesson Plans ❑ Exam ❑ PowerPoints Other ______________

Craft / Level: Copyright Date:

Module ID Number / Title:

Section Number(s):

Description:

Recommended Correction:

Your Name:

Address:

Email: Phone:

CT44_8-17

Inspect, Test, and Maintain Temperature Transmitters for Hazardous Liquid Leak Detection

OVERVIEW

Probably the most common and useful physical parameter to measure in many processes is temperature. Temperature affects numerous physical properties, such as density, pressure, and volume. A liquid may also experience changes in its ability to flow (viscosity) with temperature changes. Monitoring process temperature is therefore a critical task. Various instruments measure temperature using a number of different methods. This module describes these instruments and discusses the procedures used for inspecting, testing, and maintaining them.

Trainees with successful module completions may be eligible for credentialing through the NCCER Registry. To learn more, go to **www.nccer.org** or contact us at 1.888.622.3720. Our website has information on the latest product releases and training, as well as online versions of our *Cornerstone* magazine and Pearson's product catalog.

Your feedback is welcome. You may email your comments to **curriculum@nccer.org**, send general comments and inquiries to **info@nccer.org**, or fill in the User Update form at the back of this module.

This information is general in nature and intended for training purposes only. Actual performance of activities described in this manual requires compliance with all applicable operating, service, maintenance, and safety procedures under the direction of qualified personnel. References in this manual to patented or proprietary devices do not constitute a recommendation of their use.

CT40_8-17

Inspect, Test, and Maintain Temperature Transmitters for Hazardous Liquid Leak Detection

Objectives

When you have completed this module, you will be able to do the following:

1. Identify and describe various types of temperature monitoring devices and how they are serviced.
   a. Identify and describe various types of temperature monitoring devices.
   b. Describe how temperature monitoring devices are serviced.
2. Describe the procedures used to inspect, test, and maintain temperature transmitters for hazardous liquid leak detection (CT44_8-17).
   a. Identify the prerequisites, competencies, and abnormal operating conditions (AOCs) associated with inspecting, testing, and maintaining temperature transmitters.
   b. Describe the procedures used to inspect, test, and maintain temperature transmitters for hazardous liquid leak detection.

Performance Tasks

Under the supervision of the instructor, you should be able to do the following:

1. Inspect, test, and maintain temperature transmitters for hazardous liquid leak detection (CT44_8-17).
   - Identify potential abnormal operating conditions (AOCs) that may occur during performance of this CT, and know the appropriate actions to take in response to them.
   - Utilize the appropriate personal protective equipment according to relevant company procedures.
   - Determine the device's number and nameplate data.
   - Before beginning the calibration process, check that the test equipment has itself been certified, calibrated, and verified, and that the transmitter's input and output ranges are appropriately matched to the end device.
   - Notify the control center, local personnel, and any other personnel who might be affected by the calibration/maintenance procedure.
   - Visually inspect the device and any associated equipment.
   - Obtain an accurate reference temperature measurement from the process by using a certified thermometer.
   - Determine if transmitter calibration is required and document the current results, including all data.
   - Before beginning calibration, use the current live process temperature value to override the sensor data.
   - Disconnect the transmitter's inputs from the sensor and attach the test equipment.
   - Use the test equipment to supply simulated sensor data to the transmitter.
   - Disconnect the test equipment and restore the unit's normal connections. Terminate the override temperature value previously established.
   - Determine if transmitter calibration is required and perform any sensor trim adjustments.
   - Perform all steps required to resume normal operation.
   - Notify the control center, local personnel, and any affected personnel that the unit has been returned to normal operation.
   - Complete appropriate documentation as required by operator's procedures.

Trade Terms

Abnormal operating conditions (AOCs)
Alarm
Calibrate
End device
Fieldbus
Multifunction calibrator
Range and span
Resistance temperature detector (RTD)
Thermocouple
Thermowell
Transmitter

Industry Recognized Credentials

If you are training through an NCCER-accredited sponsor, you may be eligible for credentials from NCCER's Registry. The ID number for this module is CT44_8-17. Note that this module may have been used in other NCCER curricula and may apply to other level completions. Contact NCCER's Registry at 1.888.622.3720 or go to **www.nccer.org** for more information.

Contents

1.0.0 Temperature Monitoring Devices ... 1
1.1.0 Temperature Monitoring ... 1
1.1.1 Temperature Measuring Technologies ... 1
1.1.2 RTDs ... 2
1.1.3 Thermocouples ... 2
1.1.4 Thermocouple Wire ... 4
1.1.5 Temperature Transmitters ... 5
1.2.0 Servicing Temperature Monitoring Devices ... 5
1.2.1 Troubleshooting RTDs ... 6
1.2.2 Troubleshooting Thermocouples ... 7
1.2.3 Troubleshooting Transmitters ... 7
2.0.0 Working With Temperature Transmitters ... 9
2.1.0 Prerequisites, Competencies, and Abnormal Operating Conditions ... 10
2.1.1 Prerequisites and Competencies ... 10
2.1.2 Abnormal Operating Conditions ... 10
2.2.0 Inspecting, Testing, and Maintaining Temperature Transmitters ... 10

Figures and Tables

Figure 1 A thermowell. ... 2
Figure 2 Typical resistance temperature detector in a thermowell ... 3
Figure 3 The Seebeck effect ... 3
Figure 4 Various thermocouple joints. ... 4
Figure 5 Sheathed thermocouple. ... 4
Figure 6 A smart temperature transmitter. ... 7
Figure 7 A multifunction calibrator. ... 7
Figure 8 Temperature curves for various thermocouples. ... 8
Figure 9 Handheld communicator for smart transmitters. ... 8

Table 1 Symbols for Indicating Positive and Negative ... 5
Table 2 Table 2A ISA Color Coding of Thermocouple Wire (1 of 3) ... 6
Table 2 Table 2B ISA Color Coding of Thermocouple Wire (2 of 3) ... 6
Table 2 Table 2C ISA Color Coding of Thermocouple Wire (3 of 3) ... 6

Section One

1.0.0 Temperature Monitoring Devices

Objective

Identify and describe various types of temperature monitoring devices and how they are serviced.

a. Identify and describe various types of temperature monitoring devices.
b. Describe how temperature monitoring devices are serviced.

Trade Terms

Alarm: An alert from the control system that a process parameter has moved out of its normal range. Alarms can be audible, visual, or both.

Calibrate: The task of adjusting an instrument so that its output signal is representative of the non-adjustable input signal at any given point. Calibration necessarily includes using calibrated instruments traceable to a national standards body.

End device: The device that is the ultimate recipient of a transmitter's information. Common end devices include SCADA systems, PLCs, DCSs, and flow computers.

Fieldbus: A simple digital communications network and protocol set designed to connect instruments, controllers, and other intelligent devices together. There are several different fieldbus standards.

Multifunction calibrator: Refers to an instrument used to calibrate a wide variety of electronic devices. Some multifunction calibrators also have pneumatic calibration capabilities.

Range and span: Two properties of an instrument. Range refers to the minimum and maximum values that the instrument can measure. Span refers to the difference between the instrument's upper and lower measurement limits.

Resistance temperature detector (RTD): A temperature sensor that expresses temperature by a corresponding electrical resistance.

Thermocouple: A temperature sensor composed of junctions of dissimilar metals. Thermocouples express temperature by a corresponding voltage.

Thermowell: A thermally conductive penetration into a pipeline that allows a temperature sensor to read the process temperature while remaining physically isolated from the process material.

Transmitter: A device that conditions a sensor's raw signal value to a form suitable for the control or monitoring system to which it's connected. Transmitters are based on a number of different underlying technologies.

Temperature is essentially a measurement of the energy of motion (kinetic energy) of the particles that make up a substance. If something puts more energy into the particles, they move faster and the substance's temperature rises. Cooling a substance removes energy from the particles, so they move more slowly and the substance's temperature drops.

1.1.0 Temperature Monitoring

Temperature must be monitored, because it affects product volume, equipment performance, and many other aspects of pipeline operation. Accurate calculation of volume requires temperature information, as well as pressure and flow rate. Sensors are placed in the pipeline at strategic locations to monitor the product temperature. Other temperature sensors are placed in critical locations such as pumps, bearings, and equipment where temperature cannot exceed set limits without serious damage occurring. An alarm is included in these circuits. Data is transmitted to the control center, where the host computer displays the ongoing readings on meters and records the data for calculation and verification.

1.1.1 Temperature Measuring Technologies

Temperature is usually measured by some kind of temperature-detecting element, commonly known as a sensor. The resistance temperature detector (RTD) and thermocouple are the most common sensors used in the pipeline industry. Each represents temperature through a change in an electrical quantity (resistance or voltage).

The raw output from the sensor must be conditioned before it can be sent to a display, recorder, controller, or computer system. This is accomplished by a temperature **transmitter**. The transmitter conditions the signal, applies filtering and amplification, and generates an appropriate output signal for the system that the sensor is driving.

Temperature sensors may detect the process temperature in a number of different ways. With liquids flowing through pipelines, it is common to use a **thermowell** (*Figure 1*). A thermowell is a tubular device that penetrates the pipeline at an angle to the fluid flow. The temperature sensor itself is inserted into the thermowell, which physically (but not thermally) isolates it from the process. A thermally conductive compound, gel, or liquid is often used with the temperature sensor to improve heat transfer between it and the thermowell. Doing this makes the sensor more responsive to changes in process temperature. The thermal transfer material must be rated for the expected process temperature, however, or it could break down or change its physical properties. These changes, in turn, may lead to inaccurate temperature readings or poor sensor response.

1.1.2 RTDs

An RTD represents temperature by a specific electrical resistance value. As temperature changes, resistance goes up or down in response. RTDs contain a wire wrapped around a support that provides good thermal contact. The wire is glued in place, typically with varnish, and inserted into a sheath. The RTD is then attached to a mechanical fitting that allows it to be installed in a thermowell. A loading spring is included to help the RTD make positive contact with the thermowell. *Figure 2* shows a typical RTD in a protecting well.

One advantage of the RTD is that it can be located at a considerable distance from the indicating or recording device. This greater distance may result in error, however, due to the additional resistance of the longer wires. The effect of this resistance increases as the distance from the indicating or recording device increases. Wire resistance is negated if a three-wire or four-wire measurement bridge is used.

1.1.3 Thermocouples

A thermocouple converts temperature into an equivalent voltage. Temperature changes are represented by an increase or decrease in the sensor's output voltage. Thermocouples rely on a physics

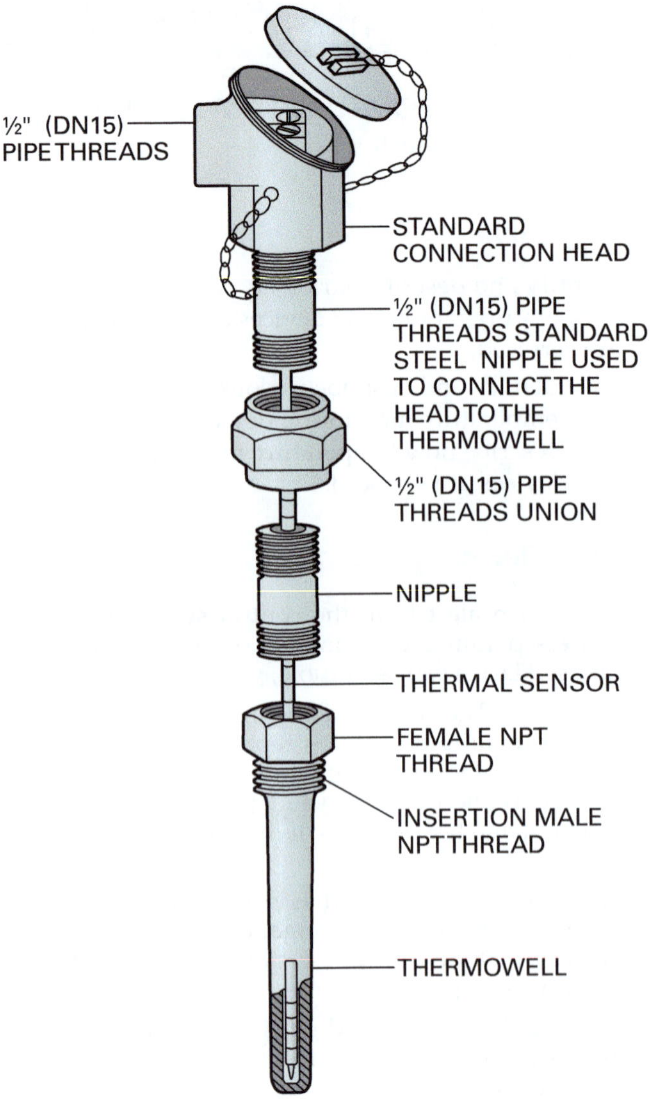

Figure 1 A thermowell.

principle called the Seebeck effect in which the temperature difference between the junctions of two different metals creates a voltage difference (*Figure 3*).

Thermocouples are available for use within the approximate limits of –300°F to +3,200°F (–184°C to +1,760°C). No single thermocouple meets every need, but each possesses characteristics desirable for selected applications. Thermocouples are classified into two groups identified as noble metals and base metals. The noble metals are the following:

- *Type S (10% rhodium/90% platinum versus platinum)* – Used for covering the international temperature scale range from 1,166°F to 1,945°F (630°C to 1,063°C). It is chemically inert and stable at high temperatures in oxidizing atmospheres. It is widely used as a standard for calibration of base metal thermocouples.

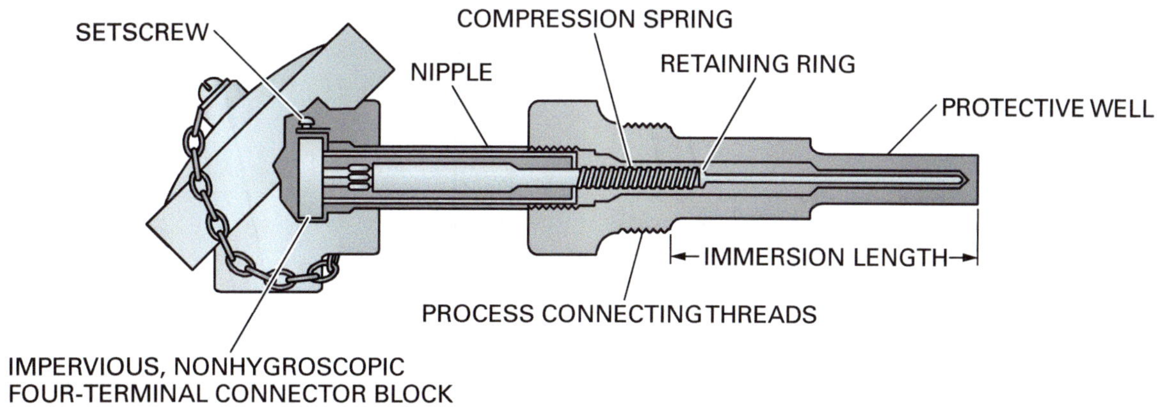

Figure 2 Typical resistance temperature detector in a thermowell.

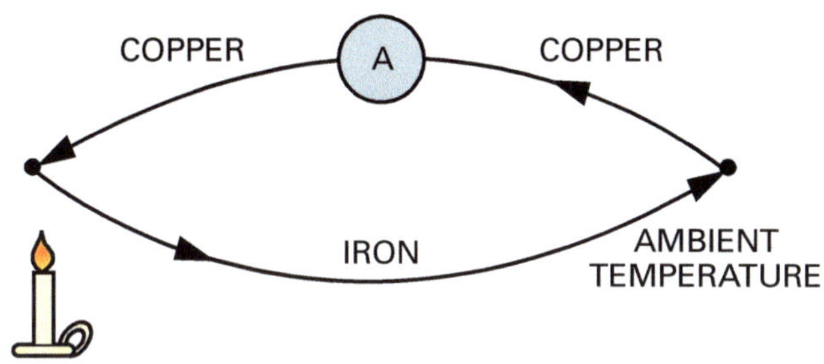

Figure 3 The Seebeck effect.

- *Type R (13% rhodium/87% platinum versus platinum)* – Similar to Type S but produces a different voltage for a given temperature.
- *Type B (30% rhodium/70% platinum versus 6% rhodium/70% platinum)* – Similar to Types S and R, but has somewhat greater physical strength and stability and can withstand somewhat higher temperatures.

The base metal types are the following:

- *Type T (copper versus constantan)* – Used over the temperature range of –300°F to +700°F (–184°C to +371°C). Constantan is an alloy of copper and nickel. Its primary use is in measuring subzero temperatures. It has superior corrosion resistance in moist atmospheres.
- *Type J (iron versus constantan)* – Used over the temperature range of –200°F to +1,400°F (–129°C to +760°C). It exhibits good stability at 1,400°F (760°C) in nonoxidizing atmospheres.
- *Type K (chromel versus alumel)* – Used over the temperature range of –200°F to +2,300°F (–129°C to +1,260°C) and is more resistant to oxidation than other base metal combinations. Chromel is an alloy of nickel and chromium. Alumel is an alloy of nickel, manganese, aluminum, and silicon. This combination must be protected against reducing atmospheres. Alternate cycling between oxidizing and reducing atmospheres is particularly destructive.
- *Type E (chromel versus constantan)* – Produces the highest thermoelectric output of any conventional thermocouple. It is used up to 1,400°F (760°C) and exhibits a high degree of calibration stability at temperatures not exceeding 1,000°F (538°C).

Various fabrication methods are used in the manufacturing of thermocouples. The factors that are considered in determining the correct fabrication process include the types of material used, the thermal and chemical environment to which the thermocouple is to be subjected, and the electrical insulation and mechanical strength requirements. Analysis of these factors determines how to join the dissimilar metals and what type of insulation and protection tube to use.

The primary requirement when joining thermocouple wires is to maintain good thermal and electrical conductivity without damaging the metallurgical properties of the wires. Several joining methods are illustrated in *Figure 4*. Method (A) is employed in the manufacturing of Type J and K thermocouples using 8 AWG through 14 AWG wire. It uses the resistance-welding technique. Method (B) is known as a butt weld. It is used to join 8 AWG through 20 AWG wires in Type E, J, and K thermocouples. This joint is also made with resistance welding. The twisted wires shown in method (C) are either arc welded or gas welded. This manufacturing method can be used in the fabrication of Types E, J, K, and T thermocouples using 8 AWG through 28 AWG gauge wire. Finally, method (D) illustrates the V-joint, which may be arc, gas, or TIG welded. It is used to join Types B, R, and S thermocouples, as well as all types previously discussed.

There are several insulating materials that can be used to cover the sensing wires in order to prevent electrical shorts and to provide abrasion protection. Some materials, such as PTFE (Teflon®) and polyvinyl chloride (PVC), are used because

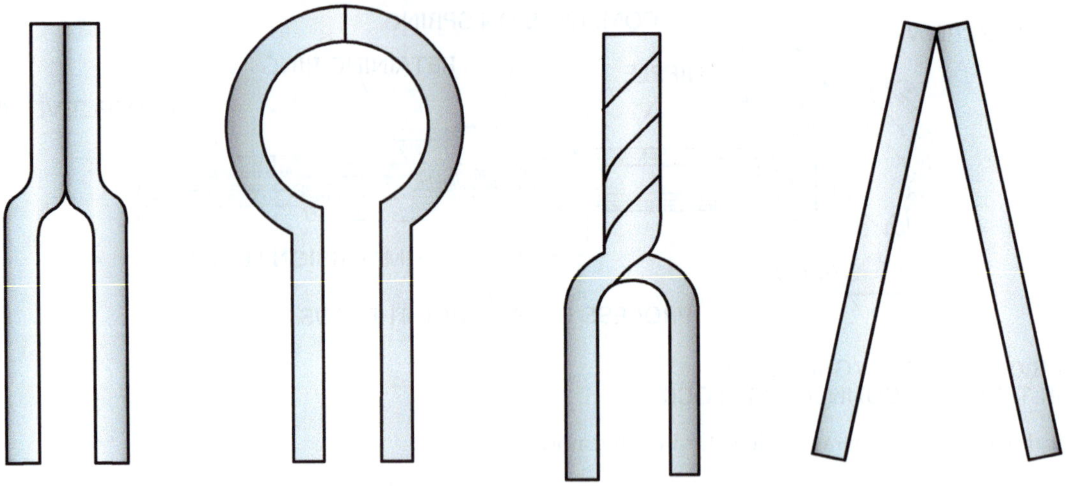

Figure 4 Various thermocouple joints.

of their natural resistance to absorbing moisture. Certain fiber materials may also be used when impregnated with wax, resin, or a similar substance. These impregnated fibers should not be used in applications where temperature exceeds 400°F (204°C), as this would vaporize the wax or resin and leave the thermocouple wire unprotected. For high-temperature applications, materials like fibrous silica and fiberglass are among the best of the non-ceramic types. These insulators are capable of withstanding operating temperatures up to 1,200°F (649°C). If temperatures greater than 1,200°F (649°C) will be measured, a ceramic insulator must be used. Ceramic insulators can operate in environments exceeding 2,500°F (1,371°C).

The insulator is typically covered by a protective tube used to prevent mechanical damage to the sensing element. The tube also provides a layer of separation between the element and any harmful environmental conditions. Materials used for tube construction are various metals, such as carbon steel, stainless steel, and high nickel alloys. Ceramics and combinations of metals and ceramics, called *cermets*, are also used. While metals provide adequate protection against mechanical damage, they are sometimes unsuitable due to their susceptibility to corrosion and because of their tendency to become porous at high temperatures. Ceramic or cermet tubes are used at temperatures exceeding approximately 2,100°F (1,149°C) because of their high strength, high density, and resistance to corrosion.

Another type of construction that has become increasingly popular is the sheathed thermocouple (*Figure 5*). This device consists of a matched pair of thermocouple wires surrounded by a metal case. The inside of the case is filled with a non-compacted ceramic insulating material. The metal is then reduced in diameter by a mechanical process that tightly compacts the insulation around the thermocouple wires. This type of construction has the most desirable characteristics of the previously discussed types. In addition, it is easier and less costly to fabricate.

1.1.4 Thermocouple Wire

Pipeline E&I technicians must be able to recognize the thermocouple type and the polarity of thermocouple wires in order to ensure that thermocouple circuits are properly installed. Technicians also should be familiar with both the letter designations and the color codes used to identify thermocouple leads and thermocouple extension wires. To make this task easier, the International Society of Automation (ISA) has developed standard *MC96.1 Temperature Measurement Thermocouples,* to establish uniformity in the designation of thermocouple wires.

Letter designations have been used for many years to identify thermocouple types. Although

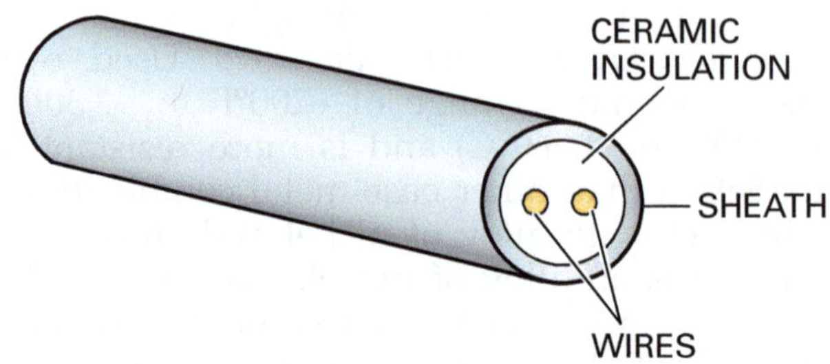

Figure 5 Sheathed thermocouple.

the letter designation is often associated with a particular thermocouple material, the letter designation applies only to the temperature-EMF relationship and not to the material. Other materials having the same temperature-EMF relationship and better physical properties for a particular application may be given the same letter designation. The types and the materials used to define the temperature-EMF relationships were listed earlier in this module.

Polarity is also indicated by letter designations. In the letter designation, P represents the positive wire and N represents the negative wire. Extension wires are frequently used with thermocouple circuits. They are designated by the letter X. *Table 1* lists the lead designations for thermocouples.

The use of color codes to indicate the value and polarity of electronic components has been an accepted practice for many years. Thermocouple manufacturers also use color coding to designate both the type and the polarity of thermocouples and thermocouple extension wires. For thermocouples and thermocouple extension wires, the color red designates the negative wire. This red-negative designation is endorsed by the ISA for all thermocouple types. *Table 2A, Table 2B,* and *Table 2C* summarize the practice of color coding leads of thermocouples and thermocouple extension leads.

CAUTION

Color codes for thermocouples and thermocouple extension wires are not standardized among all countries. Materials manufactured outside the United States may not conform to the color codes that you're familiar with. For example, red may signify positive rather than negative in some color code systems. Always confirm that the device or wire you're using is what you think it is before placing it into service.

Another common method of designating the positive and negative polarity of a thermocouple wire is by the order in which the material of the leads in the thermocouple's name is written or spoken. The first wire named is positive, and the second is negative. For example, in the Type J thermocouple (iron versus constantan), the iron wire is positive and the constantan wire is negative. The indicated polarity of the thermocouple leads apply for conditions where the measuring junction is at a higher temperature than the reference junction.

Table 1. Symbols for Indicating Positive and Negative

Thermocouple Wire		
Type	**Positive**	**Negative**
B	BP	BN
E	EP	EN
J	JP	JN
K	KP	KN
R	RP	RN
S	SP	SN
T	TP	TN

1.1.5 Temperature Transmitters

Temperature transmitters come in a wide variety of styles and have one of several possible underlying technologies. Pneumatic transmitters are an older technology found in legacy applications and with some natural gas pipelines. These rely on pressurized air or another gas to transmit information between devices. Because they are entirely mechanical, they have numerous components that can fail or require adjustment.

Electronic temperature transmitters are much more common today and come in a variety of styles. Traditional analog electronic transmitters have been around for many years and use conventional analog electronic components. They typically communicate temperature as a 4–20 mA current signal, although other values are possible as well. Analog transmitters require periodic adjustments and calibration. Some have replaceable components.

Finally, many installations have upgraded to smart electronic transmitters (*Figure 6*). These contain an embedded microprocessor that converts the sensor data to a digital form and manipulates it mathematically. These units have the advantage that calibration is greatly simplified and adjustments such as range and span are far less likely to change over time. A smart transmitter will communicate with the end device either over a standard 4–20 mA current loop or over a digital fieldbus of some type.

1.2.0 Servicing Temperature Monitoring Devices

Temperature sensors require relatively little service since they have no moving parts. Generally, if they fail, they must be discarded and replaced. Temperature transmitters, on the other hand, may have adjustments or replaceable components, depending on their underlying technology. The following section addresses some general troubleshooting principles.

Table 2. A ISA Color Coding of Thermocouple Wire (1 of 3)

Duplex-Insulated Thermocouple Wire					
Thermocouple			Color of Insulation		
Type	Positive	Negative	Overall	Positive	Negative
E	EP	EN	Brown	Purple	Red
J	JP	JN	Brown	White	Red
K	KP	KN	Brown	Yellow	Red
T	TP	TN	Brown	Blue	Red

Table 2. B ISA Color Coding of Thermocouple Wire (2 of 3)

Single-Conductor Extension Wire				
Thermocouple			Color of Insulation	
Type	Positive	Negative	Positive	Negative
B	BPX	BNX	Gray	Red-Gray Trace
E	EPX	ENX	Purple	Red-Purple Trace
J	JPX	JNX	White	Red-White Trace
K	KPX	KNX	Yellow	Red-Yellow Trace
R or S	SPX	SNX	Black	Red-Black Trace
T	TPX	TNX	Blue	Red-Blue Trace

Table 2. C ISA Color Coding of Thermocouple Wire (3 of 3)

Single-Conductor Insulator Extension Wire				
Thermocouple			Color of Insulation	
Type	Positive	Negative	Positive	Negative
B	BPX	BNX	Gray	Red-Gray Trace
E	EPX	ENX	Purple	Red-Purple Trace
J	JPX	JNX	White	Red-White Trace
K	KPX	KNX	Yellow	Red-Yellow Trace
R or S	SPX	SNX	Black	Red-Black Trace
T	TPX	TNX	Blue	Red-Blue Trace

1.2.1 Troubleshooting RTDs

When examining the RTD, you must follow company safety procedures. Safety procedures vary greatly depending on the location of the unit. Obviously, procedures are different for checking an immersion RTD on a mainline pump compared to a bearing RTD on a driver. If the mainline RTD cannot be bypassed, the pipeline might have to be shut down to access the device.

In order to test the RTD, the pipeline E&I technician needs to have the manufacturer's manual for the exact RTD in question and needs to follow the manufacturer's recommendations for the test lead connection and the appropriate readings. Particular meters for the job are specified in the manufacturer's manual. Replacement is the standard procedure for a defective RTD. Great care should be used when handling the new RTD and the bridge circuit (if applicable). Although they are very simple in construction, they are expected to be highly accurate when in operation. A general outline for replacing an RTD includes the following steps; note that the following steps are not a complete procedure, merely an overview:

Step 1 Obtain a replacement RTD. Confirm that you have the correct replacement by checking its accuracy at room temperature using an appropriate meter or calibrator.

Step 2 Disconnect associated wiring.

Step 3 Remove the RTD from the thermowell using the appropriate tool, noting the location of all seals and other parts.

Step 4 Install new seals and associated parts on the new RTD as applicable.

Step 5 If required by the application, coat the RTD with an appropriate thermal transfer compound or fill vertical thermowells with a thermal transfer fluid.

Step 6 Insert and install the RTD in the thermowell.

Step 7 Tighten according to the manufacturer's recommendations.

Step 8 Reconnect the thermowell wiring.

Step 9 Confirm that the RTD is sending reasonable readings to its attached transmitter, controller, or display.

1.2.2 Troubleshooting Thermocouples

Along with letter designations and color coding, multimeter readings can also be used to help identify thermocouples. Generally speaking, a multimeter set to measure resistance will indicate if the temperature sensing device is, in fact, a thermocouple. A resistance measurement taken with a multimeter should display a very low value—typically less than 1Ω.

A multifunction calibrator (*Figure 7*) and a calibrated heat source can be used to identify an unknown thermocouple. A number of manufacturers such as Fluke®, Druck®, and Transmation® offer these devices. When certainty is required, this approach is the best and most reliable.

A multimeter can be used to narrow down the possibilities, although with less certainty than a multifunction calibrator. Each type of thermocouple produces a specific voltage at a specific temperature. *Figure 8* shows temperature curves for seven different types of thermocouples. The curves are plotted on a graph of output voltages (in millivolts) versus temperatures between 32°F and 3,632°F (0°C and 2,000°C). As the table indicates, a Type J thermocouple produces about 28 mV at 932°F (500°C). By measuring the voltage of a thermocouple at a specific temperature and comparing that reading to a chart or graph, the specific type of thermocouple can likely be identified with a reasonable degree of accuracy.

1.2.3 Troubleshooting Transmitters

Since temperature transmitters can be based on widely differing technologies, there is no general troubleshooting procedure for them. Instead, pipeline E&I technicians should acquire the appropriate service manual for the unit in question and follow its instructions. Similarly, the tools required to troubleshoot, adjust, or calibrate a transmitter will depend on the technology. Pneumatic units usually require a pneumatic calibrator and pressure source. Analog electronic transmitters require a multifunction calibrator. Smart device requirements vary widely depending upon the brand. Some can be adjusted or calibrated directly through their front panel display and controls, while others may require a handheld communicator (*Figure 9*). Transmitters that are part of a larger intelligent control system may be able to be diagnosed remotely as well. In each case, following the manufacturer's guidelines is critical to success.

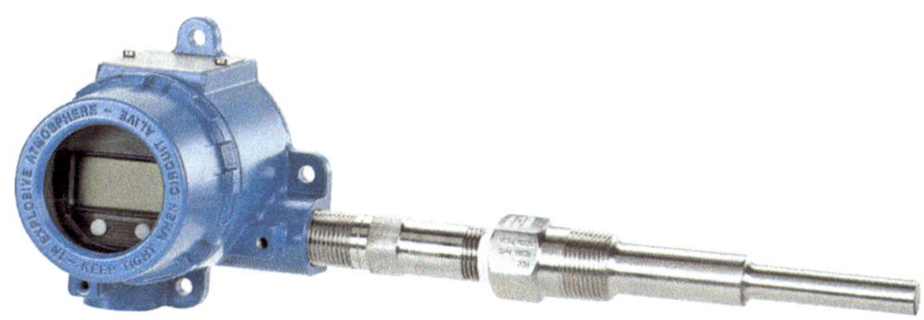

Figure 6 A smart temperature transmitter.

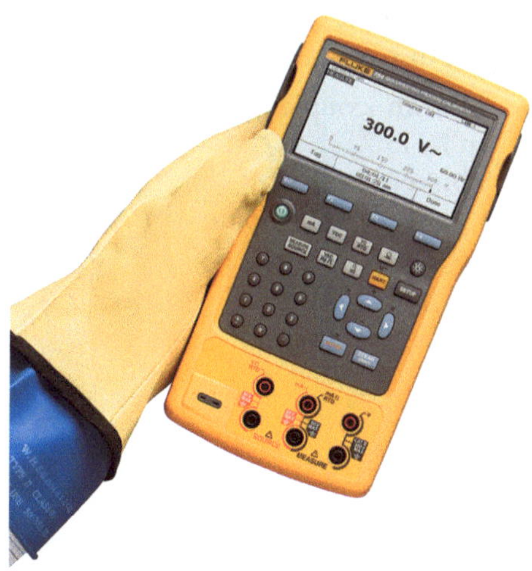

Figure 7 A multifunction calibrator.

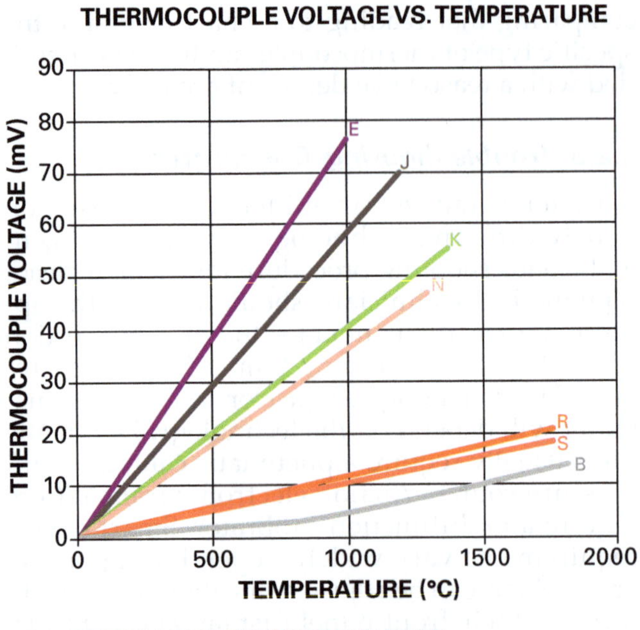

NOTE: LETTERS INDICATE THERMOCOUPLE TYPE

Figure 8 Temperature curves for various thermocouples.

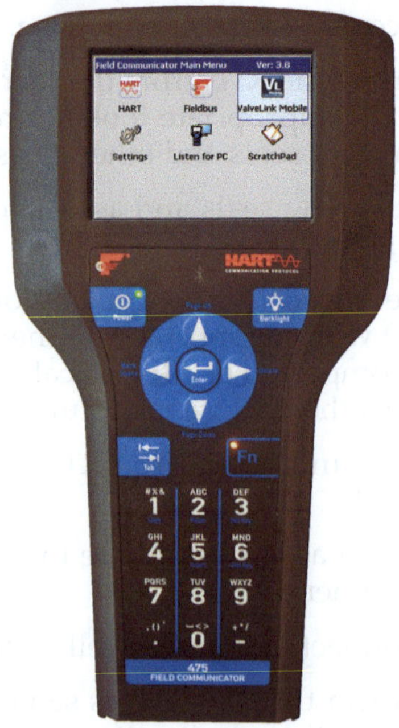

Figure 9 Handheld communicator for smart transmitters.

Additional Resources

It is assumed that you will learn to perform these procedures under the direct supervision of your crew leader, foreman, or instructor. For independent study, refer to the recommended Additional Resources at the back of this module.

1.0.0 Section Review

1. Which kind of temperature sensor expresses temperature as an electrical resistance?
   a. A thermocouple
   b. An RTD
   c. A bimetallic strip
   d. A gas bulb thermometer

2. Troubleshooting a smart temperature transmitter may require the technician to use a(n) _____.
   a. wattmeter
   b. pneumatic capsule
   c. IP converter
   d. handheld communicator

Section Two

2.0.0 Working With Temperature Transmitters

Objective

Describe the procedures used to inspect, test, and maintain temperature transmitters for hazardous liquid leak detection (CT44_8-17).

a. Identify the prerequisites, competencies, and abnormal operating conditions (AOCs) associated with inspecting, testing, and maintaining temperature transmitters.
b. Describe the procedures used to inspect, test, and maintain temperature transmitters for hazardous liquid leak detection.

Performance Task

Inspect, test, and maintain temperature transmitters for hazardous liquid leak detection (CT44_8-17).

- Identify potential abnormal operating conditions (AOCs) that may occur during performance of this CT, and know the appropriate actions to take in response to them.
- Utilize the appropriate personal protective equipment according to relevant company procedures.
- Determine the device's number and nameplate data.
- Before beginning the calibration process, check that the test equipment has itself been certified, calibrated, and verified, and that the transmitter's input and output ranges are appropriately matched to the end device.
- Notify the control center, local personnel, and any other personnel who might be affected by the calibration/maintenance procedure.
- Visually inspect the device and any associated equipment.
- Obtain an accurate reference temperature measurement from the process by using a certified thermometer.
- Determine if transmitter calibration is required and document the current results, including all data.
- Before beginning calibration, use the current live process temperature value to override the sensor data.
- Disconnect the transmitter's inputs from the sensor and attach the test equipment.
- Use the test equipment to supply simulated sensor data to the transmitter.
- Disconnect the test equipment and restore the unit's normal connections. Terminate the override temperature value previously established.
- Determine if transmitter calibration is required and perform any sensor trim adjustments.
- Perform all steps required to resume normal operation.
- Notify the control center, local personnel, and any affected personnel that the unit has been returned to normal operation.
- Complete appropriate documentation as required by operator's procedures.

Given temperature's importance to both the process and the equipment, keeping temperature transmitters and their associated sensors working correctly is crucial. The following material discusses inspecting, testing, and maintaining these devices.

Trade Terms

Abnormal operating conditions (AOCs): Conditions within a pipeline system, identified by the operator, that may indicate (1) a malfunction of a component or (2) a deviation from normal operations that may indicate a condition exceeding design limits or result in a hazard to persons, property, or the environment.

2.1.0 Prerequisites, Competencies, and Abnormal Operating Conditions

Personnel must possess a variety of skills in order to successfully inspect, test, and maintain temperature transmitters. The various operations require a series of steps, which a pipeline E&I technician must be competent to perform. Personnel must also be aware of the particular abnormal operating conditions (AOCs) that may occur during the procedure and know the proper responses to them. Personnel must also understand the company's safety policies and procedures, including the proper use of personal protective equipment (PPE).

2.1.1 Prerequisites and Competencies

Personnel must possess the following basic knowledge and skills prior to inspecting, testing, and maintaining temperature transmitters:

- Test equipment skills and knowing how to operate the following test equipment correctly:
  - Handheld communicators
  - Multifunction calibrators
  - RTD simulators
  - Multimeters
  - Decade boxes
  - Certified thermometers
- Skill and safety with electrical systems
- Familiarity with the control system and its instrumentation, including alarm, control, indication, and recording functions

2.1.2 Abnormal Operating Conditions

Personnel inspecting, testing, and maintaining temperature transmitters must be able to identify potential abnormal operating conditions (AOCs) and know the proper response to each, should they occur. For example, if the device is not functioning or is not communicating with the end device, the pipeline E&I technician must alert the control center or other appropriate personnel to take the requisite corrective action.

2.2.0 Inspecting, Testing, and Maintaining Temperature Transmitters

The following steps describe the procedure for inspecting, testing, and maintaining temperature transmitters:

Step 1 Identify potential abnormal operating conditions that may occur, and know the appropriate actions to take in response to them.

Step 2 Utilize the appropriate personal protective equipment according to relevant company procedures.

Step 3 Determine the device's number and nameplate data in order to verify that it is the correct unit to calibrate/maintain and that any data coming from the unit will be properly identified as such.

Step 4 Before beginning the calibration process, check that the test equipment has itself been certified, calibrated, and verified. Some technicians will not use test equipment until it has been tested on a known source, even if its calibration is current and properly documented.

Step 5 Check the appropriate company documentation to confirm that the transmitter's input and output ranges are appropriately matched to the end device.

Step 6 Also before beginning the process, notify the control center, local personnel, and any other personnel who might be affected by the calibration/maintenance procedure. The control center may have to provide feedback about transmitter values or alarms during the process. It may also have to override automatic shutdowns or safety systems to protect the process from interruption. These events are often known as Critical Interlock and Trips (CIATs). They usually require higher authorization to override. Pipeline E&I technicians should be aware of these rules and know what they're allowed to do.

Step 7 Visually inspect the device and any associated equipment. Check for mechanical condition, corrosion, leakage, electrical connection problems, or problems with associated electronics.

Step 8 Obtain an accurate reference temperature measurement from the process by using a certified thermometer. Insert the thermometer into an appropriate thermowell or test point. The process product must be flowing and stabilized.

Step 9 Determine if transmitter calibration is required. To do this, check the reference temperature value against the transmitter output as well as the end device reading. If the values are not within an acceptable tolerance of each other (as specified by company documentation), calibration is needed.

Step 10 Document the current results, including all data.

Step 11 Before beginning calibration, use the current live process temperature value to override the sensor data. This value will keep the control system operating normally during the calibration process. Follow company procedure.

Step 12 Disconnect the transmitter's inputs from the sensor and attach the test equipment. Confirm that the range values determined in Step 5 are correct.

Step 13 Use the test equipment to supply simulated sensor data to the transmitter. Adjust the transmitter according to the manufacturer's and company's guidelines. Repeat this step as often as necessary to bring the transmitter into proper calibration.

Step 14 Disconnect the test equipment and restore the unit's normal connections.

Step 15 Terminate the override temperature value established in Step 9.

Step 16 Repeat Step 7 and perform any sensor trim adjustments required. Follow the manufacturer's and company's documentation for specific guidelines.

Step 17 Document all procedures performed as well as the current data collected at the time that the procedure was completed. Normally, you will also need to include copies of the paperwork documenting the calibration status of the test instruments used.

Step 18 Perform all steps required to resume normal operation.

Step 19 Notify the control center, local personnel, and any affected personnel that the unit has been returned to normal operation.

Step 20 Complete appropriate documentation as required by operator's procedures.

Additional Resources

It is assumed that you will learn to perform these procedures under the direct supervision of your crew leader, foreman, or instructor. For independent study, refer to the recommended Additional Resources at the back of this module.

2.0.0 Section Review

1. With which instrument should the pipeline E&I technician be skilled when obtaining a process reference temperature?
   a. A thermocouple heat block
   b. A multifunction calibrator
   c. An RTD simulator
   d. A certified thermometer

2. What must be done before taking a temperature transmitter offline to keep the control system working correctly?
   a. Enter an override value
   b. Disable the alarm
   c. Set SCADA to ignore mode
   d. Insert a calibrator into the circuit

SUMMARY

Keeping temperature transmitters working reliably is key to having accurate information about pipeline product. Temperature transmitters are also important for monitoring various pieces of equipment to prevent heat-related damage. While temperature sensors are relatively simple and require little maintenance, temperature transmitters can require periodic calibration and adjustment. This module summarized the various maintenance issues and provided a basic overview of the procedure for inspecting, testing, and maintaining these devices.

Review Questions

1. In order to improve sensor responsiveness, a thermowell is often filled with _____.
   a. water
   b. mineral oil
   c. thermal transfer material
   d. dry nitrogen gas

2. What kind of temperature sensor is made up of two different metals?
   a. A thermocouple
   b. An RTD
   c. A mercury vapor capsule
   d. An IGBT

3. A thermocouple's electrical resistance should _____.
   a. vary with temperature
   b. be less than 1Ω
   c. be at least 1,000Ω
   d. be extremely large at room temperature

4. Which of the following is a test instrument that the pipeline E&I technician should be familiar with when maintaining temperature transmitters?
   a. A certified hydrometer
   b. A power factor meter
   c. A multifunction calibrator
   d. A capacitance box

5. In order to know that the data being read is coming from the correct temperature transmitter, the pipeline E&I technician must first _____.
   a. talk to the control room
   b. perform a fieldbus scan
   c. ping the transmitter with a handheld communicator
   d. determine the device number and nameplate data

6. During calibration, what does the connected test instrument supply to the transmitter?
   a. Calibration offsets
   b. Simulated sensor data
   c. Ramped waveforms
   d. A modulated sine wave

Trade Terms Quiz

Fill in the blank with the correct term that you learned from your study of this module.

1. A(n) ________ is a temperature sensor composed of junctions of dissimilar metals. They express temperature by a corresponding voltage.
2. The device that is the ultimate recipient of a transmitter's information is called a(n) ________. Common types of these devices include SCADA systems, PLCs, DCSs, and flow computers.
3. A(n) ________ refers to an instrument used to calibrate a wide variety of electronic devices. Some of these instruments also have pneumatic calibration capabilities.
4. An alert from the control system that a process parameter has moved out of its normal range that can be audible, visual, or both is called a(n) ________.
5. A(n) ________ is a simple digital communications network and protocol set designed to connect instruments, controllers, and other intelligent devices together. It has several different standards.
6. A thermally conductive penetration into a pipeline that allows a temperature sensor to read the process temperature while remaining physically isolated from the process material is called a(n) ________.
7. ________ is the property of an instrument that refers to the minimum and maximum values that the instrument can measure.
8. ________ is the property of an instrument that refers to the difference the instruments upper and lower measurement limits.
9. Adjusting an instrument so that its output signal is representative of the non-adjustable input signal at any given point is to ________. It necessarily includes using adjusted instruments traceable to a national standards body.
10. A temperature sensor that expresses temperature by a corresponding electrical resistance is called a(n) ________.
11. Conditions within a pipeline system that may indicate (1) a malfunction of a component or (2) a condition exceeding design limits or resulting in a hazard to persons, property, or the environment are called ________.
12. A device that conditions a sensor's raw signal value to a form suitable for the control or monitoring system to which it's connected is called a(n) ________. These devices are based on a number of different underlying technologies.

Trade Terms

Abnormal operating conditions (AOCs)
Alarm
Calibrate
End device
Fieldbus
Multifunction calibrator
Range and span
Resistance temperature detector (RTD)
Thermocouple
Thermowell
Transmitter

Trade Terms Introduced in This Module

Abnormal operating conditions (AOCs): Conditions within a pipeline system, identified by the operator, that may indicate (1) a malfunction of a component or (2) a deviation from normal operations that may indicate a condition exceeding design limits or result in a hazard to persons, property, or the environment.

Alarm: An alert from the control system that a process parameter has moved out of its normal range. Alarms can be audible, visual, or both.

Calibrate: The task of adjusting an instrument so that its output signal is representative of the non-adjustable input signal at any given point. Calibration necessarily includes using calibrated instruments traceable to a national standards body.

End device: The device that is the ultimate recipient of a transmitter's information. Common end devices include SCADA systems, PLCs, DCSs, and flow computers.

Fieldbus: A simple digital communications network and protocol set designed to connect instruments, controllers, and other intelligent devices together. There are several different fieldbus standards.

Multifunction calibrator: Refers to an instrument used to calibrate a wide variety of electronic devices. Some multifunction calibrators also have pneumatic calibration capabilities.

Range and span: Two properties of an instrument. Range refers to the minimum and maximum values that the instrument can measure. Span refers to the difference between the instrument's upper and lower measurement limits.

Resistance temperature detector (RTD): A temperature sensor that expresses temperature by a corresponding electrical resistance.

Thermocouple: A temperature sensor composed of junctions of dissimilar metals. Thermocouples express temperature by a corresponding voltage.

Thermowell: A thermally conductive penetration into a pipeline that allows a temperature sensor to read the process temperature while remaining physically isolated from the process material.

Transmitter: A device that conditions a sensor's raw signal value to a form suitable for the control or monitoring system to which it's connected. Transmitters are based on a number of different underlying technologies.

Additional Resources

The following recommended resources can provide additional helpful information related to the requirements for this covered task:

API Recommended Practice 1161, Third Edition, © 2014 American Petroleum Institute. API Publishing Services, 1220 L Street, NW, Washington, DC 20005. Distributed by Thompson Reuters (Scientific) LLC, **www.techstreet.com**.

49 *CFR* 192, **www.ecfr.gov**.

49 *CFR* 195, **www.ecfr.gov**.

The Condensed Handbook of Measurement and Control, N. E. Battikha. Third Edition. 2006. Research Triangle Park, NC: The International Society of Automation.

NCCER Module AOCCC-17, *Abnormal Operating Conditions Control Center*

NCCER Module AOCFG-17, *Abnormal Operating Conditions Field and Gas*

NCCER Module 60102-02, *Pipeline E&I Safety*

NCCER Module 64202-02, *Use of Meters and Test Equipment*

NCCER Module CT44_3-17, *Inspect, Test, and Maintain Flow Computer for Hazardous Liquid Leak Detection*

NCCER Module CT63_1-17, *Start-up of a Liquid Pipeline (Field)*

NCCER Module CT63_2-17, *Shutdown of a Liquid Pipeline (Field)*

NCCER Module CT63_4-17, *Locally Operate Valves on a Liquid Pipeline System*

The following websites offer resources for products and training:

The International Society of Automation, **www.isa.org**

Figure Credits

© Volodymyr Khotenko/Dreamstime.com, Module Opener

Emerson Process Management, Figures 6 and 9

Fluke Corporation, Figure 7

Section Review Answer Key

Answer	Section Reference	Objective
Section One		
1. b	1.1.2	1a
2. d	1.2.3	1b
Section Two		
1. d	2.1.1	2a
2. a	2.2.0	2b

NCCER CURRICULA — USER UPDATE

NCCER makes every effort to keep its textbooks up-to-date and free of technical errors. We appreciate your help in this process. If you find an error, a typographical mistake, or an inaccuracy in NCCER's curricula, please fill out this form (or a photocopy), or complete the online form at **www.nccer.org/olf**. Be sure to include the exact module ID number, page number, a detailed description, and your recommended correction. Your input will be brought to the attention of the Authoring Team. Thank you for your assistance.

Instructors – If you have an idea for improving this textbook, or have found that additional materials were necessary to teach this module effectively, please let us know so that we may present your suggestions to the Authoring Team.

NCCER Product Development and Revision

13614 Progress Blvd., Alachua, FL 32615

Email: curriculum@nccer.org

Online: www.nccer.org/olf

❑ Trainee Guide ❑ Lesson Plans ❑ Exam ❑ PowerPoints Other ____________________

Craft / Level: Copyright Date:

Module ID Number / Title:

Section Number(s):

Description:

Recommended Correction:

Your Name:

Address:

Email: Phone:

CT55_0-17

Maintain Fixed Gas Detection Equipment

Overview

There are many types of analytical equipment located on a typical pipeline. These instruments work with Supervisory Control and Data Acquisition (SCADA) systems to monitor and analyze properties of the pipeline contents, as well as to detect the presence or absence of specific elements in or around the pipeline equipment. Analytical equipment can be used to test atmospheres and as personal protective devices to indicate when hazardous conditions exist. This module describes analytical equipment and explains how gas detection equipment is maintained.

Trainees with successful module completions may be eligible for credentialing through the NCCER Registry. To learn more, go to **www.nccer.org** or contact us at 1.888.622.3720. Our website has information on the latest product releases and training, as well as online versions of our *Cornerstone* magazine and Pearson's product catalog.

Your feedback is welcome. You may email your comments to **curriculum@nccer.org**, send general comments and inquiries to **info@nccer.org**, or fill in the User Update form at the back of this module.

This information is general in nature and intended for training purposes only. Actual performance of activities described in this manual requires compliance with all applicable operating, service, maintenance, and safety procedures under the direction of qualified personnel. References in this manual to patented or proprietary devices do not constitute a recommendation of their use.

CT55_0-17

MAINTAIN FIXED GAS DETECTION EQUIPMENT

Objectives

When you have completed this module, you will be able to do the following:

1. Explain the purpose of analyzers in the pipeline industry and identify common types of analytical equipment.
   a. Explain the purpose of analyzers in the pipeline industry.
   b. Identify the types of analyzers commonly found in a pipeline setting.
2. Describe the operation and maintenance of hydrogen sulfide and sulfur analyzers.
   a. Describe the operation of hydrogen sulfide and sulfur analyzers.
   b. Describe how to use and maintain hydrogen sulfide and sulfur analyzers.
3. Describe the operation and maintenance of chromatographs.
   a. Describe the operation of chromatographs.
   b. Describe how to use and maintain chromatographs.
4. Describe the operation and maintenance of other types of pipeline analyzers.
   a. Describe the operation and maintenance of moisture analyzers.
   b. Describe the operation and maintenance of continuous emissions monitoring systems.
   c. Describe the operation and maintenance of centrifuges.
5. Describe the operation and maintenance of fixed gas detection equipment.
   a. Describe the operation of fixed gas detection equipment.
   b. Identify the prerequisites, competencies, and abnormal operating conditions (AOCs) associated with maintaining fixed gas detection equipment.
   c. Describe the procedures used to maintain fixed gas detection equipment.

Performance Task

Under the supervision of the instructor, you should be able to do the following:

1. Maintain fixed gas detection equipment (CT55_0-17).
   - Identify potential abnormal operating conditions (AOCs) that may occur during performance of this CT, and know the appropriate actions to take in response to them.
   - Utilize the appropriate personal protective equipment according to relevant company procedures.
   - Make appropriate notifications prior to beginning the inspection.
   - Verify proper operations, including the mounting, filters, cable connections, sensor placement, and backup power supply.
   - Check setpoints and alarms.
   - Calibrate gas detector (if needed).
   - Verify sensing elements and replace if necessary.
   - Make appropriate notifications upon completion of the inspection.
   - Complete appropriate documentation as required by operator's procedures.

Trade Terms

Abnormal operating conditions (AOCs)
Chemiluminescence
Fluorescence
Hygrometer
Joule-Thomson effect
Lower explosive limit (LEL)
Photomultiplier tube
Relative centrifugal force (RCF)

Industry Recognized Credentials

If you are training through an NCCER-accredited sponsor, you may be eligible for credentials from NCCER's Registry. The ID number for this module is CT55_0-17. Note that this module may have been used in other NCCER curricula and may apply to other level completions. Contact NCCER's Registry at 888.622.3720 or go to **www.nccer.org** for more information.

Contents

1.0.0 Analyzer Overview ... 1
1.1.0 The Purpose of Analyzers ... 1
1.2.0 Types of Analyzers ... 1
1.2.1 Fixed Analyzers ... 1
1.2.2 Portable Analyzers ... 2
1.2.3 Personal Analyzers ... 2
2.0.0 Hydrogen Sulfide and Sulfur Analyzers ... 4
2.1.0 Operation of Hydrogen Sulfide and Sulfur Analyzers ... 4
2.1.1 Hydrogen Sulfide and Sulfur Testing Methods ... 4
2.1.2 Lead Acetate Reaction Method ... 5
2.1.3 X-Ray Fluorescence (XRF) Method ... 5
2.1.4 Flame Photometric Detection Method ... 6
2.1.5 Chemiluminescence Method ... 8
2.2.0 Using and Maintaining Hydrogen Sulfide and Sulfur Analyzers ... 8
2.2.1 Maintaining H2S and Sulfur Analyzers ... 9
2.2.2 Calibrating H2S and Sulfur Analyzers ... 9
2.2.3 Repairing H2S and Sulfur Analyzers ... 10
2.2.4 Installing H2S and Sulfur Analyzers ... 11
3.0.0 Chromatographs ... 12
3.1.0 Chromatograph Operation ... 12
3.1.1 Sample Introduction Units ... 12
3.1.2 Fractionating/Capillary Columns ... 12
3.1.3 Detectors ... 13
3.1.4 Recording Devices ... 13
3.2.0 Using and Maintaining Chromatographs ... 13
3.2.1 Maintaining Chromatographs ... 13
3.2.2 Calibrating Chromatographs ... 14
3.2.3 Repairing Chromatographs ... 15
3.2.4 Installing Chromatographs ... 15
4.0.0 Other Pipeline Analyzers ... 17
4.1.0 Moisture Analyzers ... 17
4.1.1 Maintaining Moisture Analyzers ... 17
4.1.2 Calibrating Moisture Analyzers ... 18
4.1.3 Repairing Moisture Analyzers ... 19
4.1.4 Installing Moisture Analyzers ... 19
4.2.0 Continuous Emissions Monitoring Systems ... 20
4.3.0 Centrifuges ... 20
4.3.1 Troubleshooting the Centrifuge ... 20
4.3.2 Sample Pre-Heater ... 21

5.0.0 Fixed Vapor and Combustible Gas Detectors 21
5.1.0 Operation of Fixed Vapor and Combustible Gas Detectors 22
5.1.1 Catalytic Detectors 23
5.1.2 Semiconductor Detectors 23
5.1.3 Infrared Detectors 23
5.2.0 Prerequisites, Competencies, and Abnormal Operating Conditions 24
5.2.1 Prerequisites and Competencies 24
5.2.2 Abnormal Operating Conditions 24
5.3.0 Using and Maintaining Fixed Vapor and Combustible Gas Detectors 25
5.3.1 Maintaining Fixed Vapor and Combustible Gas Detectors 25
5.3.2 Calibrating Fixed Vapor and Combustible Gas Detectors 26
5.3.3 Repairing Fixed Vapor and Combustible Gas Detectors 27
5.3.4 Installing Fixed Vapor and Combustible Gas Detectors 27

Figures

Figure 1 Fixed analyzer 1
Figure 2 Portable analyzer 2
Figure 3 Personal analyzer 2
Figure 4 Lead acetate tape H2S analyzer flow diagram 5
Figure 5 Lead acetate tape sulfur analyzer flow diagram 6
Figure 6 Tape-type analyzer 6
Figure 7 Typical XFR sulfur analyzer flow diagram 7
Figure 8 Flame photometric detector 7
Figure 9 Typical chemiluminescence method process flow 8
Figure 10 Typical chemiluminescence sulfur analyzer 8
Figure 11 Gas chromatograph 12
Figure 12 Block diagram of typical chromatograph 13
Figure 13 Moisture analyzer 17
Figure 14 Centrifuge tubes used in the grind out 21
Figure 15 Panel-mounted combustible gas detector 23
Figure 16 Catalytic sensors 23
Figure 17 Semiconductor detector 24
Figure 18 Infrared detector 24
Figure 19 Infrared detector operation 25
Figure 20 Effect of air currents 28

SECTION ONE

1.0.0 ANALYZER OVERVIEW

Objective

Explain the purpose of analyzers in the pipeline industry and identify common types of analytical equipment.

a. Explain the purpose of analyzers in the pipeline industry.
b. Identify the types of analyzers commonly found in a pipeline setting.

Analytical equipment provides information used for maintaining product quality, pipeline integrity, and pipeline safety. The efficiency and accuracy of any analytical device depends on well-trained personnel who can properly use and interpret the results supplied by the equipment. This section provides an overview of common types of analyzers used in the pipeline industry.

1.1.0 The Purpose of Analyzers

There are various substances that must be detected, sampled, and analyzed in the pipeline environment. Analyzers are designed to identify the presence of specific substances, and some can determine the quantity of a detected substance as well. In certain situations, detection is a critical function that alerts personnel to specific hazards. In other cases, analyzers are used to assess the quality of the product.

This module explains the operation of the following types of pipeline field analytical equipment:

- Hydrogen sulfide and sulfur analyzers
- Chromatographs
- Moisture analyzers
- Continuous emissions monitoring systems
- Centrifuges
- Vapor detectors
- Combustible gas detectors

1.2.0 Types of Analyzers

There are several types and uses of analytical equipment with which the typical pipeline E&I technician must be familiar. Pipeline analytical equipment falls into three major categories: fixed analyzers, portable analyzers, and personal analyzers.

1.2.1 Fixed Analyzers

Fixed analyzers (*Figure 1*) are installed in specific operating areas, such as stations and pump rooms, to provide analytical data from samples taken from the process stream or to provide analytical data from atmospheric samples around the equipment.

Fixed process analyzers are typically installed near the pipeline and are designed to periodically remove a sample amount of the product from the pipeline for analysis. These analyzers are used to record product quality data and to transmit this data to the control center. The control center uses this information for quality and process control purposes.

Fixed analyzers are used to continuously monitor the atmosphere in pipeline facilities and around the pipeline equipment for hazardous situations. These analyzers provide monitoring and personnel protection capabilities for facilities and equipment. Various types of sensors are

Figure 1 Fixed analyzer.

used to monitor for oxygen deficiency, fire, or any buildup of hazardous vapor or combustible gas. These sensors provide atmospheric monitoring data to the control center and will notify the controller of alarm conditions.

1.2.2 Portable Analyzers

Portable analyzers (*Figure* 2) are typically small, mobile, multi-gas detectors used by technicians to detect the presence of hazardous atmospheric conditions prior to entering an area, or for monitoring atmospheric conditions while maintenance is being performed in the area. These devices are typically battery-powered. They are designed to detect several hazardous atmospheric conditions, such as oxygen deficiency or the presence of combustible or toxic gases. Most portable analyzers provide data logging. If the analyzer detects the presence of a hazardous atmosphere, an audible alarm will be sounded and a visual alarm will be activated to alert the technician.

1.2.3 Personal Analyzers

Personal analyzers (*Figure* 3) are small, pocket-sized monitors that are worn by technicians to detect the presence of hazardous gases or vapors. These analyzers are typically designed to detect a specific substance and are clipped onto the exterior of the technician's clothing. They can provide a simple indication of the target substance or provide an emergency alert if the atmosphere is immediately dangerous to life and health (IDLH).

To ensure that the atmosphere being detected is the same as that which the technician is breathing, these personal detectors should be kept as close to the breathing zone as possible. Replacement sensors as well as sensors for different gas types are available from the manufacturer.

Figure 2 Portable analyzer.

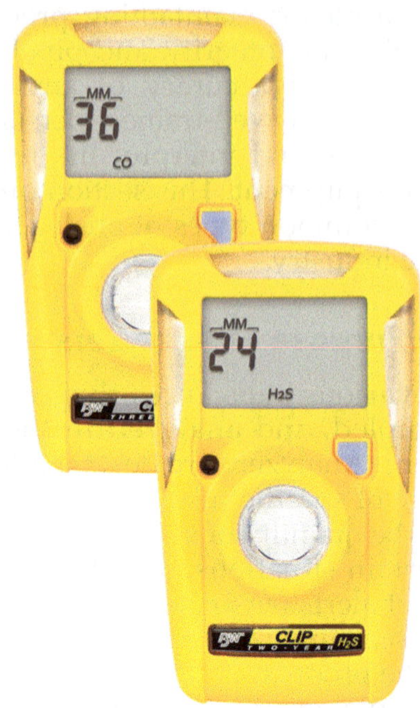

Figure 3 Personal analyzer.

Additional Resources

It is assumed that you will learn to perform these procedures under the direct supervision of your crew leader, foreman, or instructor. For independent study, refer to the recommended Additional Resources at the back of this module.

1.0.0 Section Review

1. One important purpose of analyzers is to alert pipeline personnel to _____.
   a. theft
   b. defects
   c. safety hazards
   d. sabotage

2. Personal analyzers are usually sensitive to a specific type of _____.
   a. gas
   b. liquid
   c. radiation
   d. light

Section Two

2.0.0 Hydrogen Sulfide and Sulfur Analyzers

Objective

Describe the operation and maintenance of hydrogen sulfide and sulfur analyzers.

a. Describe the operation of hydrogen sulfide and sulfur analyzers.
b. Describe how to use and maintain hydrogen sulfide and sulfur analyzers.

Trade Terms

Chemiluminescence: The emission of light as a result of a chemical reaction. Chemiluminescence may be used to identify specific substances by the wavelengths of the light emitted.

Fluorescence: The tendency for some substances to emit light (or other EM radiation) when exposed to light (or other EM radiation) of a higher wavelength.

Photomultiplier tube: A highly sensitive light detection sensor used in certain kinds of instrumentation. Photomultipliers can amplify a light signal by a factor of 100 million.

The pipeline industry uses hydrogen sulfide and sulfur analyzers for product quality and process control purposes. The presence of hydrogen sulfide (H_2S) and sulfur (S) in the pipeline product determines the grade of the product and may necessitate the use of appropriate equipment and procedures in handling the product. Hydrogen sulfide analyzers and sulfur analyzers are almost identical in operation, and both substances can be detected using the same analytical methods.

2.1.0 Operation of Hydrogen Sulfide and Sulfur Analyzers

Hydrogen sulfide is a flammable, colorless, gaseous compound of hydrogen and sulfur, which, in small amounts, has the odor of rotten eggs. It is formed naturally by the bacterial decomposition of organic animal and vegetable matter. It is found in crude oil, natural gas, sewers, and swamps and is created by a number of industrial processes.

Hydrogen sulfide is extremely corrosive and poisonous. It can cause damage to the eyes, skin, breathing passages, and lungs, attacking and paralyzing the central nervous system, especially the parts that control the lungs and heart. In large amounts after a period of time, it deadens the sense of smell, making it even more dangerous as personnel may believe that it isn't present since they can't smell it.

WARNING!

Hydrogen sulfide is an extremely dangerous compound, and exposure is deadly!

Sulfur is a common component in most petroleum, but the amount of sulfur must be determined to make the correct decisions about the product processing. The sulfur analyzers located on the pipeline provide that information. Low sulfur content is desirable in petroleum products. High sulfur content in a petroleum product makes it corrosive, environmentally hazardous when burned, and a serious safety hazard.

Most analyzer software is designed to operate with computer interfaces and notebook computers. The alarm setpoints and software parameters can be customized in the field, and self-calibration and fault reporting are typically a part of the programming. Modern units are capable of communication with the control center during system upsets. Typical pipeline hydrogen sulfide analyzers still use the old-style tape machine to detect the presence of the gas. Improvements have been made to these analyzers, including some analyzers that can sample two streams at once.

Many of the methods for testing and detecting concentrations of hydrogen sulfide and sulfur are standardized by the ASTM International. Most analyzer manufacturers build their detection and analysis systems to meet the ASTM standards for these methods. New methods of detection are typically submitted to the ASTM for testing and standardization.

2.1.1 Hydrogen Sulfide and Sulfur Testing Methods

There are several methods for analyzing H_2S and sulfur content in petroleum products. ASTM standards cover some of these testing methods, but others have yet to be standardized. The common sulfur testing methods used in the pipeline industry that are covered in this module include:

- Lead acetate reaction method
- X-ray **fluorescence** method
- Flame photometric detection method
- **Chemiluminescence** method

2.1.2 Lead Acetate Reaction Method

The lead acetate reaction method of H_2S and sulfur detection has been one of the most widely used methods in the pipeline industry during the past four decades. This method is outlined in *ASTM D4084-94 (1999) Standard Test Method for Analysis of Hydrogen Sulfide in Gaseous Fuels (Lead Acetate Reaction Method)* and in *ASTM D2420-91 (2001) Standard Test Method for Analysis of Hydrogen Sulfide in Liquefied Petroleum (LP) Gases (Lead Acetate Reaction Method).* Today, this method is less commonly used and largely has been superseded by the X-ray fluorescence method.

The lead acetate reaction method (*Figure 4*) involves exposing a section of specially constructed tape to a sample. The reaction rate, indicated by the darkening of the tape, is measured by an optical sensor.

The tape is white in color, infused with lead acetate, and is typically spooled from a supply roll to a take-up roll. As the measurement cycle begins, the tape is advanced to a new, unexposed section. Immediately, when the tape stops in the detection area, the optical sensor makes an initial reading of the tape color and stores this information for future reference. Simultaneously, the sample gas that has passed through the humidifier is passed either over or through the tape located in the confined detection area. The optical sensor takes readings approximately every second until the tape becomes too dark to provide further analysis.

The section of tape exposed to the sample will darken with the formation of lead sulfide. This darkening will occur at different rates depending on the amount of H_2S present in the sample. The degree of darkening of the color is not important during the measurement, but the rate at which the darkening occurs is very important.

Readings from the optical sensor are digitized and analyzed using the detector's internal software. Detector output may be in traditional 4–20 mA analog form or may be sent out over a digital smart communications system. This signal output is used for data acquisition of sulfur content and also for alarms when the readings exceed the alarm setpoint.

By adapting the H_2S lead acetate reaction process using a conversion furnace, the total concentration measurements are allowed to go down to approximately 10 parts per billion (ppb). In this detection method (*Figure 5*), the sample gas is passed through both a furnace and humidifier before coming in contact with the tape located in the confined detection area. *Figure 6* shows a typical lead acetate tape H_2S and sulfur analyzer.

The lead acetate method has a number of disadvantages, most of them associated with the treated tape. Jamming can result in faulty readings. The tape can run out and not be replaced in a timely manner, leading to failed readings. And the humidifier can run dry, again causing erroneous readings.

2.1.3 X-Ray Fluorescence (XRF) Method

Fluorescence technology can be used to measure the amount of any fluorescent molecule in very small volumes of liquid, or it can be used for chemical analysis of larger volumes of liquid. A

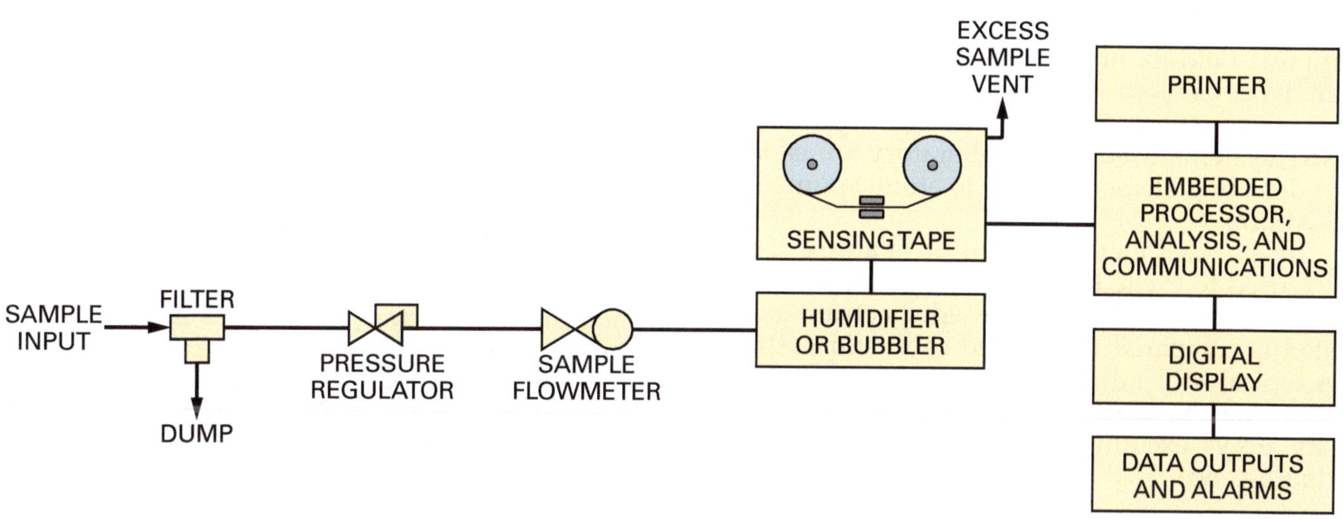

Figure 4 Lead acetate tape H_2S analyzer flow diagram.

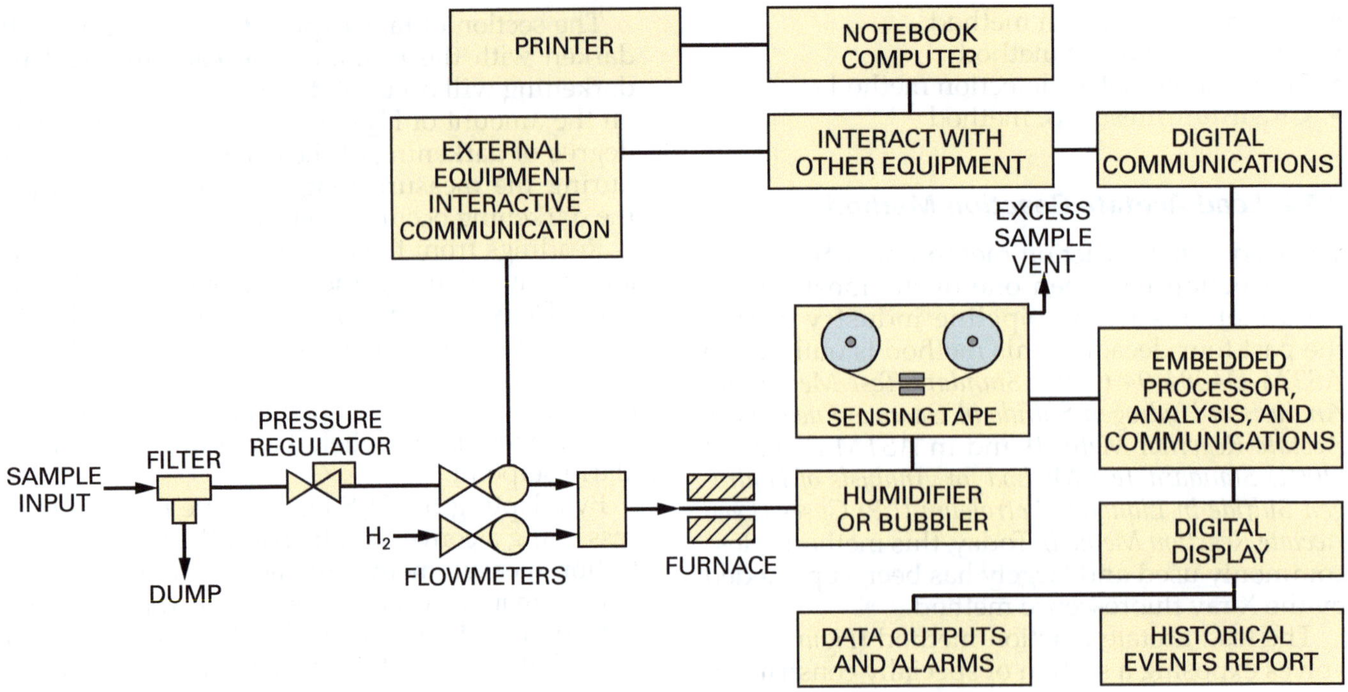

Figure 5 Lead acetate tape sulfur analyzer flow diagram.

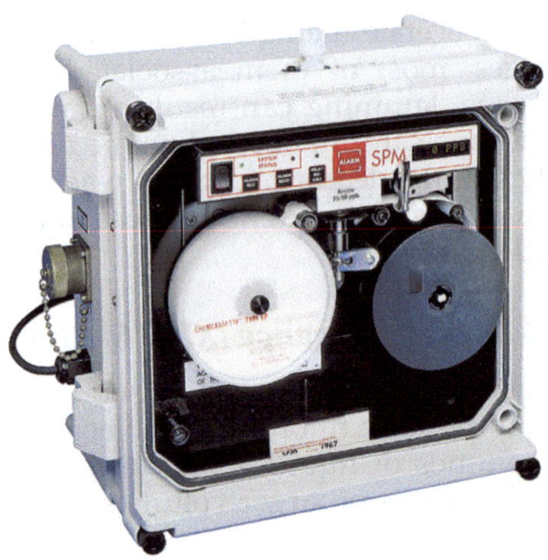

Figure 6 Tape-type analyzer.

typical field-mounted X-ray fluorescence (XRF) analyzer can detect any of the elements between atomic numbers 16 (sulfur) through 92 (uranium). XRF systems used in the laboratory are capable of detecting elements down to atomic number 11 (sodium).

In the XRF method of sulfur analysis, X-rays are used to excite the molecules in the sample to fluorescence, and the processed spectrum generated is compared to a stored calibration setting. A typical XRF sulfur analyzer (*Figure 7*) works by exposing the pipeline test sample to high-intensity X-rays, exciting the elements in the sample to fluorescence.

When the element sulfur is exposed to these X-rays, it will emit EM radiation at energy levels that are unique to sulfur. Once a sample is excited to fluorescence, mechanical filtering materials located between the X-rays and the detector block the unwanted X-ray energies and allow the others to pass through to the detector. The detector, along with its associated circuitry, is used to collect and convert the X-rays that pass through into electronic data. This electronic data is used to determine the energy and intensity emitted from the sample. The spectrum formed by the X-ray energy levels, when sorted into channels, allows identification of the sulfur content of the sample.

Due to the potentially hazardous nature of X-rays, many jurisdictions require pipeline E&I technicians to obtain appropriate training, certification, and licenses before they are allowed to use or service XRF analyzers. Technicians must be familiar with local regulations and have the proper qualifications before working with this technology.

2.1.4 Flame Photometric Detection Method

In this detection method (*Figure 8*), the flame photometric detector (FPD) uses a hydrogen-rich flame to reduce sulfur compounds to the elemental S_2 species that produces a blue chemiluminescent emission in the visible light spectrum. This blue light emission is then passed through a narrow band optical filter to a photomultiplier tube.

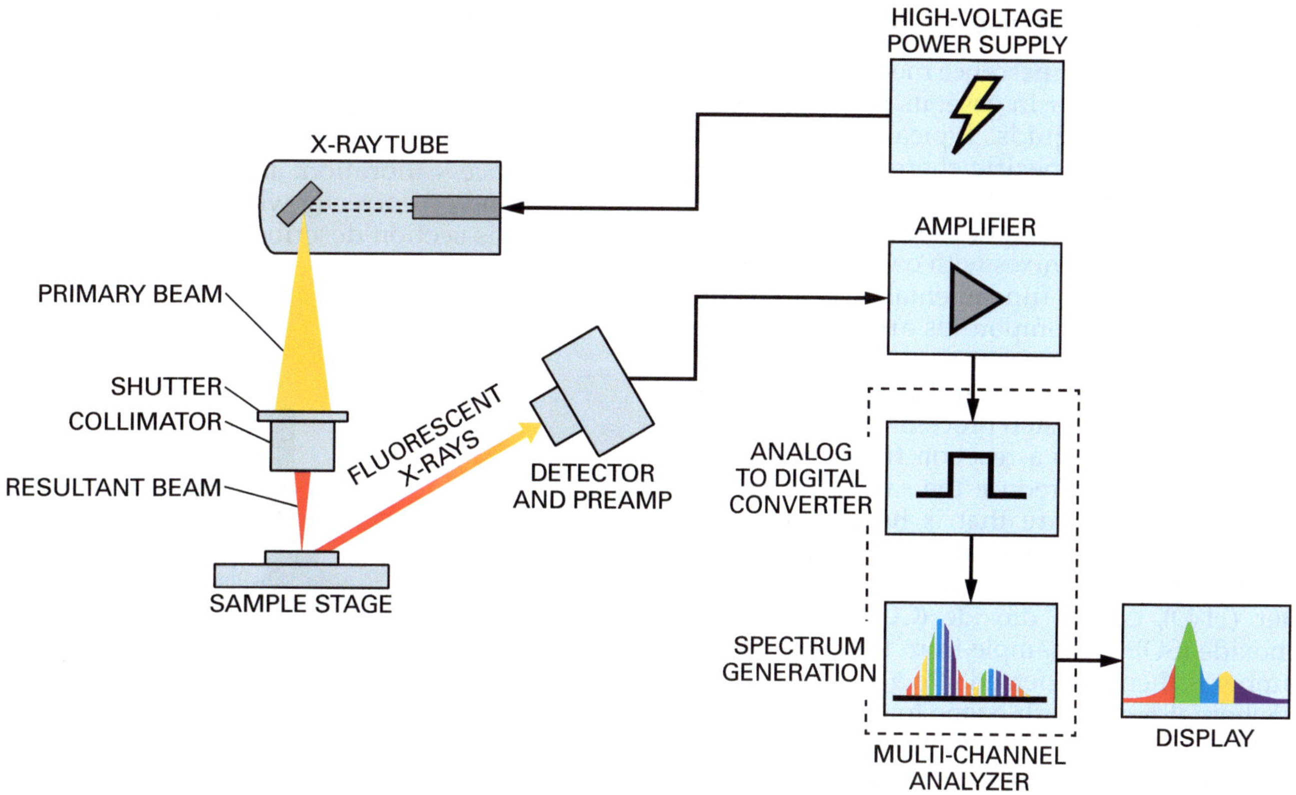

Figure 7 Typical XFR sulfur analyzer flow diagram.

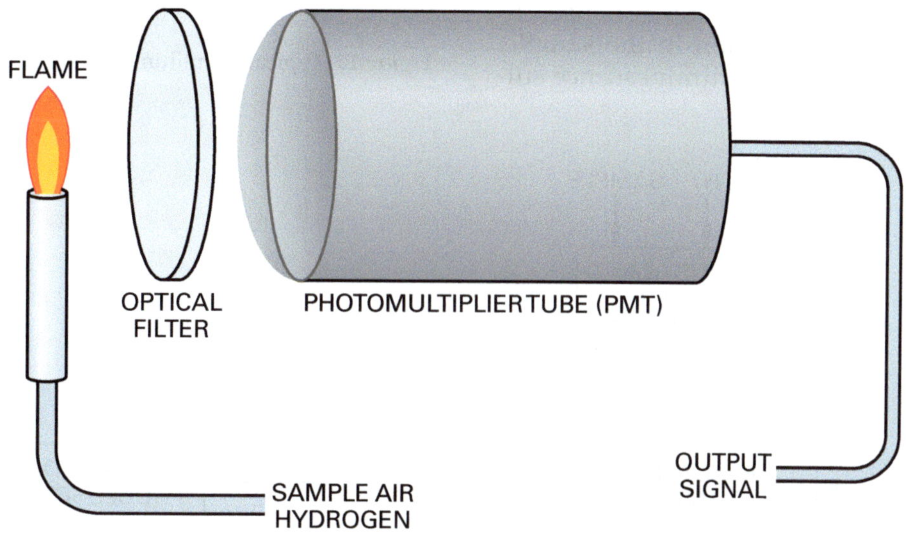

Figure 8 Flame photometric detector.

The signal output from the photomultiplier tube is used to calculate the sulfur concentration in the sample.

The flame photometric method of sulfur detection can be used to detect sulfur levels less than 500 parts per million (ppm) in pipeline samples. Some FPD devices are capable of measuring sulfur levels down to 50 parts per billion (ppb).

In order to reduce condensation buildup in the detector, the FPD block is maintained at a constant temperature of approximately 100°F. To increase the service life of the photomultiplier tube, it is maintained at a cooler operating temperature.

A quenching effect can be caused by hydrocarbons entering the detector, resulting in false flameouts or, possibly, real flameouts. These hydrocarbons can cause erroneous readings and, over time, cause a gradual buildup of carbon deposits in the detector. This buildup reduces the amount of light entering the photomultiplier tube and reduces the detector's sensitivity.

2.1.5 Chemiluminescence Method

The sulfur chemiluminescence method is another way to analyze sulfur in gaseous fuels, gasoline, and other refined liquids. Typical measurement limits for a sulfur-specific chemiluminescence detector are 10 ppm to 5,000 ppm. In the chemiluminescence method, light produced when partially burned sulfur mixes with ozone in a partial vacuum provides the fundamental measurement. All sulfur-bearing compounds are measured as a single peak reading that represents the total sulfur content. Performing sulfur chemiluminescence involves a two-step process (*Figure 9*).

The first step uses a reaction furnace or combustion chamber to reduce the sample in an air and hydrogen mixture that is held under vacuum. The sulfur components in the combustion chamber are oxidized into sulfur dioxide (SO_2), water (H_2O), carbon dioxide (CO_2), and sulfur monoxide (SO). The sample from the combustion chamber is then transferred into a reaction chamber where it reacts with ozone to generate sulfur dioxide and light. The light generated by this reaction is measured by a photomultiplier tube, and the output signal generated is linearly proportional to the quantity of sulfur contained in the sample. The detector output signal is used to calculate the sulfur concentration in the sample. *Figure 10* shows a typical chemiluminescence sulfur analyzer.

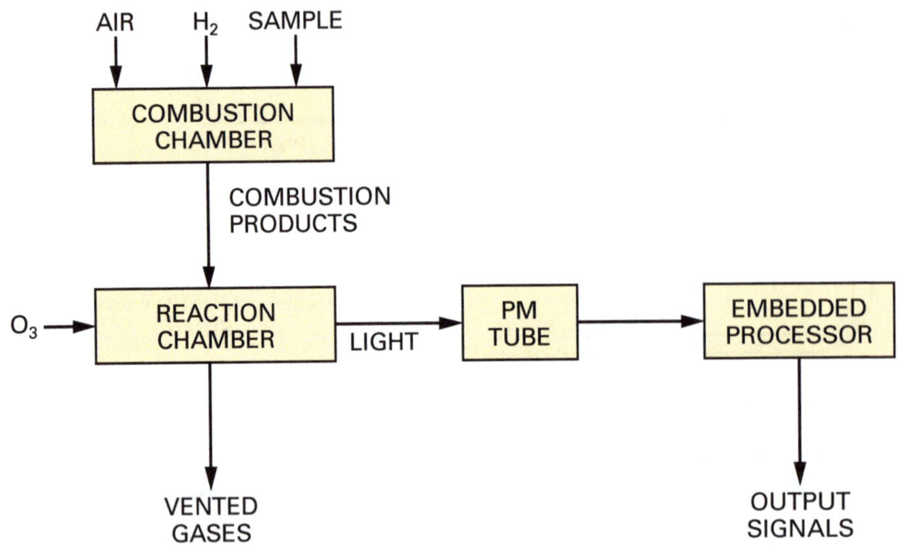

Figure 9 Typical chemiluminescence method process flow.

2.2.0 Using and Maintaining Hydrogen Sulfide and Sulfur Analyzers

Hydrogen sulfide and sulfur analyzers require a certain amount of routine care in the form of maintenance, calibration, and repair. They must also be handled correctly when installed and operated. This section describes some of these processes.

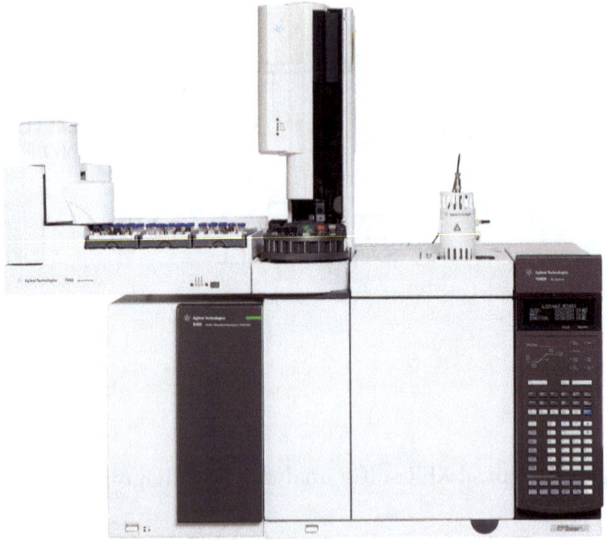

Figure 10 Typical chemiluminescence sulfur analyzer.

2.2.1 Maintaining H2S and Sulfur Analyzers

Periodic maintenance is required to ensure the proper operation and detection ability of the analyzer. For example, the lead acetate H_2S or sulfur analyzer requires periodic removal and replacement of the tape supply in the sensing unit. This must be performed based on usage rates for the tape. The lenses and optical sensor also require periodic cleaning.

Typical maintenance procedures for these analyzers include the following steps:

Step 1 Notify the operations control center and/or affected personnel before performing any maintenance on the system.

Step 2 If the system is currently in the alarm condition, do not enter the area until you have confirmed that it is safe. Don't assume a false alarm due to system malfunction. Treat alarms as genuine until you can confirm otherwise.

Step 3 Ensure that the operations controlled by the analyzer are clearly understood.

Step 4 Ensure that the sensor mounting is secure.

Step 5 Ensure that the sensor filter is clear of water, oil, dust, and paint that might clog it.

Step 6 Check for tubing leaks. If any are found, correct the problem and perform a bubble test to verify that the problem is fixed.

Step 7 Check all cable connections for tightness or possible damage.

Step 8 Ensure that the system is supplied with a stable carrier gas.

Step 9 Check that all sensor placements are up-to-date with the plant processes.

Step 10 Ensure that the system has a backup power supply. Since many backup systems rely on batteries, be sure to schedule periodic battery testing and replacement to ensure constant and reliable operation.

Step 11 Verify the data to SCADA and local readouts. The data may be digital or analog.

Step 12 Document your work using the appropriate forms and following company standard procedure. Include a copy of the bottled gas certification to confirm that the gas was correct and not past its expiration date. If a certificate is not available, include a photo of the bottle label.

2.2.2 Calibrating H2S and Sulfur Analyzers

Most modern H_2S and sulfur analyzers require periodic testing and calibration to maintain proper operation. Company policy will require periodic inspection and calibration to ensure that the analyzers are operating as designed, and most companies have detailed procedures on how to perform equipment testing.

Modern technology has improved the troubleshooting and testing of these devices. Local and remote fault indications that are produced by the analyzer provide immediate information on the status of the analyzer's operation. Text displays and indicators located on the analyzer provide fault messages and prompts on how to perform specific procedures, such as calibration.

In most cases, the calibration frequency is determined by company policies or governmental agency regulations. Testing and calibration of any detector must be performed based on the manufacturer's recommended practices. Typically, calibration of an H_2S or sulfur detector will involve either the use of a sample test medium with a known H_2S or sulfur content, or it will involve an H_2S gas generator. This calibration sample is then used to verify the accuracy of the H_2S or sulfur detector measurements.

Test mediums are specifically designed to provide the proper calibration of the analyzer. Some analyzers are automatically calibrated at regular time intervals, and others have the calibration mode activated by another means. When calibrating an H_2S or sulfur analyzer using a reference sample, always take great care, since the sample is highly toxic. This level of caution is especially crucial when working indoors, where gas can easily accumulate to toxic levels. Pipeline E&I technicians should wear an appropriate personal monitor when performing this task to alert them to hazardous situations.

The specific control features and operating parameters for the analyzers can typically be modified in the field by connecting to and interfacing with the analyzer's software through a notebook computer. Additionally, the analyzer's data logs can be accessed, uploaded, or downloaded for analysis or recordkeeping. Some typical steps that may be performed for testing and calibration include the following:

Step 1 Notify the operations control center and/or affected personnel before performing any tests or calibrations on the system.

Step 2 Ensure that the operations controlled by the analyzer are clearly understood.

Step 3 Determine the proper test methods, and verify the calibration status and date of any test equipment. Test methods include the following:

- *Bench condition testing (BCT)* – Isolated completely from the system
- *Static condition testing* (SCT) – In place but isolated electrically and mechanically from process operations
- *Dynamic condition testing* (DCT) – In place and not isolated electrically and mechanically from process operation.

CAUTION

If performing dynamic condition testing, exercise extra caution in applying a setpoint to a device to ensure that the calibration process does not cause an upset or abnormal effect on the system.

Step 4 If necessary, remove the analyzer from service following company policy.

Step 5 Check the alignment of the optics, and clean with a soft cloth. Lenses must be clean and free of dust, dirt, and smears.

Step 6 Balance the resistance of the detecting and reference photocells.

Step 7 Check the analyzer zero reading and the differentiator gain.

NOTE

Steps 6 and 7 will depend upon the brand of analyzer. Many analyzers perform these adjustments automatically.

Step 8 Prepare a calibration gas sample, and introduce the sample into the analyzer.

Step 9 Perform the analyzer test using an approved test medium. If a dynamic condition test is performed, check with the operations control center to see if the analyzer output was received properly.

Step 10 Check the calibration readings, and ensure that they are within the acceptable calibration range. Repeat the calibration check if the first reading is out of range.

Step 11 Return the analyzer to service according to company policy.

Step 12 Notify the operations control center that testing is completed.

Step 13 Document your work using the appropriate forms and following company standard procedure. Include a copy of the bottled gas certification to confirm that the gas was correct and not past its expiration date. If a certificate is not available, include a photo of the bottle label.

2.2.3 Repairing H2S and Sulfur Analyzers

Due to the simple modular designs and built-in diagnostics of H_2S and sulfur analyzers, most repairs can be performed by pipeline E&I technicians. Repairs mainly involve swapping parts or removing the analyzer from service and sending it back to the vendor for repair.

When performing any type of repair on H_2S or sulfur analyzers, the manufacturer's maintenance procedures must be followed. Many analyzers are supplied with warranties, and if an analyzer is not maintained and repaired according to the manufacturer's instructions, the warranty may be voided. Before beginning any repair work on a particular analyzer, check to see if the warranty is still in effect. If the analyzer is still under warranty, any repair work must be performed in an authorized manner. In some cases, the entire analyzer may have to be removed from service and returned to the manufacturer.

If an analyzer is to be repaired by company personnel, they must be trained and qualified. They must also make sure that the manufacturer's instructions are followed in detail. If the analyzer must be removed for service and is not replaced with a spare, company policy may require a Management of Change (MOC) to ensure that the process protection and/or personnel safety is not compromised.

2.2.4 Installing H2S and Sulfur Analyzers

The decision to remove and replace an analyzer will depend on the process, and not all situations will warrant an immediate equipment swap. There are no standard procedures for H_2S or sulfur analyzer installation because the optimum analyzer location will vary for each application, and because installation on the pipeline is typically by engineering design.

Understanding how the equipment is being used in the process is very important. Each installation must be evaluated. Proper installation is critical to the operation of the analyzer and the accuracy of its measurements. The mounting site of an analyzer should be as close as possible to the sampling point. Prior to the installation of an H_2S or sulfur analyzer, refer to the manufacturer's recommendations on proper installation of the components. In general, locate an analyzer:

- Where it will be away from sources of vibration.
- Where it will be away from sources of heat (verify the temperature range of the analyzer).
- Where it will not be in direct sunlight or icy winds.
- Where it will be easy to access for calibration and maintenance.
- Where it will not be contaminated by water buildup or coated with contaminating substances.

These basic steps should be followed for the installation of H_2S and sulfur analyzers:

Step 1 Notify the operations control center and/or affected personnel before performing any installations on the system.

Step 2 Ensure that the operations controlled by the analyzer are clearly understood.

Step 3 Connect sample lines, vent lines, and drain lines. Sample supply tubing should run up-grade to prevent water accumulation and possible blockage from occurring.

Step 4 Make signal and power connections at the terminal block. Signal connections will depend on the technology involved (digital or analog). Do not connect power until all other connections are made. Follow the manufacturer's recommendations on how to connect wiring to the detector.

Step 5 Before applying power to the detector, check all wiring connections and replace all covers.

Step 6 After initially powering up the detector, allow the detector unit to stabilize before performing calibrations. Confirm that it is receiving the proper voltage.

Step 7 Document your work using the appropriate forms and following company standard procedure.

Additional Resources

It is assumed that you will learn to perform these procedures under the direct supervision of your crew leader, foreman, or instructor. For independent study, refer to the recommended Additional Resources at the back of this module.

2.0.0 Section Review

1. Which H_2S/sulfur analyzer method may require the pipeline E&I technician to obtain special training and licensing to operate or service?
   a. Lead acetate reaction
   b. X-ray fluorescence
   c. Flame photometric detection
   d. Chemiluminescence

2. Which form of analyzer testing is performed in place but with the analyzer isolated from the process?
   a. Static condition testing
   b. Dynamic condition testing
   c. Isolated condition testing
   d. Bench condition testing

Section Three

3.0.0 Chromatographs

Objective

Describe the operation and maintenance of chromatographs.

a. Describe the operation of chromatographs.
b. Describe how to use and maintain chromatographs.

A chromatograph can be used to determine the concentration of specific elements in vapors or liquids. The term *chromatography* is used to describe the physical method of separating samples based on the differences in the absorption rate of the substances in the sample.
In experiments performed in 1906, it was found that when a solution of mixed colored pigments was allowed to filter through a column of firmly packed pulverized calcium carbonate, the individual pigments passed down through the column at different rates and could be separated into different color bands. This result was called a *chromatogram*. A modern variation of this principle can be found in paper chromatography, where compounds in solution migrate at different speeds across a sheet of porous paper.

Liquid and gas chromatographs (*Figure 11*) are typically used in the pipeline industry for custody transfer measurement of the energy content and quality of the product. Both liquid and gas chromatographs operate by similar methods. Gas chromatographs separate product samples by passing the sample mixture and a carrier gas (typically helium, hydrogen, or nitrogen) through a column packed with a suitable fractionating agent. As it passes through the column, the fractionating agent causes the various components to be retained for different periods of time. As the sample components emerge from the column, a detection device records the times of emergence, and quantities of the various components can be detected and measured.

Using a specific column and fractionating agent in a system that maintains constant pressure, carrier flow rate, and column and agent temperature, the retention times can all be held constant. Therefore, the time that elapses between sample injection and component detection allows specific components in the sample to be measured. The following sections describe chromatograph equipment and how it operates.

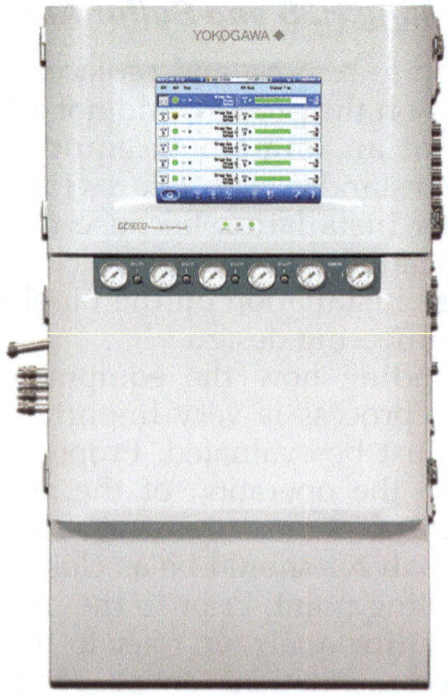

Figure 11 Gas chromatograph.

3.1.0 Chromatograph Operation

The chromatograph is comprised of four major components: the sample introduction units, fractionating/capillary column, detector, and recording device. *Figure 12* shows a block diagram of a typical chromatograph.

3.1.1 Sample Introduction Units

Several types of valves are used by equipment manufacturers for sample injection and also to switch fractionating columns. The sample valves are designed to inject specific volumes of the sample into the fractionating column. Due to the difficulty in handling liquid samples, they are often vaporized prior to entering the sampling valve.

3.1.2 Fractionating/Capillary Columns

Depending on the model and the type of substances the chromatograph must analyze, it will have either a fractionating or capillary column. In either case, the column is the heart of the system. A fractionating column is usually constructed from a copper or stainless steel tube about $\frac{1}{4}$ inch (6-mm) in diameter and from 1 to 50 feet (0.3 to 15 m) long. Adsorption-type fractionating col-

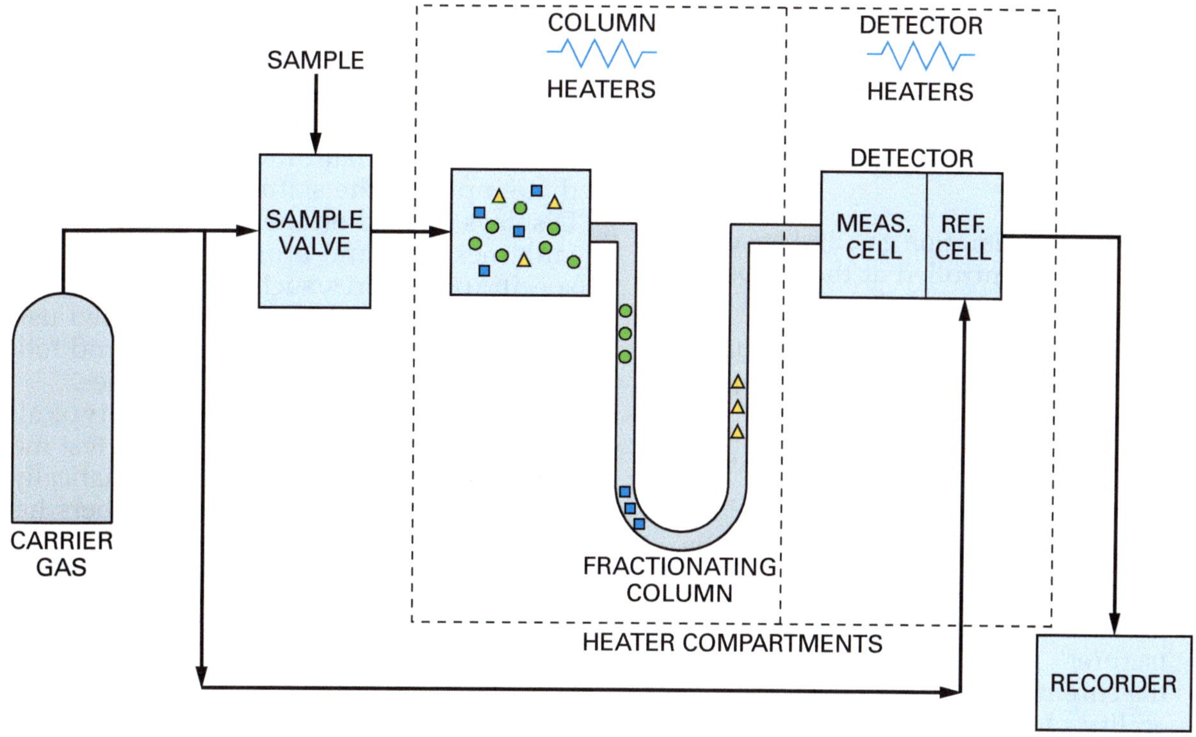

Figure 12 Block diagram of typical chromatograph.

umns are packed with a fine mesh material, such as charcoal, silica gel, or activated alumina. Partition-type fractionating columns are packed with an inert material, such as firebrick or diatomaceous earth, and coated with a nonvolatile liquid called a *partitioner.*

A capillary column is quite different in structure. It consists of tubes with an inside diameter of 0.01-inch (0.3-mm) and a length of 150 to 1,000 feet (46 to 305 m), coiled into a compact helix. The sample introduced into such a column must be kept small in size, but the column is highly selective and has a relatively high speed of response—as fast as a few seconds compared to many minutes for adsorption and partition-type fractionating columns.

3.1.3 Detectors

Detectors may be of several types. Differential thermal-conductivity cells are the most popular because they are relatively inexpensive, sensitive, and stable. They may be comprised of either matched resistance wires or thermistors arranged in a bridge circuit. Other types are the hydrogen flame ionization detector, argon ionization detector, and catalytic combustion detector. The detector causes a recording instrument to indicate a concentration measurement based on the magnitude of the peak reading.

3.1.4 Recording Devices

Recording devices include chart recorders and computers of various types. Detector readings are analyzed using software, and the results are recorded and stored for future reference. Data can also be supplied to the control center through the data acquisition system.

3.2.0 Using and Maintaining Chromatographs

Gas chromatographs require a certain amount of routine care in the form of maintenance, calibration, and repair. They must also be handled correctly when installed and operated. This section describes some of these processes.

3.2.1 Maintaining Chromatographs

Periodic maintenance may be required to ensure the proper operation and detection ability of a chromatograph. Due to the use of printed circuit boards and other components that could be damaged from handling, little maintenance is recommended unless the chromatograph is malfunctioning. When it is necessary to handle a printed circuit board, always take precautions against electrostatic discharge (ESD) damage.

Typical maintenance procedures include the following steps:

Step 1 Verify that the proper component concentrations have been entered in the component data table by checking calibration certificates.

Step 2 Check the calibration gas, and ensure that it is being controlled at the proper temperature, pressure, or flow.

Step 3 To ensure repeatable results, allow the calibration gas sufficient purge time.

Step 4 Visually check the chromatograph for baseline stability and peak separation.

Step 5 Ensure that retention times logged on the chromatograph were entered correctly.

Step 6 Verify the response factors, and check repeated analysis reports against the manufacturer's specified repeatability range. If the chromatograph does not show repeatability, check alarms and perform diagnostics to repair the unit.

Step 7 Check that the calibration gas percentages from the analyzed results match the certification sheet supplied by the calibration gas supplier.

Step 8 Visually monitor the chart and/or printer record to ensure that the programmed functions are performed automatically and that the analyzer is operating normally.

Step 9 Check the chart and/or printer paper supply, and replenish as needed.

Step 10 Check the pressure in the supply gas cylinders, and maintain a record of gas usage and estimated operating times.

Step 11 Check to make sure the carrier gas indications and air pressure indications are showing the proper setpoint.

Step 12 Check sample pressures and flow rates. Make sure to allow proper purging between analyses.

Step 13 Retain documentation for future reference including chromatograms, calibration reports, and repeatability results.

Step 14 Verify the data to SCADA and local readouts. Data may be digital or analog.

Step 15 Document your work using the appropriate forms and following company standard procedure.

3.2.2 Calibrating Chromatographs

Most modern chromatographs require little or no actual testing. Technology has improved the troubleshooting and testing of these devices. Local and remote fault indications provide immediate input on the status of the chromatograph. Displays located on the chromatograph provide fault messages and prompts on how to perform specific procedures, such as calibration. Testing of chromatographs should be performed using the manufacturer's approved method and following all company standards and guidelines.

Calibration of chromatographs is typically performed using a company-approved test medium. Some chromatographs are automatically calibrated at regular time intervals; others have the calibration mode activated by other means.

Control features for the chromatographs can be modified in the field by connecting to and interfacing through a notebook computer. Additionally, data logs can be accessed and downloaded for analysis. Typical analyzer testing includes performing field calibrations per company and manufacturer specifications, as follows:

Step 1 Notify the operations control center and/or affected personnel before performing any tests or calibrations on the system.

Step 2 Ensure that the operations controlled by the chromatograph are clearly understood.

Step 3 Determine the proper test methods, and verify calibration status and date of any test equipment. Test methods include the following:

- *Bench condition testing (BCT)* – Isolated completely from the system
- *Static condition testing* (SCT) – In place but isolated electrically and mechanically from process operations
- *Dynamic condition testing* (DCT) – In place and not isolated electrically and mechanically from process operations

CAUTION

If dynamic condition testing, exercise extra caution in applying a setpoint to a device to ensure that the calibration process does not cause an upset or abnormal effect on the system.

Step 4 If necessary, remove the chromatograph from service following company policy.

Step 5 Prepare a calibration sample, and introduce the calibration sample into the chromatograph.

Step 6 Perform the chromatograph test using an approved test medium. If a dynamic condition test is performed, check with the operations control center to see if the chromatograph output was received properly.

Step 7 Check the calibration readings, and ensure that the readings are within the acceptable calibration range. Repeat the calibration check if the first reading is out of range.

Step 8 Return the chromatograph to service following company policy.

Step 9 Notify the operations control center that testing is completed.

Step 10 Document your work using the appropriate forms and following company standard procedure.

3.2.3 Repairing Chromatographs

Due to the simple modular designs and built-in diagnostics of chromatographs, most repairs can be performed by pipeline E&I technicians. When troubleshooting and repairing a typical chromatograph, use the following general procedure:

Step 1 Check that the flow gauges are properly set.

Step 2 Check that the flows are balanced.

Step 3 Check that the temperature is properly set and maintained.

Step 4 Check that the timing sequence is properly set.

Step 5 Perform leak checks on system components.

Step 6 Check balance setup.

Step 7 Perform valve flow tests.

Step 8 Perform valve pressure tests.

Step 9 Perform bench calibrations.

Step 10 Perform program editing.

Step 11 Document your work using the appropriate forms and following company standard procedure.

When performing any type of repair on a chromatograph, the manufacturer's maintenance procedures must be followed. Many chromatographs are supplied with a warranty, and if a chromatograph is not maintained and repaired according to the manufacturer's instructions, the warranty may be voided. Before beginning any repair work on a particular chromatograph, check to see if the warranty is still in effect. If the chromatograph is still under warranty, any repair work must be performed in an authorized manner.

In some cases, the entire chromatograph may have to be removed from service and returned to the manufacturer. If a chromatograph is to be repaired by company personnel, they must be trained and qualified. They must also make sure that the manufacturer's instructions are followed in detail.

3.2.4 Installing Chromatographs

Installation of chromatographs on the pipeline is typically by engineering design. Proper installation is critical to the operation of the chromatograph and the accuracy of the measurements. Prior to installation of a chromatograph, refer to the manufacturer's recommendations on proper installation of the components.

The following steps describe the general procedure for a typical chromatograph installation:

Step 1 Notify the operations control center and/or affected personnel before performing any installations on the system.

Step 2 Ensure that the operations controlled by the chromatograph are clearly understood.

Step 3 Identify the proper mounting location for the chromatograph equipment. Be sure to provide adequate space for future maintenance or repairs.

Step 4 Wire the chromatograph equipment. Be sure to follow applicable electrical codes and consult the latest edition of the *National Electrical Code®* (*NEC®*). In some locations, the presence of hazardous substances may require the use of specialized electrical fittings in order to meet code and create a safe working environment.

Step 5 Install the sampling system, carrier gas system, and vent lines.

Step 6 Recheck all wiring connections to ensure proper installation, and then power up the chromatograph.

Step 7 Perform leak checks on the chromatograph equipment and components.

Step 8 Adjust the carrier gas supply, and prepare the system for purging.

Step 9 Allow the chromatograph temperature to stabilize, and perform the system purge.

Step 10 Enter initial programming information, such as date and time.

Step 11 Perform calibration procedures using an approved calibration medium.

Step 12 Put the chromatograph into service.

Step 13 Document your work using the appropriate forms and following company standard procedure.

Additional Resources

It is assumed that you will learn to perform these procedures under the direct supervision of your crew leader, foreman, or instructor. For independent study, refer to the recommended Additional Resources at the back of this module.

3.0.0 Section Review

1. Which type of chromatography column can be extremely long?
   a. Capillary
   b. Adsorption
   c. Partition
   d. Glass-filled

2. The final step of chromatograph calibration is to _____.
   a. notify the control center
   b. re-activate the chromatograph
   c. confirm SCADA readings
   d. document your work

SECTION FOUR

4.0.0 OTHER PIPELINE ANALYZERS

Objective

Describe the operation and maintenance of other types of pipeline analyzers.

a. Describe the operation and maintenance of moisture analyzers.
b. Describe the operation and maintenance of continuous emissions monitoring systems.
c. Describe the operation and maintenance of centrifuges.

Trade Terms

Hygrometer: An instrument used to determine the amount of moisture present (humidity) in a gas such as air.

Joule-Thomson effect: A cooling that occurs when a gas at a high pressure suddenly drops to a much lower pressure and expands. Such cooling can cause vapors (water or hydrocarbons) mixed with the gas to condense. Named for nineteenth-century British scientists James Joule and William Thomson.

Relative centrifugal force (RCF): A measure of the force generated in a centrifuge. A function of the centrifuge's rotational speed (rpm) measured in gees.

In addition to the analyzers already examined, a typical pipeline setting will usually include other analyzers or tools used to assess the product. This section describes other technologies that the pipeline E&I technician may be assigned to maintain.

4.1.0 Moisture Analyzers

Moisture analyzers (*Figure 13*) are used to detect the presence of moisture in the products transferred in the pipeline. Moisture analysis is very important in the pipeline industry because controlling moisture reduces corrosion, equipment breakdowns, and freeze-ups. Moisture content will vary from product to product.

There are many different instruments that are used to monitor moisture content, and there are various methods used to control moisture. In the pipeline industry, high levels of moisture in the product can cause maintenance problems and measurement inaccuracies.

Figure 13 Moisture analyzer.

When water in the product combines with either H_2S or with CO_2, corrosive acids are formed. These acids accelerate wear on pipeline equipment and components. Additionally, moisture can combine with liquid hydrocarbons in the product and freeze up, causing flow restrictions or blockages. Moisture in the product can also freeze due to the **Joule-Thomson effect** when rapid expansion occurs, such as at an orifice plate.

There are two basic methods of moisture analysis used in the pipeline industry—direct measurement and indirect measurement. The direct measurement technique involves the use of a chilled-mirror **hygrometer** to accurately measure the temperature at which condensate forms on a chilled surface in the presence of a gas. Indirect measurement techniques use electronic moisture analyzers that rely on changes in resistance or capacitance to indicate the amount of moisture present.

4.1.1 Maintaining Moisture Analyzers

Periodic maintenance is required to ensure the proper operation and detection ability of a moisture analyzer. The following steps describe general maintenance procedures for typical moisture analyzers:

> **WARNING!**
> Always confirm that the unit is located in a safe area before opening the lines or the unit itself. If a hazardous atmosphere is present, the consequences could be deadly.

Step 1 Check the sample line for leaks.

Step 2 Check the condition of flow meters, aspirators, regulators, valves, gauges, and traps. Repair or replace any damaged parts.

Step 3 Check that all printed circuit boards are secure in their connectors and that all cabling and plugs are secure.

Step 4 Check that all relays are secure.

Step 5 Check for signs of overheating in the power supply, on the printed circuit boards, and in the wiring.

Step 6 Check for excessive accumulation of dust, dirt, or debris, and blow clean with a clean, dry, low-pressure air supply.

> **WARNING!**
>
> Never use canned air that contains a flammable propellant if the circuit is powered up. A fire or explosion could result.

Step 7 Refill the moisture generator reservoir.

Step 8 Replace filters.

Step 9 Refill dryer or trap.

Step 10 Replace the molecular sieve and contaminant trap at prescribed intervals.

Step 11 Replace the molecular sieve desiccant at prescribed intervals.

Step 12 Perform leak testing of moisture analyzer components.

Step 13 Perform field calibrations.

Step 14 Check valve solenoid and heater operation.

Step 15 Change analyzer programmed operating parameters.

Step 16 Verify the data to SCADA and local readouts. Data may be digital or analog.

Step 17 Document your work using the appropriate forms and following company standard procedure.

4.1.2 Calibrating Moisture Analyzers

All moisture analyzers require periodic testing and calibration. In most cases, the frequency of calibration is determined by company policies or custody transfer contracts. Testing and calibration of any moisture analyzer must be performed based on the manufacturer's recommended practices.

Typically, calibration of a moisture analyzer will involve the use of a sample with a known moisture content. This calibration sample is then used to verify the accuracy of the moisture analyzer measurements. Typical analyzer testing includes performing automatic system diagnostic tests, leak tests, and field calibrations per company and manufacturer specifications.

Typical steps that may be performed for testing and calibration include the following:

Step 1 Notify the operations control center and/or affected personnel before performing any tests or calibrations on the system.

Step 2 Ensure that the operations controlled by the moisture analyzer are clearly understood.

Step 3 Determine the proper test methods, and verify calibration status and date of any test equipment. Test methods include the following:

- *Bench condition testing (BCT)* – Isolated completely from the system
- *Static condition testing* (SCT) – In place but isolated electrically and mechanically from process operations
- *Dynamic condition testing* (DCT) – In place and not isolated electrically and mechanically from process operations

> **CAUTION**
>
> If dynamic condition testing, exercise extra caution in applying a setpoint to a device to ensure that the calibration process does not cause an upset or abnormal effect on the system.

Step 4 If necessary, remove the moisture analyzer from service following company policy.

Step 5 Prepare a calibration sample, and introduce the sample into the analyzer.

Step 6 Perform the analyzer test using an approved test medium. If a dynamic condition test is performed, check with the operations control center to see if the analyzer output was received properly.

Step 7 Check the calibration readings, and ensure that the readings are within the acceptable calibration range. Repeat the calibration check if the first reading is out of range.

Step 8 Return the analyzer to service following company policy.

Step 9 Notify the operations control center that testing is completed.

Step 10 Document your work using the appropriate forms and following company standard procedure.

4.1.3 Repairing Moisture Analyzers

When performing any type of repair on a moisture analyzer, the analyzer manufacturer's maintenance procedures must be followed. Many moisture analyzers are supplied with a warranty, and if a moisture analyzer is not maintained and repaired according to the manufacturer's instructions, the warranty may be voided. Before beginning any repair work on a particular moisture analyzer, check to see if the warranty is still in effect. If the moisture analyzer is still under warranty, any repair work must be performed in an authorized manner.

In some cases, the entire moisture analyzer may have to be removed from service and returned to the manufacturer. If an analyzer is to be repaired by company personnel, they must be trained and qualified. They must also make sure that the manufacturer's instructions are followed in detail.

4.1.4 Installing Moisture Analyzers

Moisture analyzers are installed on the pipeline per engineering design. Proper installation is critical to the operation of the moisture analyzer and the accuracy of the measurements. Before installing a moisture analyzer, refer to the manufacturer's recommendations on proper installation of the components. Typical installation procedures include the following steps:

Step 1 Notify the operations control center and/or affected personnel before performing any installations on the system.

Step 2 Ensure that the operations controlled by the moisture analyzer are clearly understood.

Step 3 Identify the proper mounting location for the moisture analyzer control unit and field unit. Be sure to provide adequate space for future maintenance or repairs.

Step 4 Wire the moisture analyzer equipment. Be sure to follow applicable electrical codes (consult the latest edition of the *NEC®*). In some locations, hazardous substances may necessitate the use of specialized electrical fittings in order to meet code and create a safe working environment.

Step 5 Install the sampling system, reference gas system, and vent lines.

Step 6 Recheck all wiring connections to ensure proper installation, and then power up the moisture analyzer.

Step 7 Perform leak checks on the moisture analyzer equipment and components.

Step 8 Adjust the gas supply flow and backpressure settings, and prepare the system for operation.

Step 9 Allow the moisture analyzer temperature to stabilize, and perform the system purge.

Step 10 Enter initial programming information, such as date and time.

Step 11 Perform calibration procedures using an approved calibration medium.

Step 12 Put the moisture analyzer into service.

Step 13 Document your work using the appropriate forms and following company standard procedure.

4.2.0 Continuous Emissions Monitoring Systems

The Clean Air Act Amendments of 1990 and other regulatory pressures have provided incentives for the advancement of continuous emission monitoring systems (CEMS). A typical CEMS consists of several equipment components, both software and hardware, used to sample, analyze, measure, and permanently record emissions in a pipeline process. Many of the analyzers discussed in this module are incorporated into continuous emission monitoring systems to perform analysis of the pipeline processes. There are several types of CEMS including in-situ, straight extractive, dilution extractive, and parameter.

In-situ systems allow personnel to take measurements in place, without transporting a sample to another location for analysis. In-situ CEMS methods provide measurement over a short distance and are used for opacity measurement and stack flow. The in-situ type can be one of two methods—the point method or the path method. With the point method, an electro-optical or electrochemical sensor provides the measurement. With the path method, light or sound is transmitted across the sampling area, and the interaction is referenced back to specific gas characteristics.

Extractive methods involve removing a representative sample and then transporting the sample to a CEMS shelter for analysis. This method involves using a probe, sample lines, filters, condensers, pumps, and analyzers. The extractive method can be one of two types—the straight extractive method or dilution extractive method. Both of these methods use similar analytical techniques for measuring the content of the sample, including infrared, ultraviolet, chemiluminescent, flame ionization detection, and gas chromatography.

Parameter methods of CEMS involve the use of specific physical measurements, such as temperature, pressure, and flow, as well as other process measurements to predict emissions. These methods do not use any analytical equipment for measuring the content of the sample.

4.3.0 Centrifuges

Centrifuges are used to separate product samples in order to calculate sediment and water content. Samples that have been collected by the automatic sampling unit are mixed well to give a composite sample of the entire batch. A number of samples are withdrawn from the composite sample. Equal portions are placed in graduated centrifuge tubes, placed in a centrifuge, and spun at a high rate of speed (called the *grind out*).

Centrifugal force causes the heavier items, sediment and water, to separate out of the petroleum and migrate to the end of the tube. By reading the scale on the centrifuge tube, a calculation can be made of the sediment and water content of the product. The true volume of product can be calculated by subtracting the percentage of sediment and water from the total volume of the batch or tank. Sediment and water tests normally apply to unrefined products such as crude oil. *Figure 14* shows centrifuge tubes used in the grind out.

A centrifuge is an electric motor-driven unit with special hinged fittings that allow the tubes to swing to a horizontal plane as the unit speed increases. Some centrifuges hold the tubes at a fixed angle. The speed, coupled with the angle of the tubes, causes the sediment and water to migrate to the end of the tube where it can be measured. The basis for the operation of a centrifuge is its ability to develop high rpm. Some units are portable and may be battery powered, but most plug in to a regular electrical outlet.

4.3.1 Troubleshooting the Centrifuge

Occasional troubleshooting may be required when using a centrifuge. Some indications of a problem include the following:

- Failure to start up
- Slow acceleration
- Low speed
- Excessive heat
- Odor of hot wiring/insulation
- Vibration
- Excessive or unusual noise

In some cases, problems with a centrifuge could be caused by broken power supply wires or connectors. Switches often fail. If a problem with speed is suspected and the centrifuge does not have a tachometer, a portable digital tachometer should be secured to make certain that it is reaching the specified rpm. A factory-supplied speed chart should be used along with the tachometer to ensure that the unit is reaching a **relative centrifugal force (RCF)** of at least 500 gees. If it is not, the unit must be adjusted, repaired, or replaced.

Many problems with a centrifuge involve internal parts, usually inside the motor. In this case, the unit needs to be referred to a shop or returned to the manufacturer for service.

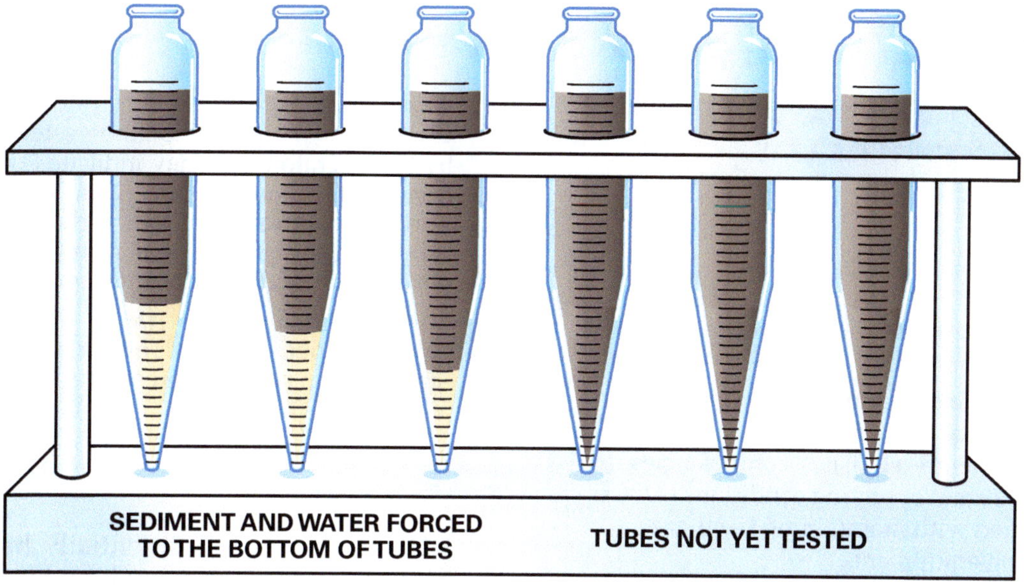

Figure 14 Centrifuge tubes used in the grind out.

4.3.2 Sample Pre-Heater

Samples must be heated to 140°F (60°C) before testing. Some centrifuge machines are supplied with a heater that is sufficient to keep the sample's temperature above the required level while it's spinning in the machine. Failure of this part of the equipment involves testing the heating element and heat-control device as directed by the operator's manual.

Whether it is an internal or separate unit, it must have the ability to heat and maintain 140°F (60°C) ±5°F (±3°C) for the duration of the test. Wiring and connectors should be examined for a broken connection. Some units may be a closed module type that must be discarded and replaced. A centrifuge repair may have to be referred to a shop, or the unit may have to be returned to the manufacturer for service.

Additional Resources

It is assumed that you will learn to perform these procedures under the direct supervision of your crew leader, foreman, or instructor. For independent study, refer to the recommended Additional Resources at the back of this module.

4.0.0 Section Review

1. When water combines with gases like H_2S and CO_2, it can form _____.
   a. ice crystals
   b. pipe blockages
   c. corrosive acids
   d. flammable mixtures

2. The continuous emissions monitoring system that takes measurements right at the pipeline is the _____.
   a. in-situ
   b. straight extractive
   c. dilution extractive
   d. parameter

3. A centrifuge used for grind out should be able to reach an RCF of at least _____.
   a. 50 gees
   b. 100 gees
   c. 500 gees
   d. 1,000 gees

SECTION FIVE

5.0.0 FIXED VAPOR AND COMBUSTIBLE GAS DETECTORS

Objective

Describe the operation and maintenance of fixed gas detection equipment.

a. Describe the operation of fixed gas detection equipment.
b. Identify the prerequisites, competencies, and abnormal operating conditions (AOCs) associated with maintaining fixed gas detection equipment.
c. Describe the procedures used to maintain fixed gas detection equipment.

Performance Task

Maintain fixed gas detection equipment (CT55_0-17).

- Identify potential abnormal operating conditions (AOCs) that may occur during performance of this CT, and know the appropriate actions to take in response to them.
- Utilize the appropriate personal protective equipment according to relevant company procedures.
- Make appropriate notifications prior to beginning the inspection.
- Verify proper operations, including the mounting, filters, cable connections, sensor placement, and backup power supply.
- Check setpoints and alarms.
- Calibrate gas detector (if needed).
- Verify sensing elements and replace if necessary.
- Make appropriate notifications upon completion of the inspection.
- Complete appropriate documentation as required by operator's procedures.

Trade Terms

Abnormal operating conditions (AOCs): Conditions within a pipeline system, identified by the operator, that may indicate (1) a malfunction of a component or (2) a deviation from normal operations that may indicate a condition exceeding design limits or result in a hazard to persons, property, or the environment.

Lower explosive limit (LEL): The lowest concentration of a gas or vapor in air that will ignite if an ignition source is present.

There are several potentially hazardous liquids, vapors, and combustible gases associated with pipelines. Therefore, maintaining the safety of personnel and equipment is very important. Fixed vapor and combustible gas detectors are used to detect the presence of hazardous vapors or atmospheres around pipeline equipment and buildings. They are able to detect a buildup of specific gases or vapors during normal pipeline operations or in the event of an accidental release. They can also be used to determine the extent of an accidental release.

Fixed vapor detectors and combustible gas detectors may be wall installations or panel-mounted (*Figure 15*). Many are online and are used at stations and pump rooms constantly to monitor for the presence of hazardous vapors. These detectors communicate with the SCADA system to trigger an alarm at the control center if the vapor concentrations exceed the alarm setpoint. When a hazardous vapor alarm is received, the operator in the control center will respond to the alarm, shut down operations for that station, and contact a pipeline E&I technician to investigate the alarm.

Since explosive ranges of flammable liquids and gases are very wide, the detectors are calibrated to the **lower explosive limit (LEL)**. Flammable gas and vapor concentrations must be kept below the LEL to maintain safety. Most fixed vapor and gas detectors are commonly calibrated

Figure 15 Panel-mounted combustible gas detector.

to a range of the LEL. For example, a detector calibrated to detect ethane at 0 to 100 percent LEL may initiate a warning at 20 percent of the LEL and an alarm condition at 40 percent of the LEL.

5.1.0 Operation of Fixed Vapor and Combustible Gas Detectors

Several methods are used to detect the presence of hazardous or combustible vapors. Some of these methods include catalytic reaction, semiconductor, and infrared absorption techniques.

5.1.1 Catalytic Detectors

Catalytic detectors (*Figure 16*) use a catalyst that is incorporated into an electrically heated sensor containing a temperature-sensitive device. If a flammable gas is present, the catalyst causes it to be continuously oxidized to carbon dioxide and water. The temperature sensor measures the heat generated during oxidation of the flammable gas, and this measurement is used to determine the concentration of the gas.

Newer types of catalyst detectors use a catalytic bead encapsulated in a highly porous bead of alumina. Others use two catalytic beads, one as a detection bead and one as a reference bead. Both beads are identical, but the reference bead is

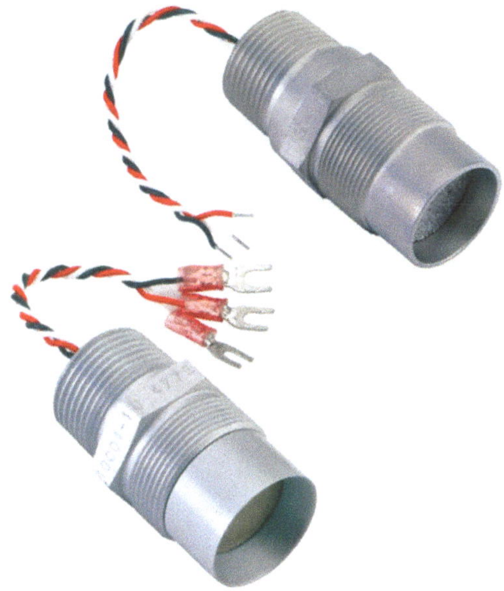

Figure 16 Catalytic sensors.

coated to prevent gas from oxidizing the catalyst. This reference bead will respond to changes of temperature, pressure, and humidity in the same manner as the detection bead, and provides a more accurate reference based on the testing conditions. These catalytic bead-type detectors have a higher resistance to corruption, or poisoning, that can result from a buildup on the catalyst.

5.1.2 Semiconductor Detectors

Semiconductor detectors (*Figure 17*) are solid-state sensors based on the use of metal oxide semiconductors. Similar to catalytic detectors, semiconductor detectors rely on the absorption of the gas on a heated oxide surface. As the flammable gas is absorbed on the surface, a change in the electrical conductivity of the metal oxide reflects the amount of flammable gas present in the surrounding atmosphere. Semiconductor detectors are used mainly where low concentrations of flammable gas indicate alarm conditions.

5.1.3 Infrared Detectors

Infrared detectors (*Figure 18*) use infrared radiation to detect the presence of combustible gases. All flammable gas molecules absorb radiation in

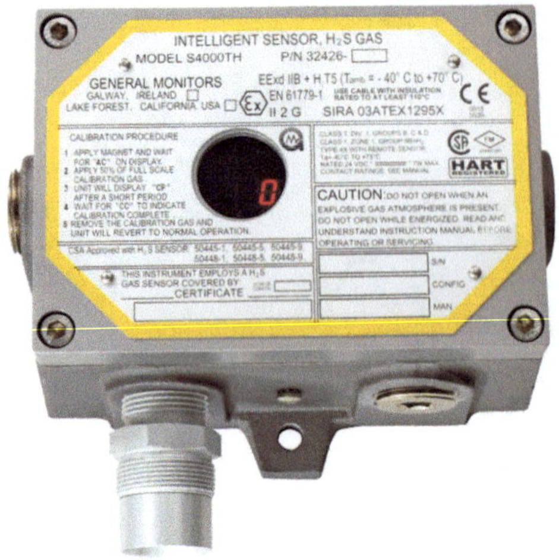

Figure 17 Semiconductor detector.

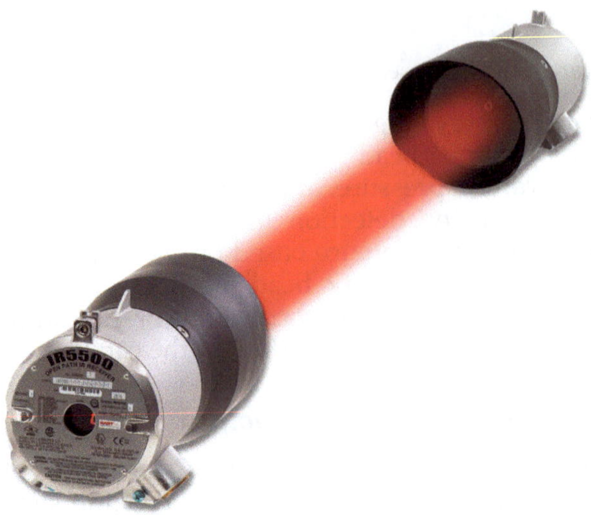

Figure 18 Infrared detector.

the infrared region of the electromagnetic spectrum. The specific absorption spectra produced that are characteristic of flammable gases act as a fingerprint that can be identified by a detector. As infrared radiation is passed through a gas or vapor sample (*Figure 19*), the molecules in the sample absorb part of the radiation, and the radiation that emerges through the sample is measured by a detector that evaluates the gas concentration.

There are two basic types of infrared systems that are used to detect vapors and combustible gases. The dispersive infrared detector uses a prism or grating to disperse the radiation from the source. This allows only the specific band of radiation that is absorbed by the target flammable gas to pass through to the detector.

In the non-dispersive infrared detectors, either all of the radiation produced is passed through the sample, or an infrared filter is used to narrow the radiation band that is passed through the sample. If all the radiation is passed through the sample, the target gas is selected by the detector. If a filter is used to narrow the radiation band, then the target gas is selected by the filter.

An advantage of using infrared gas detection is that the signal from the instrument is a function only of the concentration of the gas to be measured. It is also less subject to interference from other gases, and can be highly specific for the gas to be measured.

5.2.0 Prerequisites, Competencies, and Abnormal Operating Conditions

Personnel must possess a variety of skills in order to successfully maintain fixed gas detector equipment. The various operations require a series of steps, which the technician must be competent to perform. Personnel must also be aware of the particular **abnormal operating conditions (AOCs)** that may occur during the procedure and know the proper responses to them.

5.2.1 Prerequisites and Competencies

Personnel must possess the following basic knowledge and skills prior to maintaining fixed gas detector equipment:

- Thorough knowledge of company safety policies and procedures, and the proper use of personal protective equipment
- Thorough knowledge of the gas detection equipment being inspected
- An understanding of the proper electrical supply to the gas detection equipment
- Ability to test the gas detection equipment using a sample gas
- Ability to calibrate the gas detection equipment to alarm at the specified concentration
- Ability to properly document the inspection

5.2.2 Abnormal Operating Conditions

Personnel maintaining fixed gas detection equipment must be able to identify AOCs and know the proper response to each, should they occur. The following are some potential AOCs, and the appropriate action personnel should take in response to them:

- *Electric shock occurs* – Make appropriate notifications, including calling 911.
- *Component failure occurs, or detector doesn't operate as intended* – Notify appropriate personnel to make repairs.

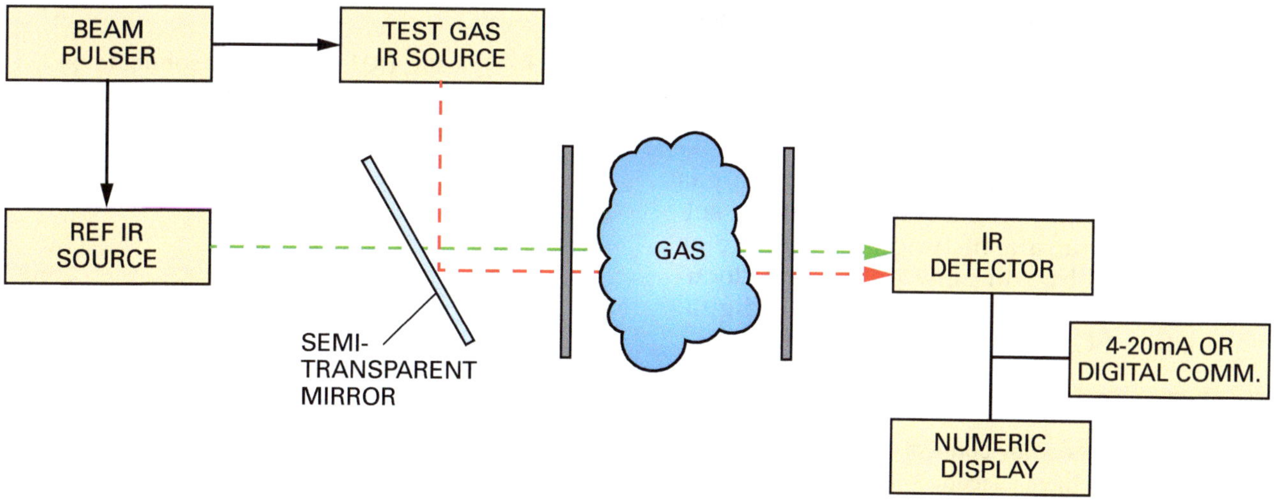

Figure 19 Infrared detector operation.

- *Unexpected hazardous liquid or carbon dioxide leak is encountered* – Isolate the leak if it is safe to do so. Evacuate the area and allow it to ventilate. When safety backup arrives, make appropriate repairs if it is safe to do so.
- *Fire or explosion occurs* – Notify 911 emergency personnel and make appropriate company notifications. Secure the area if it is safe to do so.

5.3.0 Using and Maintaining Fixed Vapor and Combustible Gas Detectors

Fixed vapor and combustible gas detectors require a certain amount of routine care in the form of maintenance, calibration, and repair. They must also be handled correctly when installed and operated. This section describes some of these procedures.

5.3.1 Maintaining Fixed Vapor and Combustible Gas Detectors

When performing maintenance on fixed vapor and combustible gas detection equipment, the procedures must be performed without disturbing the process. Safety of the pipeline and personnel must always be the first consideration. Even when proper precautions are taken, an abnormal operating condition (AOC) may result. Pipeline E&I technicians should be familiar with the AOCs that might occur and know how to respond to these conditions.

Periodic maintenance is required to ensure the proper operation and detection ability of fixed vapor and combustible gas detectors. Some regulatory agencies specify maintenance and testing intervals. If applicable, consult these guidelines as well. The following general procedure outlines typical items that should be checked during maintenance activities.

WARNING!

Always obtain and use the appropriate PPE for each task. Only qualified individuals may make electrical connections and operate equipment. Always verify the absence of voltage and current before making connections. Follow the manufacturer's instructions for the equipment being installed.

Follow these steps to maintain fixed vapor and combustible gas detection equipment:

Step 1 Identify potential abnormal operating conditions that may occur during performance of this CT, and know the appropriate actions to take in response to them.

Step 2 Utilize the appropriate personal protective equipment according to relevant company procedures.

Step 3 Notify the operations control center and/or affected personnel so that they are aware of the task activities in order to ensure the safety of the pipeline and of personnel.

Step 4 Ensure that the sensor mounting is secure.

Step 5 Ensure the sensor filter is clear of water, oil, dust, and paint that might clog it.

Step 6 Check all cable connections for tightness and possible damage.

Step 7 Ensure that the sensor placement is up to date with the plant processes.

Step 8 Verify that vapor and combustible gas detection equipment setpoints, alarm setpoints, and alarm functions operate as specified by company policy and documentation. Verify the analog and digital data to SCADA and/or local readouts.

> **CAUTION**
>
> Some detectors may connect to a PLC or a DCS and can have cascade shutdown effects on the process. Before performing any tests, be completely familiar with the system and the equipment.

Step 9 Calibrate the vapor and combustible gas detection equipment according to manufacturer's procedures and company policy.

Step 10 Replace sensing elements as needed.

Step 11 Ensure that the system has a backup power supply. Since many backup systems rely on batteries, be sure to schedule periodic battery testing and replacement to ensure constant operation.

Step 12 Alert nearby personnel and the operations control center when the procedure is completed in order to maintain safety of the pipeline personnel and to ensure uninterrupted operation.

Step 13 Complete all necessary documentation as required by company policy; properly file documentation.

5.3.2 Calibrating Fixed Vapor and Combustible Gas Detectors

All fixed vapor and combustible gas detectors require periodic testing and calibration. In most cases, the frequency of calibration is determined by company policies. Testing and calibration of any fixed vapor or combustible gas detector must be performed based on the manufacturer's recommended practices. Typically, calibration of fixed vapor and combustible gas detectors will involve the use of a sample with a known flammable gas content. This calibration sample is then used to verify the accuracy of the detector measurements.

> **NOTE**
>
> - Typically, DOT requires personnel to be operator qualified on CT55_0-17 to calibrate and test vapor and combustible gas detectors. Verify requirements and your qualifications before performing any of these tasks.

The general procedure for calibration includes the following steps:

Step 1 Notify the operations control center and/or affected personnel before performing any tests or calibrations on the system.

Step 2 Ensure that the operations controlled by the fixed vapor or combustible gas detector are clearly understood.

Step 3 Determine the proper test methods, and verify calibration status and date of any test equipment. Test methods include:

- *Bench condition testing (BCT)* – Isolated completely from the system
- *Static condition testing* (SCT) – In place but isolated electrically and mechanically from process operations
- *Dynamic condition testing* (DCT) – In place and not isolated electrically and mechanically from process operations

> **CAUTION**
>
> If dynamic condition testing, exercise extra caution in applying a setpoint to a device to ensure that the calibration process does not cause an upset or abnormal effect on the system.

Step 4 If necessary, remove the detector from service following company policy. If the detector must be removed for service and is not replaced with a spare, company policy may require a Management of Change (MOC) to ensure that the process protection and/or personnel safety is not compromised. In some cases, the process may have to be shut down.

Step 5 Prepare a calibration sample, and introduce the calibration sample into the detector.

Step 6 Perform the detector test using an approved test medium. If a dynamic condition test is performed, check with the operations control center to see if the detector output was received properly.

Step 7 Check the calibration readings, and ensure that the readings are within the acceptable calibration range. Repeat the calibration check if the first reading is out of range.

Step 8 Return the detector to service following company policy.

Step 9 Notify the operations control center that testing is completed.

Step 10 Document your work using the appropriate forms and following company standard procedure.

5.3.3 Repairing Fixed Vapor and Combustible Gas Detectors

When performing any type of repair on fixed vapor and combustible gas detectors, the equipment manufacturer's maintenance procedures must be followed. Many vapor and combustible gas detectors are supplied with warranties, and if a detector is not maintained and repaired according to the manufacturer's instructions, the warranty may be voided. Before beginning any repair work on a particular vapor or combustible gas detector, check to see if the warranty is still in effect. If the detector is still under warranty, any repair work must be performed in an authorized manner.

In some cases, the entire detector may have to be removed from service and returned to the manufacturer. If a vapor or combustible gas detector is to be repaired by company personnel, they must be trained and qualified. They must also make sure that the manufacturer's instructions are followed in detail.

NOTE

Typically, DOT requires personnel to be operator qualified on CT55_0-17 to repair vapor and combustible gas detectors. Verify requirements and your qualifications before performing any repairs.

5.3.4 Installing Fixed Vapor and Combustible Gas Detectors

There are no standard procedures for fixed vapor and combustible gas detector installation because the optimum sensor location will vary for each application. Vapor and combustible gas detectors are typically installed per engineering design. Each installation must be evaluated, and proper installation is critical to the operation of the equipment and the accuracy of its measurements. Prior to installation of a detector, refer to the manufacturer's recommendations on proper installation of the components.

As a general guideline, locate the detector:

- Where prevailing air currents would contain the maximum concentration of gas (*Figure 20*).
- Where it will be easy to access for calibration.
- Where it will be away from sources of heat (verify the temperature range of the detector).
- Where it will be away from radio transmitters and similar equipment.
- Where it will be protected from excessive wind, dust, water, or vibration.
- Where it will not be contaminated by water buildup or coated with contaminating substances.
- Where it will be near possible sources of a leak. Depending upon the density and other characteristics of the target gas, the detector may have to be mounted above or below this point.

WARNING!

Always obtain and use the appropriate PPE for each task. Only qualified individuals may make electrical connections and operate equipment. Always verify the absence of voltage and current before making connections. Follow the manufacturer's instructions for the equipment being installed.

Use the general installation procedure provided in the following steps:

Step 1 Notify the operations control center and/or affected personnel before performing any installations on the system.

Step 2 Ensure that the operations controlled by the detector are clearly understood.

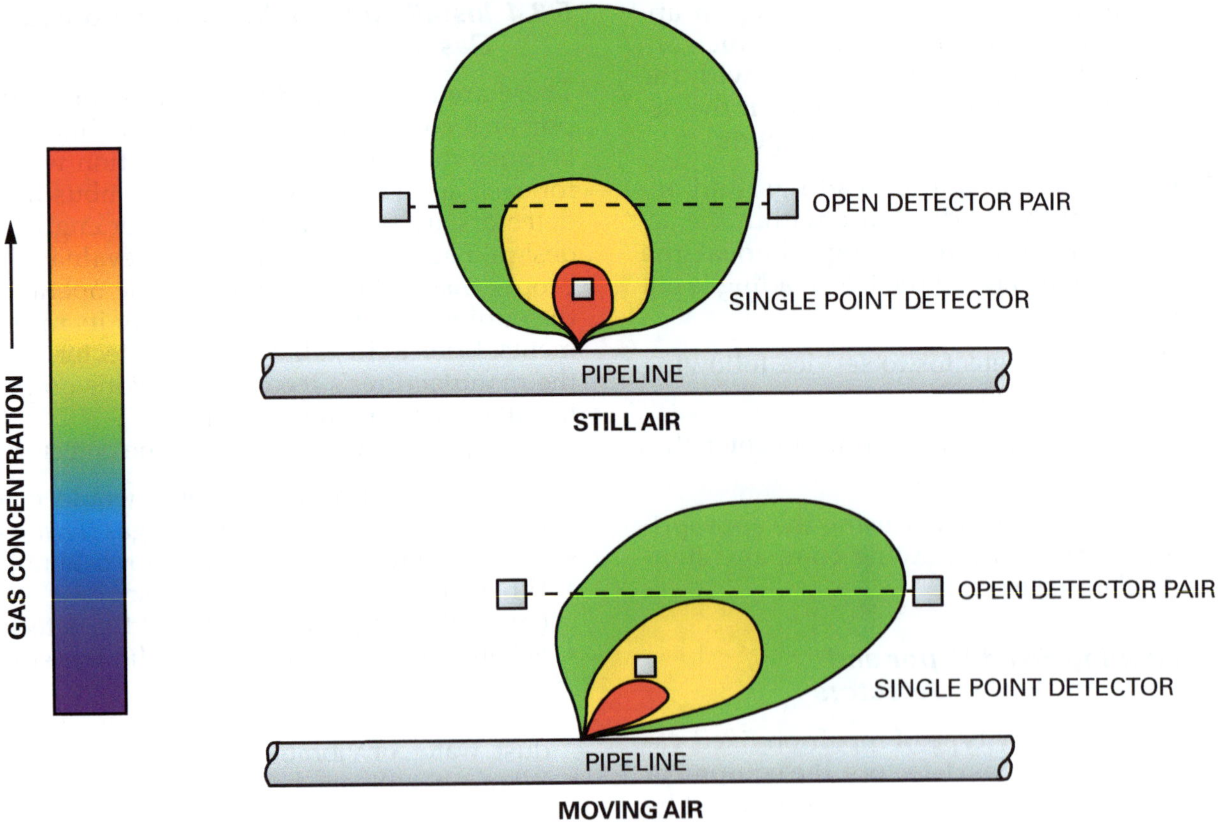

Figure 20 Effect of air currents.

Step 3 Make signal and power connections at the terminal block. Signal connections will depend on the technology involved (digital or analog). Power supplies for detectors vary, but 24 VDC and 120 VAC are common. Other supplies may be used, however. Confirm correct voltage with the manufacturer's manual. Do not connect power until all other connections are made. Follow the manufacturer's recommendations on how to connect wiring to the detector. Also, be sure to follow *NEC*® guidelines for wiring in hazardous locations since many detectors will be located near potential hazards.

Step 4 Before applying power to the detector, check all wiring connections and replace all covers.

Step 5 After initially powering up the detector, allow the detector unit to stabilize before performing calibrations.

Step 6 Document your work using the appropriate forms and following company standard procedure.

Additional Resources

It is assumed that you will learn to perform these procedures under the direct supervision of your crew leader, foreman, or instructor. For independent study, refer to the recommended Additional Resources at the back of this module.

5.0.0 Section Review

1. The combustible gas detector that measures a change in the electrical conductivity of a metal oxide is the _____.
   a. catalytic
   b. semiconductor
   c. infrared
   d. chromatic

2. If an unexpected gas release occurs during the maintenance of fixed gas detection equipment, the proper first response is to _____.
   a. call 911
   b. isolate the leak if it is safe to do so, then evacuate the area
   c. document the incident according to company procedure
   d. d.notify personnel who can make repairs

3. Before beginning a repair on fixed gas detection equipment, first determine the existence and status of _____.
   a. the last calibration
   b. the repair history log
   c. a manufacturer's warranty
   d. electrical codes

SUMMARY

Analytical equipment is essential in all phases of pipeline operation to maintain the integrity of the pipeline. Many types of analyzers are available in both permanently installed and portable models. Maintaining and repairing analyzers requires training and skill. This module described some of the analyzers used in the pipeline industry and explained how to maintain fixed gas detection equipment.

Review Questions

1. Pipeline analytical equipment provides information for maintaining product quality, pipeline integrity, and pipeline _____.
   a. viability
   b. economy
   c. safety
   d. productivity

2. Analyzers can determine the type of specific substances in the pipeline environment, and often the _____.
   a. danger level
   b. quantity
   c. toxicity
   d. reactivity

3. Analyzers installed at specific operating areas to provide analytical data from samples taken from the process stream or to provide analytical data from atmospheric samples around the area equipment are called _____.
   a. fixed analyzers
   b. portable analyzers
   c. personal analyzers
   d. standard analyzers

4. To ensure that the atmosphere being detected is the same as that which the technician is breathing, a clothing-mounted detector should be kept close to the _____.
   a. pipeline
   b. breathing zone
   c. lungs
   d. body midpoint

5. The presence of hydrogen sulfide and sulfur in the pipeline product determines the product _____.
   a. pressure
   b. temperature
   c. density
   d. grade

6. During the past four decades, one of the most widely used methods in the pipeline industry for sulfur detection is the _____.
   a. flame photometric method
   b. XRF method
   c. lead acetate reaction method
   d. chemiluminescence method

7. The method of sulfur detection that uses a photomultiplier tube to detect a blue light emission from the sample is the _____.
   a. flame photometric method
   b. XRF method
   c. lead acetate reaction method
   d. chemiluminescence method

8. The frequency of hydrogen sulfide detector calibration should be based on company policy and _____.
   a. whenever economically feasible
   b. NIST bulletin recommendations
   c. ASME guidelines
   d. governmental agency regulations

9. Chromatography separates substances based on different _____.
   a. boiling points
   b. flash points
   c. retention rates
   d. chromogenic constants

10. Gas chromatographs use a carrier gas such as _____.
    a. xenon
    b. helium
    c. oxygen
    d. chlorine

11. What kind of column does the fastest type of chromatograph use?
    a. Capillary
    b. Adsorption
    c. Partition
    d. Alumina

12. A chromatograph is calibrated with an approved _____.
    a. carrier gas
    b. hydrocarbon mixture
    c. test medium
    d. inert gas

13. Moisture analysis is very important in the pipeline industry because it reduces corrosion, equipment breakdowns, and _____.
    a. odorization
    b. freeze-ups
    c. power usage
    d. thaw-outs

14. A family of technologies designed to monitor a pipeline's output and its effect on the environment is known as _____.
    a. PPMS
    b. CEMS
    c. UEL
    d. LEL

15. What tool is used to separate product samples in order to calculate sediment and water content?
    a. CEMS
    b. Vapor and combustible gas detector
    c. Centrifuge
    d. Semiconductor detector

16. When equal portions of a sample are placed in graduated tubes and spun at high speed it is called the _____.
    a. grind out
    b. spin out
    c. top out
    d. wash out

17. Since explosive ranges of flammable liquids and gases are very wide, combustible gas detectors are calibrated to the _____.
    a. MMI
    b. UEL
    c. CDL
    d. LEL

18. Which type of combustible gas detector relies on the absorption of the gas on a heated oxide surface?
    a. Catalytic
    b. Infrared
    c. Semiconductor
    d. Ionization

19. Before calibrating or testing fixed gas detection equipment, determine whether _____.
    a. the control room has been notified
    b. bench calibration is mandatory
    c. DOT Operator Qualification is required
    d. the system has a backup detector

20. Typically, fixed gas detection equipment installations are _____.
    a. highly standardized
    b. performed by specialists
    c. performed during shutdowns
    d. dependent on engineering design

Trade Terms Quiz

Fill in the blank with the correct term that you learned from your study of this module.

1. Conditions within a pipeline system, identified by the operator, that may indicate (1) a malfunction of a component or (2) a deviation from normal operations that may indicate a condition exceeding design limits or result in a hazard to persons, property, or the environment are called _________.
2. The cooling that occurs when a gas at a high pressure suddenly drops to a much lower pressure and expands is known as the _________.
3. The measure of the force generated in a centrifuge, measured in gees, is the _________.
4. An instrument used to determine the amount of moisture present (humidity) in a gas such as air is called a(n) _________.
5. A highly sensitive light detection sensor used in certain kinds of instrumentation is called a(n) _________.
6. The emission of light as a result of a chemical reaction is called _________.
7. The lowest concentration of a gas or vapor in air that will ignite if an ignition source is present is known as the _________.
8. The tendency for some substances to emit light (or other EM radiation) when exposed to light (or other EM radiation) of a higher wavelength is called _________.

Trade Terms

Abnormal operating conditions (AOCs)
Chemiluminescence
Fluorescence
Hygrometer
Joule-Thomson effect
Lower explosive limit (LEL)
Photomultiplier tube
Relative centrifugal force (RCF)

Trade Terms Introduced in This Module

Abnormal operating conditions (AOCs): Conditions within a pipeline system, identified by the operator, that may indicate (1) a malfunction of a component or (2) a deviation from normal operations that may indicate a condition exceeding design limits or result in a hazard to persons, property, or the environment.

Chemiluminescence: The emission of light as a result of a chemical reaction. Chemiluminescence may be used to identify specific substances by the wavelengths of the light emitted.

Fluorescence: The tendency for some substances to emit light (or other EM radiation) when exposed to light (or other EM radiation) of a higher wavelength.

Hygrometer: An instrument used to determine the amount of moisture present (humidity) in a gas such as air.

Joule-Thomson effect: A cooling that occurs when a gas at a high pressure suddenly drops to a much lower pressure and expands. Such cooling can cause vapors (water or hydrocarbons) mixed with the gas to condense. Named for James Joule and William Thomson.

Lower explosive limit (LEL): The lowest concentration of a gas or vapor in air that will ignite if an ignition source is present.

Photomultiplier tube: A highly sensitive light detection sensor used in certain kinds of instrumentation. Photomultipliers can amplify a light signal by a factor of 100 million.

Relative centrifugal force (RCF): A measure of the force generated in a centrifuge. It's a function of the centrifuge's rotational speed (rpm) measured in gees.

Additional Resources

The following recommended resources can provide additional helpful information related to the requirements for this covered task:

API Recommended Practice 1161, Third Edition, © 2014 American Petroleum Institute. API Publishing Services, 1220 L Street NW, Washington, DC 20005. Distributed by Thompson Reuters (Scientific) LLC, **www.techstreet.com**.

NCCER Module AOCCC-17, *Abnormal Operating Conditions Control Center.*

NCCER Module AOCFG-17, *Abnormal Operating Conditions Field and Gas.*

NCCER Module 62105-14, *Introduction to Pipeline Documents.*

NCCER Module 64202-02, *Use of Meters and Test Equipment.*

NCCER Module 66101-14, *Introduction to the Pipeline Industry.*

Instrument Engineers Handbook, Volume 1: Process Measurement and Analysis. Béla G. Lipták (editor). Fourth Edition. 2003. Boca Raton, FL: CRC Press.

Instrumentation Reference Book. Walt Boyes (editor). Fourth Edition. 2009. Oxford, UK: Butterworth-Heinemann.

Measurement and Instrumentation Principles. Alan S. Morris. Third Edition. 2001. Oxford, UK: Butterworth-Heinemann.

The Condensed Handbook of Measurement and Control. N.E. Battikha. Third Edition, 2006. Research Triangle Park, NC: The International Society of Automation.

The following websites provide resources for products and training:

The International Society of Automation, **www.isa.org.**

Figure Credits

Soundsnaps/Shutterstock.com, Module Opener

Courtesy of Honeywell International Inc., Figures 1–3, 6

© Agilent Technologies, Inc. 2015 Reproduced with Permission, Courtesy of Agilent Technologies, Inc., Figure 10

Yokogawa Corporation of America, Figure 11

General Monitors, Figures 15–18

MEECO, Inc., Figure 13

Section Review Answer Key

Answer	Section Reference	Objective
Section One		
1. c	1.1.0	1a
2. a	1.2.3	1b
Section Two		
1. b	2.1.3	2a
2. a	2.2.2	2b
Section Three		
1. a	3.1.2	3a
2. d	3.2.2	3b
Section Four		
1. c	4.1.0	4a
2. a	4.2.0	4b
3. c	4.3.1	4c
Section Five		
1. b	5.1.2	5a
2. b	5.2.2	5b
3. c	5.3.3	5c

NCCER CURRICULA — USER UPDATE

NCCER makes every effort to keep its textbooks up-to-date and free of technical errors. We appreciate your help in this process. If you find an error, a typographical mistake, or an inaccuracy in NCCER's curricula, please fill out this form (or a photocopy), or complete the online form at **www.nccer.org/olf**. Be sure to include the exact module ID number, page number, a detailed description, and your recommended correction. Your input will be brought to the attention of the Authoring Team. Thank you for your assistance.

Instructors – If you have an idea for improving this textbook, or have found that additional materials were necessary to teach this module effectively, please let us know so that we may present your suggestions to the Authoring Team.

NCCER Product Development and Revision

13614 Progress Blvd., Alachua, FL 32615

Email: curriculum@nccer.org

Online: www.nccer.org/olf

❑ Trainee Guide ❑ Lesson Plans ❑ Exam ❑ PowerPoints Other ____________

Craft / Level: Copyright Date:

Module ID Number / Title:

Section Number(s):

Description:

Recommended Correction:

Your Name:

Address:

Email: Phone: